Introduction to the
Physical Properties of Soil

Contributors :
Hussein Othmanli,
Xiaoning Zhao, *et al.*

KOROS PRESS LIMITED
London, UK

Introduction to the Physical Properties of Soil
Contributors : Hussein Othmanli *and* Xiaoning Zhao, *et al.*

Published by Koros Press Limited

www.korospress.com

United Kingdom

Copyright 2016

Printed in 2017 for Sale in the Indian Subcontinent

Introduction to the Physical Properties of Soil

ISBN: 978-1-78163-557-5

British Library Cataloguing in Publication Data
A CIP record for this book is available from the British Library

Exclusively distributed by CBS Publishers & Distributors Pvt. Ltd.

Sales & Distribution Rights only for India, Pakistan, Bangladesh, Sri Lanka, Nepal and Bhutan.This book is not to be sold outside these territories.

PREFACE

Soil is a thin layer of material on the Earth's surface in which plants have their roots. It is made up of many things, such as weathered rock and decayed plant and animal matter. Soil is formed over a long period of time.

Soil Formation takes place when many things interact, such as air, water, plant life, animal life, rocks, and chemicals. The inorganic components of soil are weathered rock, air, water and minerals. The organic matter is the decomposing fragments of plants and animals. The spaces between the small particles that make up the soil are filled with air or water. Living plants and animals live in the soil and improve aeration and drainage. Some organisms, such as bacteria, play an important role in converting plant foods or nutrients, *e.g.* nitrogen, into a form that plants can use to grow. Important plant foods include nitrogen (helps leaves and stems grow), phosphate (helps roots and fruits develop) and potassium (stimulates overall plant health). When plants die, they return the nutrients they initially absorbed from the soil, back to the soil, and enrich the soil. In this way soil plays a very important role in the recycling of nutrients.

Soil takes thousands of years to develop from the parent rock – 10mm of soil takes between 100 and 1000 years to form. In South Africa 1mm of soil takes about 40 years to form. The time depends on the speed of weathering (parent rock being broken down into small particles). Weathering can be physical (frost, temperature changes, salt chrystallisation), chemical (chemical action of water, oxygen, carbon dioxide and organic acids) or biological (tree roots that widen crevices and cracks).

The soil profile generally consists out of three main layers (horizons): the topsoil (100 – 200 mm deep) or darker layer, where air, water and humus allow plants to grow in; the sub-soil, a more clay-like layer which acts as a reservoir (water store) for the plants, and the bedrock or parent material, which is the underlying layer from which the first two horizons are formed. In South Africa sub-soil can be transported or residual, or both. Transported soil originates from wind, water or gravitational processes, while residual soils are the in situ (undisturbed) decomposed product of the underlying rocks. Soil horizons are set apart from other soil layers by differences in physical and chemical composition, organic structure, or a combination of those properties. Soil horizons are developed by the interactions, through time, of climate, living organisms, and the configuration of the land surface (relief).

Owing to its lucid style and presentation of advanced topics, the book will be useful to postgraduate students as also to practising soil sciences.

CONTENTS

Preface   *v*

1. INTRODUCTION TO SOIL   **1-19**

    What is the Soil?   2

    Components of Climate   2

    Soil Profile   7

    Soil Texture   8

    Soil Structure and Soil Particles   9

    Soil Slope   11

    Principal Reasons for Soil Erosion   11

    Soil Structure   14

    Organic Activity of Soil   16

2. PROPERTIES OF SOIL   **20-43**

    Origin of Soils   21

    Classification of Soils   43

3. SOIL BIODIVERSITY   **44-55**

    Soils as Components of Ecosystems   44

    Soils as Organising Centers in Ecosystems   45

    Major Soil Processes   46

    Biodiversity in Soils   47

    Micro-fauna   51

    Mesofauna   52

4. SOIL-HABITAT   **56-64**

    Soil Biota and Ecosystem Functioning   56

    What are Soil Biota   61

    What do Soil Biota do?   62

    Importance of Soil Biota   62

    What Affects Soil Biota?   63

5. **PRINCIPALS OF SOIL ORGANISMS** **65-87**

    Soil Health and Plant Health   66

    Soil Organisms   67

    Managing the Soil Biota   71

    Crop Losses from Soil Organisms   81

    The Way Ahead   85

6. **SOIL MICROBES** **88-114**

    Soil Microbes Overview   88

    Fungi   89

    Bacteria   92

    Other Microbes   92

    Soil Food Web   93

    Roles of Soil Organisms   96

    Management Effects on Soil Biota   99

    Grassing   100

    Cultivation   101

    Fertilizers   102

    Inorganic Fertilizers   102

    P and S Fertilizers   103

    N Fertilizers   104

    Lime   104

    Organic Fertilizers   104

    Plant Residue Retention   106

    Fire   108

    Tree Clearing   109

    Body Form and Lifestyle   110

7. **SOIL MICRO-ORGANISMS** **115-136**

    Introduction   115

    Role and Application   118

    Controlling the Soil Micro-Flora   122

    Classification of Soils Based on their Micro-biological Properties   127

    Control   131

    Affects of Soil Micro-organisms on Plant Health and Nutrition   133

8. **SOIL ECOSYSTEM** **137-166**

    Introduction   137

    Abiotic Soil Components   139

Soil Processes   140

Chemical Composition of Soil   142

Methods to Quantify Effects of Chemical Input on
    Abiotic Soil Characteristics   145

Methods to Assess the Potential Risk of Chemicals for
    Soil Organisms (Prognosis)   149

Methods to Assess the Impact of Soil Contamination on
    Soil Organisms (Diagnosis)   162

Conclusions   165

9. SOIL MICROBIAL ACTIVITY   **167-210**

Soil Fertility and Nutrition   167

Nitrogen Fertilizer   188

Characteristics of Phosphorus   197

10. CHEMICAL DYNAMICS OF SOIL   **211-216**

Some Basic Ideas of Soil   211

11. MECHANICS OF SOILS   **217-224**

Introduction   217

12. SOILS, WEATHERING, AND NUTRIENTS   **225-235**

Soils   225

13. SOIL EROSION AND CONSERVATION   **236-252**

Soil Conservation   236

Research on Soil Erosion and Conservation   237

Important of Soil Erosion and its Conservation   239

Soil Biology and the Biological Micro-environment   241

Management for Maintenance of Soil Biology and Nutrition   243

Soil Bulk Density Determination   244

Soil Compaction and Building   247

Soil Analysis and Soil Nutrient Management   249

Interpretation of the Soil Analysis   251

14. SOIL TEMPERATURES AND CONDITIONS   **253-263**

Temperature   253

Composition and Structure   255

Soil Resources   260

Soil Resources and Cropping Systems   261

15. WATER USE EFFICIENCY IN SALINE SOILS UNDER COTTON CULTIVATION IN THE TARIM RIVER BASIN **264-284**

Xiaoning Zhao, Hussein Othmanli, Theresa Schiller, Chengyi Zhao, Yu Sheng, Shamaila Zia, Joachim Müller and Karl Stahr

Introduction 265

Materials and Methods 267

Results 272

Discussion 276

Conclusions 279

References 280

LIST OF CONTRIBUTORS

Xiaoning Zhao

Institute of Soil Science and Land Evaluation, University of Hohenheim, Emil-Wolff-Str. 27, Stuttgart 70593, Germany; E-Mails: husseinothmanli@hotmail.com (H.O.); theresa.schiller@gmx.net (T.S.); karl.stahr@uni-hohenheim.de (K.S.)

Hussein Othmanli

Institute of Soil Science and Land Evaluation, University of Hohenheim, Emil-Wolff-Str. 27, Stuttgart 70593, Germany; E-Mails: husseinothmanli@hotmail.com (H.O.); theresa.schiller@gmx.net (T.S.); karl.stahr@uni-hohenheim.de (K.S.)

Theresa Schiller

Institute of Soil Science and Land Evaluation, University of Hohenheim, Emil-Wolff-Str. 27, Stuttgart 70593, Germany; E-Mails: husseinothmanli@hotmail.com (H.O.); theresa.schiller@gmx.net (T.S.); karl.stahr@uni-hohenheim.de (K.S.)

Chengyi Zhao

Key Laboratory of Oasis Ecology and Desert Environment, Xinjiang Institute of Ecology and Geography, Chinese Academic of Science, Urumqi 830011, China; E-Mails: zcy@ms.xjb.ac.cn (C.Z.); shengyu@ms.xjb.ac.cn (Y.S.)

Yu Sheng

Key Laboratory of Oasis Ecology and Desert Environment, Xinjiang Institute of Ecology and Geography, Chinese Academic of Science, Urumqi 830011, China; E-Mails: zcy@ms.xjb.ac.cn (C.Z.); shengyu@ms.xjb.ac.cn (Y.S.)

Shamaila Zia

Institute of Agricultural Engineering, Hohenheim University, Stuttgart 70593, Germany; E-Mails: shamailazia@googlemail.com (S.Z.); joachim.mueller@uni-hohenheim.de (J.M.)

Joachim Müller

Institute of Agricultural Engineering, Hohenheim University, Stuttgart 70593, Germany; E-Mails: shamailazia@googlemail.com (S.Z.); joachim.mueller@uni-hohenheim.de (J.M.)

Karl Stahr

Institute of Soil Science and Land Evaluation, University of Hohenheim, Emil-Wolff-Str. 27, Stuttgart 70593, Germany; E-Mails: husseinothmanli@hotmail.com (H.O.); theresa.schiller@gmx.net (T.S.); karl.stahr@uni-hohenheim.de (K.S.)

Chapter 1

INTRODUCTION TO SOIL

Soil is a thin layer of material on the Earth's surface in which plants have their roots. It is made up of many things, such as weathered rock and decayed plant and animal matter. Soil is formed over a long period of time.

Soil Formation takes place when many things interact, such as air, water, plant life, animal life, rocks, and chemicals. The inorganic components of soil are weathered rock, air, water and minerals. The organic matter is the decomposing fragments of plants and animals. The spaces between the small particles that make up the soil are filled with air or water. Living plants and animals live in the soil and improve aeration and drainage. Some organisms, such as bacteria, play an important role in converting plant foods or nutrients, *e.g.* nitrogen, into a form that plants can use to grow. Important plant foods include nitrogen (helps leaves and stems grow), phosphate (helps roots and fruits develop) and potassium (stimulates overall plant health). When plants die, they return the nutrients they initially absorbed from the soil, back to the soil, and enrich the soil. In this way soil plays a very important role in the recycling of nutrients.

Soil takes thousands of years to develop from the parent rock – 10mm of soil takes between 100 and 1000 years to form. In South Africa 1mm of soil takes about 40 years to form. The time depends on the speed of weathering (parent rock being broken down into small particles). Weathering can be physical (frost, temperature changes, salt chrystallisation), chemical (chemical action of water, oxygen, carbon dioxide and organic acids) or biological (tree roots that widen crevices and cracks).

The soil profile generally consists out of three main layers (horizons) : the topsoil (100 –200 mm deep) or darker layer, where air, water and humus allow plants to grow in; the *sub-soil*, a more clay-like layer which acts as a reservoir (water store) for the plants, and the *bedrock* or *parent material*, which is the underlying layer from which the first two horizons are formed. In South Africa sub-soil can be transported or residual, or both. Transported soil originates from wind, water or gravitational processes, while residual soils are the *in situ* (undisturbed) decomposed product of the underlying rocks. Soil horizons are set apart from other soil layers by differences in physical and chemical composition, organic structure, or a combination of those properties. Soil horizons are developed by the interactions, through time, of climate, living organisms, and the configuration of the land surface (relief).

A city, like Greater Johannesburg, is an artificial man-made environment, dependent on technology and imported energy. The system is kept at an artificial equilibrium, and depends on the surrounding soil, agriculture and natural resources to ensure its well-being and sustain economic development. Soils are unfortunately deteriorating at an alarming pace due to poor management practices.

WHAT IS THE SOIL?

Soil can be defined in a variety of ways depending on one's outlook and purpose for considering or looking at soil, and these multiple views are not incompatible with each other. One person might think of soil as a medium within which plants extract water and nutrients in order to grow, while another might consider soil as a medium for waste disposal which functions as a chemical and biological reactor to ameliorate elements harmful to the environment. Another person might alternatively consider soil to be simply material with certain physical properties, within or upon which buildings or roads will be built. Most pedologists (soil scientists) consider soil to be a natural three-dimensional entity occurring at the earth's surface with identifiable horizons (layers) which have formed as a result of five soil forming factors. Hans Jenny (1941) explained that the five soil forming factors were climate, organisms, parent material (geology), relief (or topography) and time. While each soil forming factor can be considered individually, one should keep in mind that, in reality, the factors are often interrelated and interdependent.

$$S = f(C, O, P, R, T)$$

COMPONENTS OF CLIMATE

The two major components of climate which affect soils are precipitation (mainly rainfall) and temperature. In some regions, climatic variations are seen gradually over larger distances. In some situations, however, climactic changes may be much more abrupt. Rainfall contributes moisture to the soil which permits important chemical reactions and biological activity to occur. In the eastern USA and other areas where rainfall is abundant and precipitation exceeds evapotranspiration, an excess of moisture may accumulate in the soil causing percolation and leaching of soluble chemical constituents. Depending on the degree of leaching, soils may become acid (to varying degrees). In parts of the western USA where rainfall is more limited and evapotranspiration exceeds precipitation, soluble chemical components may actually accumulate in the soil. Certain soil horizons may be especially enriched in calcium carbonate or gypsum.

Fig. Two soil profiles formed under dramatically different climates. The soil shown in profile 1a is formed under an arid climate (annual rainfall 12 in. [300 mm]) and has a zone of calcium carbonate accumulation in the Bk horizon that extends from 16 to 40 in. (40 to 100 cm). The pH in the Bk horizon is approximately 8.1. The soil shown in profile 1b formed in a humid environment (annual rainfall 44 in. [1100 mm]) where soluble weathering products are quickly leached from the soil. This soil profile has no carbonates present at all. The pH of the Bt horizon, which extends from 8 to 28 in. (20 to 70 cm), is approximately 5.5. Small photo scale increments are 4 in. (10 cm).

Temperature also affects the moisture state of the soil because higher temperatures generally lead to greater evapotranspiration. Thus the degree of leaching in a soil is related to temperature as well as rainfall. The balance between rainfall and precipitation will also affect the likelihood that a soil may experience a seasonally-high water table (where the soil porosity becomes filled with water and saturated close to the soil surface). Temperature also affects chemical reactions in the soil. The rates of most chemical reactions increase as temperatures rise. This is true in the soil as it is in the laboratory. The degradation and weathering of rocks and minerals in the soil is driven by various chemical reactions. Therefore, the weathering of soil minerals and rocks may be much more dramatic in warmer regions (*e.g.* south-eastern USA) than in cooler ones (such as the Mid-Atlantic Region). Similarly, the decomposition of soil organic matter is generally faster under warmer soil conditions.

Organisms Affect of Soil Formation

Organisms which affect soil formation are both large and small and include plants, animals, and humans. The effects of large animals (such as burrowing rodents) often are more dramatic when observed, but are probably less important than those of smaller invertebrates such as earthworms and insects which often ingest soil and help form small aggregates of finer mineral particles, and in special instances cause significant mixing of the upper soil horizons.

Fig. Soil organisms such as ants or termites can cause significant mixing of the upper horizons of the soil. Termites have mixed the soil during the formation of mounds in this Australian landscape.

All organisms living on or in the soil contribute organic matter to the soil but the contributions by plants are generally held to be much greater than that of animals. Plants contribute organic matter to the soil directly as their roots grow and then die within the soil and they also add organic matter to the surface through the senescence of above ground parts (such as stem or leaf fall). Soil animals may then help to incorporate leaves and stem parts into the soil. Different types of plants contribute different amounts of organic matter to the soil through leaves versus below ground parts. In the Great Plains, for example, grasses add a great deal of organic matter to the soil through their roots and this may extend to a depth of as much as 3 ft. (1 m). Trees, on the other hand, contribute a greater proportion of organic matter onto the soil surface. Thus, organic rich A horizons of prairie soils tend to be thicker and darker than do the A horizons of soils formed under forest vegetation. The nature of the organic debris from plants may also differ chemically. Certain plants, such as conifers and heather, tend to be very acidic and their contribution of acidic organic matter may accentuate acid weathering reactions in some soils.

Fig. The soil on the left, which has a fairly thin A horizon, was formed under forest vegetation where much of the organic matter is added to the soil surface during leaf fall. The soil on the right, which has a much thicker A horizon, was formed under prairie vegetation where much of the organic matter is added within the soil by the grass roots. Small photo scale increments are 4 in. (10 cm).

Humans often have profound effects on soil formation through alteration of soils and all five of the soil forming factors. Humans alter soil climate (hydrology) by irrigating, flooding, ponding, and draining soils. Native vegetation is altered by agriculture and forestry, which further impacts organic and nutrient cycling and hydrology. Heavy equipment is used to construct new landforms and alter relief on existing landforms. This often results in destruction of existing soil properties and the renewal of soil forming processes, as old soils are transformed into new soils.

Parent Material

Parent material is the geological material from which the soil originated. Parent materials may be residual, meaning that they are derived from the underlying hard consolidated bedrock such as granite or limestone. Parent materials may also be transported, meaning that they have been moved by some agent such as water, ice, wind, or gravity and then subsequently deposited.

These would include alluvium deposited by streams, till and drift deposited by glaciers, loess and dunes deposited by wind, and colluvium deposited downslope by gravity. Because the parent materials represent the starting point of soils, they often contribute many significant properties to the soils (as do parents contribute genetically to the characteristics of their offspring).

The particle size, mineralogy, chemistry and colour of soils may all be initially related to (inherited from) the properties of the parent material. In locations where the diversity of parent materials is great over relatively small distances, there may be a disproportionately large number of different soils which form.

Fig. The soil on the left has formed from a gray shale residuum, and the shale bedrock is evident from approximately 16 in. (40 cm) below the soil surface. The soil on the right has formed from sandy and gravelly sediments deposited by melting glaciers. Small photo scale increments are 4 in. (10 cm).

Relief and Topography of Soil

Relief and topography refer to differences in the magnitude of slope, differences in location on the landscape (summit, shoulder, backslope, footslope, etc.)

and differences in slope aspect (N, S, E, W, etc.). Both the position of a soil on the landscape, as well as the magnitude of the slope (2 per cent *vs.* 10 per cent *vs.* 30 per cent) have major effects on the degree to which a given soil is likely to be eroded (convex shoulders or steep backslopes) or is likely to receive additions of newly deposited materials (concave footslopes), or where it is likely to remain stable and experience little erosion or deposition (nearly level summits). There may also be hydrological effects caused by the location of a soil along a particular landscape.

The depth to the water table is generally not uniform across dissected landscapes. Soils occurring at particular positions on the landscape may be poorly drained; while well drained soils may be found in close proximity, but on other landscape settings. In highly dissected areas with steep slopes, the aspect of the slope may cause the soil temperatures to be different on north facing *versus* south facing slopes.

As one might expect, soils on steep north facing slopes tend to have cooler soil temperatures than corresponding south facing slopes. In general, the relief/topography factor of soil formation affects the erosion/sedimentation processes and also mediates climatic conditions (meaning that it can cause a soil to be wetter, drier, warmer or cooler than it would otherwise be).

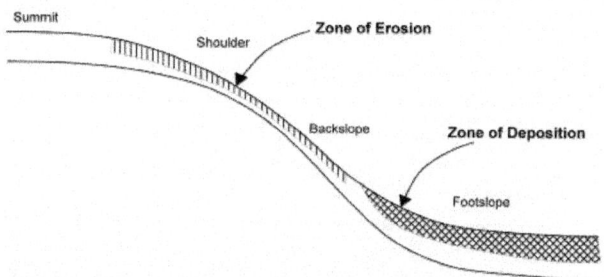

Fig. Schematic diagram of a hillslope illustrating the various components. Erosional processes are usually most dramatic along the shoulder and backslope portions of the hillslope and depositional processes are most pronounced in the footslope portion.

Time Factor of Soil

The time factor is related to the age of the soil, which must not be confused with the age of the soil parent material. The age of a soil is the amount of time which has passed since the other soil forming factors (climate, organisms, relief) first began to impact the soil parent material to begin to form soil horizons. The age of soils may range from a few years (such as in soils formed in recent landslide deposits) all the way to millions of years (in the case of soils formed on some very stable landscapes). The parent materials may be much older than the soil.

For example, relatively recent tectonic uplift and erosion may have exposed granitic bedrock of Precambrian age (approximately one billion years old), but the landscape itself may have only been exposed to soil forming processes a much shorter time – perhaps a few hundred thousand years ago. Alternatively, the age of a soil may be approximately equal to the age of a glacial till deposit in which it has formed. Determining the age of a soil is often a difficult task and in some cases there may be no clear way to accurately assess soil age.

Fig. The soil on the left is young having formed from sediments deposited in the early 19th century (*i.e.* age < 100 yrs). Note the absence of well developed horizons and the preservation of original sedimentary stratification at depths below 20 in. (50 cm). The very old soil on the right was formed in terrace deposits, which are perhaps as much as 1 million years old. This soil has a strong argillic (Bt) horizon that begins around 24 in. (60 cm) and continues down to nearly 10 ft.(3 m). Small photo scale increments are 4 in. (10 cm).

In general, younger soils tend to strongly reflect the properties of their parent materials. A soil which is one day old would look much like the soil parent material. As time goes on, soil forming processes driven by soil forming factors gradually cause changes in the parent material leading to the formation of soil horizons with distinct morphological, physical, chemical, mineralogical, and biological properties. Time then becomes the matrix within which the other soil forming factors operate.

SOIL PROFILE

A vertical cross-section through a soil typically represents a layered pattern. This section is called a "profile" and the individual layers are called "horizons". The uppermost layer includes the "surface soil" or "topsoil" and is designated the A horizon. This is the layer which is most subject to climatic and biological influence.

It is usually the layer of maximum organic accumulation, has a darker colour, and has less clay than sub-soil. The majority of plant roots and most of the soil's fertility are contained in this horizon. The next successive horizon is called the "sub-soil" or B horizon. It is a layer which commonly accumulates materials that have migrated downward from the surface.

Much of the deposition is clay particles, iron and aluminum oxides, calcium carbonate, calcium sulfate and possibly other salts. The accumulation of these materials creates a layer which is normally more compact and has more clay than the surface. This often leads to restricted movement of moisture and reduced crop yields.

The parent material (C horizon) is the least affected by physical, chemical and biological weathering agents. It is very similar in chemical composition to the original material from which the A and B horizons were formed. Parent material which has formed in its original position by weathering of bedrock is termed "residual", or called "transported" if it has been moved to a new location by natural forces. This latter type is further characterised on the basis of the kind of natural force responsible for its transportation and deposition. When water is the transporting agent, the parent materials are referred to as "alluvial" (stream deposited). This type is especially important in Oklahoma. These are often the most productive soils for agricultural crops. Wind-deposited materials are called "aeolian".

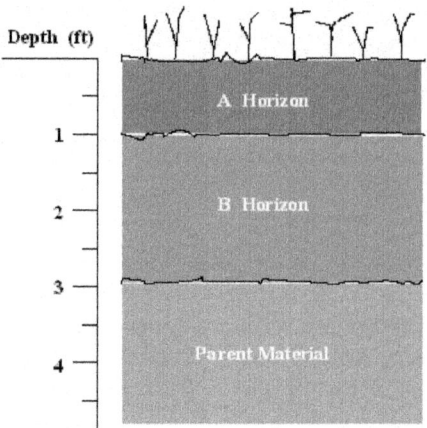

Fig. A typical soil profile.

Climate has a strong influence on soil profile development. Certain characteristics of soils formed in areas of different climates can be described. For example, soils in western Oklahoma are drier and tend to be coarser textured, less well developed and contain more calcium, phosphorus, potassium and other nutrients than do soils in the humid eastern part of the state.

The soil profile is an important consideration in terms of plant growth. The depth of the soil, its texture and structure, its chemical nature as well as the soil position on the landscape and slope of the land largely determine crop production potential. The potential productivity is vitally important in determining the level of fertilization.

SOIL TEXTURE

Soils are composed of particles with an infinite variety of sizes. The individual particles are divided by size into the categories of sand, silt and clay. Soil texture refers to the relative proportion of sand, silt and clay in the soil. Textural class is the name given to soil based on the relative amounts of sand, silt, and clay present, as indicated by the textural triangle.

Such divisions are very meaningful in terms of relative plant growth. Many of the important chemical and physical reactions are associated with the surface of the particles, and hence are more active in fine than coarse texture soils.

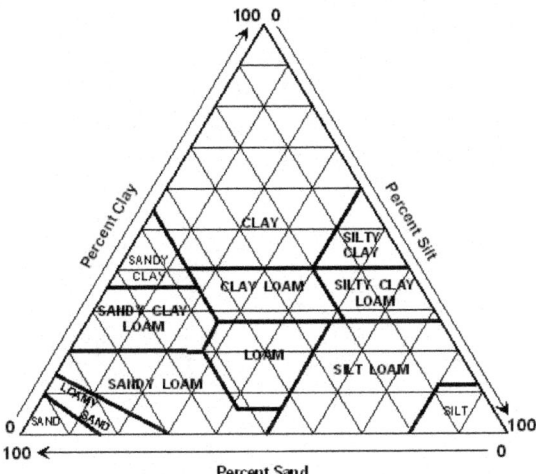

Fig. Triangle for determining soil textural classes.

A textural class description of soils can tell a lot about soil-plant interactions, since the physical and chemical properties of soils are determined largely by texture. In mineral soils, exchange capacity (ability to hold plant nutrient elements) is related closely to the amount and kind of clay in soils. Texture is a major determining factor for water holding capacity. Fine-textured soils (high percentage of silt and clay) hold more water than coarse-textured soils (sandy). Water and air movement through the finer textured soils is reduced and they can be more difficult to work. From the standpoint of plant growth, medium-textured soils, such as loams, sandy loams and silt loams, are probably the most ideal. Nevertheless, the relationships between soil textural class and soil productivity cannot be generally applied to all soils, since texture is one of the many factors that influence crop production. Check the texture of the surface and sub-soil. Normally, the surface includes the top foot of soil, but it may be shallower or deeper in certain situations. Soil below the tillage zone is called "sub-soil". It is also necessary to consider the sub-soil texture when determining productivity potentials.

SOIL STRUCTURE AND SOIL PARTICLES

Soil structure refers to the presence of aggregates of soil particles that have been bound together to form distinct shapes. Sometimes the binding or cementing is only weak, however the aggregates are much larger than individual soil particles. Soil organic matter contributes significantly as a cementing agent. Air and water movement and root penetration in the soil is related to the soil structure. The better the structure, the higher the productivity of the soil is. Size and shape of the structure units is important. When height of the structure unit is approximately equal to its width (blocky structure) we expect good air and water movement. Structure units that have greater height than width (prismatic structure) are often associated with sub-soils that swell when wet and shrink when dry, resulting in poor air and water movement. When particles have greater width than height (platy structure) water and air movement and root development in the soil is restricted, compared to a soil with desirable structure. Granular structure particularly in fine-textured soils is ideal for water penetration and air movement. Water and air move more freely through sub-soils that have blocky structure than those with platy structure. Good air and water movement is conducive to plant root development.

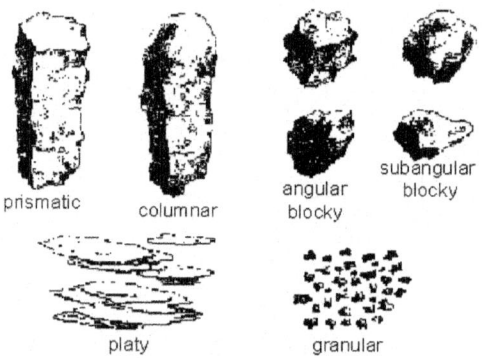

Fig. Types of soil structure.

The productivity of the soil is influenced by both surface and sub-soil texture and structure. An approximate rating for soils considering texture and structure.

Table. Soil Productivity Rating as Affected by Texture.*

Subsoil Texture	Surface Soil Texture				
	Sand	Sandy Loam	Loam	Clay Loam	Clay; Silty Clay
	-------- Percent of Maximum Productivity --------				
Sandy	50	55	65	60	55
Sandy Loam	60	70	80	75	65
Loam	70	80	95	90	75
Clay Loam	70	80	90	90	75
Clay; Silty Clay	65	70	80	80	70

Note : *Numbers represent average soil conditions.

Raise or lower the rating 10 to 20 per cent, according to whether the soil structure is more, or less, favourable than the average. If gravel occurs in the soil, lower the rating according to its effect on the productive capacity.

SOIL DEPTH

Soil depth is generally used to describe how deep roots can favourably penetrate. Soils that are deep, well drained, and have desirable texture and structure are suitable for production of most crops. For satisfactory production, most plants require considerable soil depth for root development from which to secure nutrients and water. Plants growing on shallow soils have little soil volume from which to secure water and nutrients. Depth of soil, and its capacity to hold nutrients and water, frequently determines crop yield, particularly for summer crops. Roots of most crops will extend three feet or more into favourable soil. Soils should be at least six feet deep to give maximum production. Look for materials or conditions that limit soil depth, such as hardpans, shale, coarse gravelly layers and tight impervious layers. These are almost impossible to change. On the other hand, a high water table may limit root growth, but it can usually be corrected by drainage. Soil productivity estimates on the basis of soil depth.

Table. Soil Productivity Rating as Affected by Depth.

Soil Depth Usable by Crop Roots	Relative Productivity
(Feet)	(Percent)
1	35
2	60
3	75
4	85
5	95
6	100

SOIL SLOPE

Topography of the land largely determines potential for run-off and erosion, method of irrigation, and management practices needed to conserve soil and water. Higher sloping land requires more management, labour and equipment expenditures. Table below can be used to rate land productivity based on slope. If slope varies, use steeper slopes for the rating.

Table. Soil Productivity Ratings as Affected by Slope.

Slope	Relative Productivity	
	Stable Soil	Unstable, Easily Eroded Soil
	---------------------------------- % ----------------------------------	
0-1	100	95
1-3	90	75
3-5	80	50
5-8	60	30
8-12	40	10

PRINCIPAL REASONS FOR SOIL EROSION

Principal reasons for soil erosion in Oklahoma are :

• Insufficient vegetative cover, which is usually a result of inadequate fertility to support a good plant cover,

• Growing cultivated crops on soils not suited to cultivation, and

- Improper tillage of the soil.

Soil erosion can be held to a minimum by :

- Using the soil to produce crops for which it is suitable,
- Using adequate fertilizer and lime to promote vigourous plant growth, and
- Using proven soil preparation and tillage methods.

Soils that have lost part or all their surfaces are usually harder to till and have lower productivity than non-eroded soils. To compensate for surface soil loss, more fertilization, liming and other management practices should be used.

Soil and Available Water

Plants are totally dependent upon water for growth and production. Even with well fertilized soils, limited water can greatly reduce yields. Rainfall is not always dependable in Oklahoma, and therefore, crops are dependent upon the moisture stored in the soil profile for growth and production. Soils differ in their ability to supply water to plants. Limited root zones caused by shallow soils, high water table or claypans, or extremely porous sub-soils cause drought stress in plants faster than more desirable soils. Illustrates the differences in available water in selected soil profiles. Soils with silt loam or fine sandy loam surface textures have high available water holding capacities. Differences in available water holding capacity between the soils caused by widely varying textures of the sub-soil and soil depth point out the need for knowing what is below the surface. (This kind of information is available in county soil survey manuals). During a drought, differences of two inches of available water can keep plants growing for an extra ten days during peak plant use and could be the difference between success and crop failure.

Table. Effect of Depth and Texture on Available Water for Crop Use.

Soil Name	Texture	Depth	Available Water
		---------- inches ----------	
Dennis	silt loam	0-11	1.98
	silty clay loam	11-23	2.52
	clay	23-60	5.55
	Total	60	10.05
Sallisaw	silt loam	0-10	1.80
	silt loam	10-20	1.80
	gravelly clay loam	20-40	2.80
	very gravelly clay loam	40-60	1.60
	Total	60	8.00
Shellabarger	fine sandy loam	0-16	1.92
	sandy clay loam	16-52	5.86
	fine sandy loam	52-60	0.88
	Total	60	8.66
Stephenville	fine sandy loam	0-14	1.82
	sandy clay loam	14-38	3.84
	sandstone	38+	-----
	Total	38+	5.66

Soil Fertility

Soil fertility is the soil's ability to provide essential plant nutrients in adequate amounts and proper proportions to sustain plant growth. Soil fertility is a component of soil productivity that is quite variable and strongly influenced by management. Other components of soil productivity, especially soil slope and soil depth, will be the same year after year. Together with climate, these components set the soil productivity limits, above which yields cannot be obtained even with ideal use of fertilizer. It is important to realise this and understand that added fertilizer cannot compensate for a soil that is unproductive because it is excessively stony or has a sub-soil layer that restricts normal root growth and development.

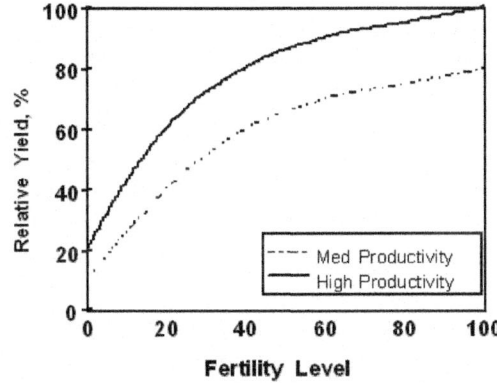

Fig. Influence of Soil Productivity on Yield Response to Fertility.

Soil Management

There are numerous other soil characteristics that can be important to soil productivity in specific areas. These include : soil drainage, soli salinity, presence of stone and/or rocks, and organic matter content. They are not major limiting factors over wide areas, and therefore, will not be discussed here. One additional factor on which soil productivity is highly dependent is soil management. This implies using the best available knowledge, techniques, materials, and equipment in crop production. The use of minimum tillage is an important management practice used to reduce the potential damage to soil structure from over-working, and for economic and fuel conservation purposes as well as to allow farming of more acres per unit of labour.

Soil conservation is a concept integrating important management practices which deserves close attention. It is estimated that annually in the U.S. four billion tons of sediment are lost from the land in run-off waters, and with it much of the natural and applied fertility. That is equivalent to the total loss of topsoil (six inches deep) from four million acres. Wind erosion is also a problem in certain areas. Management practices such as contouring, strip planting, covercropping, reduced tillage, terracing and crop residue management help to eliminate or minimise the loss of soil from water and wind erosion. Proper utilisation of crop residues can be a key management practice. Crop residues returned to the soil improve soil productivity through the addition of organic matter and plant nutrients. The organic matter also contributes to an improved physical condition of the soil, which increases water infiltration and storage and aids aeration. This is vital to crop growth.

SOIL STRUCTURE

Soil structure refers to units composed of primary particles. The cohesion within these units is greater than the adhesion among units. As a consequence, under stress, the soil mass tends to rupture along predetermined planes or zones. These planes or zones, in turn, form the boundary. Compositional differences of the fabric matrix appear to exert weak or no control over where the bounding surfaces occur. If compositional differences control the bounding surfaces of the body, then the term "concentration" is employed. The term "structural unit" is used for any repetitive soil body that is commonly bounded by planes or zones of weakness that are not an apparent consequence of compositional differences. A structural unit that is the consequence of soil development is called a ped. The surfaces of peds persist through cycles of wetting and drying in place. Commonly, the surface of the ped and its interior differ as to composition or organisation, or both, because of soil development.

Earthy *clods* and *fragments* stand in contrast to peds, for which soil forming processes exert weak or no control on the boundaries. Some clods, adjacent to the surface of the body, exhibit some re-arrangement of primary particles to a denser configuration through mechanical means. The same terms and criteria used to describe structured soils should be used to describe the shape, grade, and size of clods. Structure is not inferred by using the terms interchangeably. A size sufficient to affect tilth adversely must be considered. The distinction between clods and fragments rests on the degree of consolidation by mechanical means. Soil fragments include (1) units of undisturbed soil with bounding planes of weakness that are formed on drying without application of external force and which do not appear to have predetermined bounding planes, (2) units of soil disturbed by mechanical means but without significant re-arrangement to a denser configuration, and (3) pieces of soil bounded by planes of weakness caused by pressure exerted during examination with size and shape highly dependent on the manner of manipulation.

Some soils lack structure and are referred to as *structureless*. In structureless layers or horizons, no units are observable in place or after the soil has been gently disturbed, such as by tapping a spade containing a slice of soil against a hard surface or dropping a large fragment on the ground. When structureless soils are ruptured, soil fragments, single grains, or both result. Structureless soil material may be either single grain or massive. Soil material of single grains lacks structure. In addition, it is loose. On rupture, more than 50 per cent of the mass consists of discrete mineral particles. Some soils have simple structure, each unit being an entity without component smaller units. Others have *compound structure,* in which large units are composed of smaller units separated by persistent planes of weakness.

In soils that have structure, the shape, size, and grade (distinctness) of the units are described. Field terminology for soil structure consists of separate sets of terms designating each of the three properties, which by combination form the names for structure.

Compound Structure

Smaller structural units may be held together to form larger units. Grade, size, and shape are given for both and the relationship of one set to the other is indicated : "strong medium blocks within moderate coarse prisms," or "moderate coarse prismatic structure parting to strong medium blocky."

Extra-Structural Cracks

Cracks are macroscopic vertical planar voids with a width much smaller than length and depth. A crack represents the release of strain that is a consequence of drying. In many soils, cracks bound individual structural units. These cracks are repetitive and usually quite narrow. Their presence is part of the concept of the structure. The cracks to be discussed are the result of localised stress release which forms planar voids that are wider than the repetitive planar voids between structural units or which occur in massive or weakly structured material at relatively wide intervals. These cracks may be coextensive with crack space between structural units. If they are coextensive, the width exceeds that of the associated structural cracks. The areal percentage of such cracks, either on a vertical exposure or on the ground surface, may be measured by line-intercept methods. For taxonomic purposes, the width and depth of cracks has importance.

Four kinds of extra-structural cracks may be recognised :

- *Surface-initiated reversible cracks* form as a result of drying from the surface downward. They close after relatively slight surficial wetting and have little influence on ponded infiltration rates.

- *Surface-initiated irreversible cracks* form on near-surface water reduction from exceptionally high water content related to freeze-thaw action and other processes. The cracks do not close completely when rewet and extend through the crust formed by frost action. They act to increase ponded infiltration rates.

- *Subsurface-initiated reversible cracks* form as a result of appreciable reduction in water content from "field capacity" in horizons or layers with considerable extensibility. They close in a matter of days if the horizon is brought to the moderately moist or wetter state. They extend upward to the soil surface unless there is a relatively thick overlying horizon that is very weakly compacted (loose or very friable) and does not permit the propagation of cracks (mechanically bulked sub-zones). Such cracks importantly influence ponded infiltration rates and evaporation directly from the soil.

- *Sub-surface-initiated irreversible cracks* are the "permanent" cracks of the USDA soil taxonomy system. They have a similar origin to surface-initiated irreversible cracks, although quite different agencies are involved.

The foregoing genetic definition of cracks does not directly relate to prediction of infiltration. For such predictions, the surface connectiveness of the cracks and their depth must be specified. *Surface-connected cracks* occur at the ground surface or are covered by up to 10-15 cm of soil material that would permit the accumulation of free water at the plane that marks the top of the crack under conditions that may occur in most years. If the antecedent water state of the overlying zone were very moist, free water from 25 mm of rainfall in one hour should reach the top of the cracks. Usually the zone would have *very high* or *high* saturated hydraulic conductivity. Such sub-zones may exhibit structure or be single grain. The structure units range widely in size. The common characteristic is that the consistent units of the mass as a whole are highly discrete and the porosity of the interstices among the structural units is high. If not too thick, the *mechanically bulked* sub-zone of tilled surface horizons would be such a zone.

ORGANIC ACTIVITY OF SOIL

A mass of mineral particles alone do not constitute a true soil. True soils are influenced, modified, and supplemented by living organisms. Plants and animals aid in the development of a soil through the addition of organic matter. Fungi and bacteria decompose this organic matter into a semi-soluble chemical substance known as humus. Larger soil organisms, like earthworms, beetles, and termites, vertically redistribute this humus within the mineral matter found beneath the surface of a soil. Humus is the bio-chemical substance that makes the upper layers of the soil become dark. It is coloured dark brown to black. Humus is difficult to see in isolation because it binds with larger mineral and organic particles.

Humus provides soil with a number of very important benefits :

- It enhances a soil's ability to hold and store moisture.
- It reduces the eluviation of soluble nutrients from the soil profile.
- It improves soil structure which is necessary for plant growth.

Organic activity is usually profuse in the near surface layers of a soil. For instance, one cubic centimeter of soil can be the home to more than 1,000,000 bacteria. A hectare of pasture land in a humid mid-latitude climate can contain more than a million earthworms and several million insects. Earthworms and insects are extremely important because of their ability to mix and aerate soil. Higher porosity, because of mixing and aeration, increases the movement of air and water from the soil surface to deeper layers where roots reside. Increasing air and water availablity to roots has a significant positive effect on plant productivity. Earthworms and insects also produce most of the humus found in a soil through the incomplete digestion of organic matter.

Translocation of Soil

When water moves downward into the soil, it causes both mechanical and chemical translocations of material. The complete chemical removal of substances from the soil profile is known as leaching. Leached substances often end up in the groundwater zone and then travel by groundwater flow into water bodies like rivers, lakes, and oceans. Eluviation refers to the movement of fine mineral particles (like clay) or dissolved substances out of an upper layer in a soil profile. The deposition of fine mineral particles or dissolved substances in a lower soil layer is called illuviation.

Soil Texture

The texture of a soil refers to the size distribution of the mineral particles found in a representative sample of soil. Particles are normally grouped into three main classes : sand, silt, and clay. Describes the classification of soil particles according to size.

Table. Particle Size Ranges for Sand, Silt, and Clay.

Type of Mineral Particle	Size Range
Sand	2.0 - 0.06 millimeters
Silt	0.06 - 0.002 millimeters
Clay	less than 0.002 millimeters

Clay is probably the most important type of mineral particle found in a soil. Despite their small size, clay particles have a very large surface area relative to their volume. This large surface is highly reactive and has the ability to attract and hold positively charged nutrient ions. These nutrients are available to plant roots for nutrition. Clay particles are also somewhat flexible and plastic because of their lattice-like design. This feature allows clay particles to absorb water and other substances into their structure.

Soil pH

Soils support a number of inorganic and organic chemical reactions. Many of these reactions are dependent on some particular soil chemical properties. One of the most important chemical properties influencing reactions in a soil is pH. Soil pH is primarily controlled by the concentration of free hydrogen ions in the soil matrix. Soils with a relatively large concentration of hydrogen ions tend to be acidic. Alkaline soils have a relatively low concentration of hydrogen ions. Hydrogen ions are made available to the soil matrix by the dissociation of water, by the activity of plant roots, and by many chemical weathering reactions.

Soil fertility is directly influenced by pH through the solubility of many nutrients. At a pH lower than 5.5, many nutrients become very soluble and are readily leached from the soil profile. At high pH, nutrients become insoluble and plants cannot readily extract them. Maximum soil fertility occurs in the range 6.0 to 7.2.

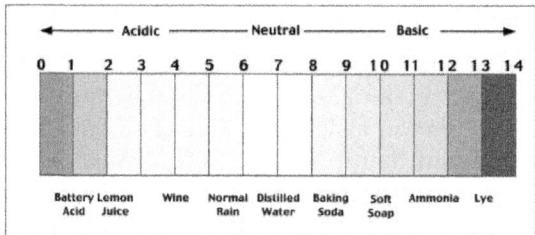

Fig. The pH Scale. A Value of 7.0 is Considered Neutral. Values Higher than 7.0 are Increasingly Alkaline or Basic. Values Lower than 7.0 are Increasingly Acidic. The Illustration Above also Describes the pH of some Common Substances.

Soil Colour

Soils tend to have distinct variations in colour both horizontally and vertically. The colouring of soils occurs because of a variety of factors. Soils of the humid tropics are generally red or yellow because of the oxidation of iron or aluminum, respectively. In the temperate grasslands, large additions of humus cause soils to be black. The heavy leaching of iron causes coniferous forest soils to be gray. High water tables in soils cause the reduction of iron, and these soils tend to have greenish and gray-blue hues. Organic matter colours the soil black. The combination of iron oxides and organic content gives many soil types a brown colour. Other colouring materials sometimes present include white calcium carbonate, black manganese oxides, and black carbon compounds.

Soil Profiles

Most soils have a distinct profile or sequence of horizontal layers. Generally, these horizons result from the processes of chemical weathering, eluviation, illuviation, and organic decomposition. Up to five layers can be present in a typical soil : O, A, B, C, and R horizons.

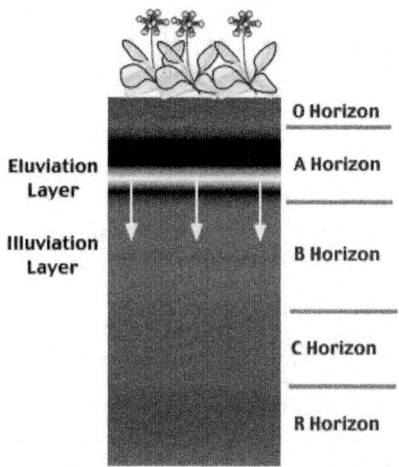

Fig. Typical Layers Found in a Soil Profile.

The O horizon is the top most layer of most soils. It is composed mainly of plant litter at various levels of decomposition and humus. A horizon is found below the O layer. This layer is composed primarily of mineral particles. The A horizon has two important characteristics : it is the layer in which humus and other organic materials are mixed with mineral particles, and it is a zone of translocation from which eluviation has removed finer particles and soluble substances, both of which may be deposited at a lower layer. Thus the A horizon is dark in colour and usually light in texture and porous. The A horizon is commonly differentiated into a darker upper horizon or organic accumulation, and a lower horizon showing loss of material by eluviation.

The B horizon is a mineral soil layer which is strongly influenced by illuviation. Consequently, this layer receives material eluviated from the A horizon. The B horizon also has a higher bulk density than the A horizon due to its enrichment of clay particles. The B horizon may be coloured by oxides of iron and aluminum or by calcium carbonate illuviated from the A horizon.

The C horizon is composed of weathered parent material. The texture of this material can be quite variable with particles ranging in size from clay to boulders. The C horizon has also not been significantly influenced by the pedogenic processes, translocation, and/or organic modification. The final layer in a typical soil profile is called the R horizon. This soil layer simply consists of unweathered bedrock.

Chapter 2

PROPERTIES OF SOIL

No department of agricultural chemistry is surrounded with greater difficulties and uncertainties than that relating to the properties of the soil. When chemistry began to be applied to agriculture, it was not unnaturally supposed that the examination of the soil would enable us to ascertain with certainty the mode in which it might be most advantageously improved and cultivated, and when, as occasionally happened, analysis revealed the absence of one or more of the essential constituents of the plant in a barren soil, it indicated at once the cause and the cure of the defect.

But the expectations naturally formed from the facts then observed have been as yet very partially fulfilled; for, as our knowledge has advanced, it has become apparent that it is only in rare instances that it is possible satisfactorily to connect together the composition and the properties of a soil, and with each advancement in the accuracy and minuteness of our analysis the difficulties have been rather increased than diminished.

Although it is occasionally possible to predicate from its composition that a particular soil will be incapable of supporting vegetation, it not unfrequently happens that a fruitful and a barren soil are so similar that it is impossible to distinguish them from one another, and cases even occur in which the barren appears superior to the fertile soil.

The cause of this apparently anomalous phenomenon lies in the fact that analysis, however minute, is unable to disclose all the conditions of fertility, and that it must be supplemented by an examination of its physical and other chemical properties, which are not indicated by ordinary experiments.

Of late years very considerable progress has been made in the investigation of the properties of the soil, and many facts of great importance have been discovered, but we are still unable to assert that all the conditions of fertility are yet known, and the practical application of those recently discovered is still very imperfectly understood.

It must not be supposed that a careful analysis of a soil is without value, for very important practical deductions may often be drawn from it, and when this is not practicable it is not unfrequently due to its being imperfect or incomplete, for it is so complex that the cases in which all the necessary details have been eliminated are even now by no means numerous.

In fact, the want of a large number of thorough analyses of soils of different kinds is a matter of some difficulty, and so soon as a satisfactory mode of investigation can be determined upon, a full examination of this subject would be of much importance.

ORIGIN OF SOILS

The constituents of the soil, like those of the plant, may be divided into the great classes of organic and inorganic. The origin of the former has been already discussed : they are derived from the decay of plants which have already grown upon the soil, and which, in various stages of decomposition, form the numerous class of substances grouped together under the name of humus.

The organic substances may therefore be considered as in a manner secondary constituents of the soil, which have been accumulated in it as the consequence of the growth and decay of successive generations of plants, while the primeval soil consisted of inorganic substances only. The inorganic constituents of the soil are obtained as the result of a succession of chemical changes going on in the rocks which protrude through the surface of the earth.

We have only to examine one of these rocks to observe that it is constantly undergoing a series of important changes. Under the influence of air and moisture, aided by the powerful agency of frost, it is seen to become soft, and gradually to disintegrate, until it is finally converted into an uniform powder, in which the structure of the original rock is with difficulty, if at all distinguishable.

The rapidity with which these changes take place is very variable; in the harder rocks, such as granite and mica slate it is so slow as to be scarcely perceptible, while in others, such as the shales of the coal formation, a very few years' exposure is sufficient for the purpose. These actions, operating through a long series of years, are the source of the inorganic constituents of all soils.

Geology points to a period at which the earth's surface must have been altogether devoid of soil, and have consisted entirely of hard crystalline rocks, such as granite and trap, by the disintegration of which, slowly proceeding from the creation down to the present time, all the soils which now cover the surface have been formed.

But they have been produced by a succession of very complicated processes; for these disintegrated rocks being washed away in the form of fine mud, or at least of minute particles, and being deposited at the bottom of the primeval seas, have there hardened into what are called sedimentary rocks, which being raised above the surface by volcanic action or other great geological forces, have been again disintegrated to yield different soils.

Thus, then, all soils are directly or indirectly derived from the crystalline rocks, those overlying them being formed immediately by their decomposition, while those found above the sedimentary rocks may be traced back through them to the crystalline rocks from which they were originally formed.

Such being the case, the composition of different soils must manifestly depend on that of the crystalline rocks from which they have been derived. Their number is by no means large, and they all consist of mixtures in variable proportions of quartz, felspar, mica, hornblende, augite, and zeolites.

With the exception of quartz and augite, these names are, however, representatives of different classes of minerals. There are, for instance, several different minerals commonly classified under the name of felspar, which have been distinguished by mineralogists by the names of orthoclase, albite, oligoclase and labradorite; and there are at least two sorts of mica, two of hornblende, and many varieties of zeolites.

Quartz consists of pure silica, and when in large masses is one of the most indestructible rocks. It occurs, however, intermixed with other minerals in small crystals, or irregular fragments, and forms the entire mass of pure sand. The four kinds of felspar which have been already named are compounds of silica with alumina, and another base which is either potash, soda, or lime.

It is obvious that soils produced by the disintegration of these minerals must differ materially in quality. Those yielded by orthoclase must generally abound in potash, while albite and labradorite, containing little or none of that element must produce soils in which it is deficient. The quality of the soil they yield is not however entirely dependent on the nature of the particular felspar which yields it, but is also intimately connected with the extent to which the decomposition has advanced.

It is observed that different felspars undergo decomposition with different degrees of rapidity but after a certain time they all begin to lose their peculiar lustre, acquire a dull and earthy appearance, and at length fall into a more or less white and soft powder.

During this change water is absorbed, and, by the decomposing action of the air, the alkaline silicate is gradually rendered soluble, and at length entirely washed away, leaving a substance which, when mixed with water, becomes plastic, and has all the characters of common clay.

In this instance the decomposition of the felspar had reached its limit, a mere trace of potash being left, but if taken at different stages of the process, variable proportions of that alkali are met with. This decomposition of felspar is the source of the great deposits of clay which are so abundantly distributed over the globe, and it takes place with nearly equal rapidity with potash and soda felspar.

It is rarely complete, and the soils produced from it frequently contain a considerable proportion of the undecomposed mineral, which continues for a long period to yield a supply of alkalies to the plants which grow on them. Mica is a very widely distributed mineral, and two varieties of it are distinguished by mineralogists, one of which is characterised by the large quantity of magnesia it contains.

Mica undergoes decomposition with extreme slowness, as is at once illustrated by the fact that its shining scales may frequently be met with entirely unchanged in the soil. Its persistence is dependent on the small quantity of alkaline constituents which it contains; and for this reason it is observed that the magnesian micas undergo decomposition less rapidly than those containing the larger quantity of potash.

Eventually, however, both varieties become converted into clay, their magnesia and potash passing gradually into soluble forms. Hornblende and augite are two widely distributed minerals, which are so similar in composition and properties that they may be considered together.

In these minerals alkalies are entirely absent, and their decomposition is due to the presence of protoxide of iron, which readily absorbs oxygen from the air, when the magnesia is separated and a ferruginous clay left.

The minerals just referred to, constitute the great bulk of the mountain masses, but they are associated with many others which take part in the formation of the soil. Of these the most important are the zeolites which do not occur in large masses but are disseminated through the other rocks in small quantity.

They are chiefly characterised by containing their silica in a soluble state, and hence may yield that substance to the plants in a condition particularly favourable for absorption. It is obvious from what has been stated that all these minerals are capable, by their decomposition, of yielding soft porous masses having the physical properties of soils, but most of them would be devoid of many essential ingredients, while not one of them would yield either phosphoric acid, sulphuric acid, or chlorine.

It has, however, been recently ascertained that certain of these minerals, or at least the rocks formed from them, contain minute, but distinctly appreciable traces of phosphoric acid, although in too small quantity to be detected by ordinary analysis; and small quantities of chlorine and sulphuric acid may also in most instances be found. Still it will be observed that most of these minerals would yield a soil containing only two or three of those substances, which, as we have already learned, are essential to the plant.

Thus, potash felspar, while it would give abundance of potash, would be but an inefficient source of lime and magnesia; and labradorite, which contains abundance of lime, is altogether deficient in magnesia and potash. Nature has, however, provided against this difficulty, for she has so arranged it that these minerals rarely occur alone, the rocks which form our great mountain masses being composed of intimate mixtures of two or more of them, and that in such a manner that the deficiencies of the one compensate those of the other.

We shall shortly mention the composition of these rocks. Granite is a mixture of quartz, felspar, and mica in variable proportions, and the quality of the soil it yields depends on whether the variety of felspar present be orthoclase or albite.

When the former is the constituent, granite yields soils of tolerable fertility, provided their climatic conditions be favourable; but it frequently occurs in high and exposed situations which are unfavourable to the growth of plants. Gneiss is a similar mixture, but characterised by the predominance of mica, and by its banded structure.

Owing to the small quantity of felspar which it contains, and the abundance of the difficulty decomposable mica, the soils formed by its disintegration are generally inferior. Mica slate is also a mixture of quartz, felspar, and mica, but consisting almost entirely of the latter ingredient, and consequently presenting an extreme infertility.

The position of the granite, gneiss, and mica slate soils in this country is such that very few of them are of much value; but in warm climates they not unfrequently produce abundant crops of grain. Syenite is a rock similar in composition to granite, but having the mica replaced by hornblende, which by its decomposition yields supplies of lime and magnesia more readily than they can be obtained from the less easily disintegrated mica.

For this reason soils produced from the syenitic rocks are frequently possessed of considerable fertility.

The series of rocks of which greenstone and trap are types, and which are very widely distributed, differ greatly in composition from those already mentioned. They are divisible into two great classes, which have received the names of diorite and dolerite, the former a mixture of albite and hornblende, the latter of augite and labradorite, sometimes with considerable quantities of a sort of oligoclase containing both soda and lime, and of different kinds of zeolitic minerals.

Generally speaking, the soils produced from diorite are superior to those from dolerite. The albite which the former contains undergoes a rapid decomposition, and yields abundance of soda along with some potash, which is seldom altogether wanting, while the hornblende supplies both lime and magnesia.

Dolerite, when composed entirely of augite and labradorite, produces rather inferior soils; but when it contains oligoclase and zeolites, and comes under the head of basalt, its disintegration is the source of soils remarkable for their fertility; for these latter substances undergoing rapid decomposition furnish the plants with abundant supplies of alkalies and lime, while the more slowly decomposing hornblende affords the necessary quantity of magnesia.

In addition to these, the basaltic rocks are found to contain appreciable quantities of phosphoric acid, so that they are in a condition to yield to the plant almost all its necessary constituents. The different rocks now mentioned, with a few others of less general distribution, constitute the whole of our great mountain masses; and while their general composition is such as has been stated, they frequently contain disseminated through them quantities of other minerals which, though in trifling quantity, nevertheless add their quota of valuable constituents to the soils.

Moreover, the exact composition of the minerals of which the great masses of rocks are composed is liable to some variety. Those which we have taken as illustrations have been selected as typical of the minerals; but it is not uncommon to find albite containing 2 or 3 per cent of potash, labradorite with a considerable proportion of soda, and zeolitic minerals containing several per cent of potash, the presence of which must of course considerably modify the properties of the soils produced from them.

They are also greatly affected by the mechanical influences to which the rocks are exposed; and being situated for the most part in elevated positions, they are no sooner disintegrated than they are washed down by the rains. A granite, for instance, as the result of disintegration, has its felspar reduced to an impalpable powder, while its quartz and mica remain, the former entirely, the latter in great part, in the crystalline grains which existed originally in the granite.

If such a disintegrated granite remains on the spot, it is easy to see what its composition must be; but if exposed to the action of running water, by which it is washed away from its original site, a process of separation takes place, the heavy grains of quartz are first deposited, then the lighter mica, and lastly the felspar.

Thus there may be produced from the same granite, soils of very different nature and composition, from a pure and barren sand to a rich clay formed entirely of felspathic debris. The sedimentary or stratified rocks are formed of particles carried down by water and deposited at the bottom of the primeval seas from which they have been upheaved in the course of geological changes.

The process of their formation may be watched at the present day at the mouths of all great rivers, where a delta composed of the suspended matters carried down by the waters is slowly formed. The nature of these rocks must therefore depend entirely on that of the country through which the river flows.

If its course runs through a country in which lime is abundant, calcareous rocks will be deposited, and if it passes through districts of different geological characters the deposit must necessarily consist of a mixture of the disintegrated particles of the different rocks the river has encountered. For this reason it is impossible to enter upon a detailed account of their composition.

It is to be observed, however, that the particles of which they are composed, though originally derived from the crystalline rocks, have generally undergone a complex series of changes, geology teaching that, after deposition, they may in their turn undergo disintegration and be carried away by water, to be again deposited.

Their composition must therefore vary not merely according to the nature of the rock from which they have been formed, but also according to the extent to which the decomposition has gone, and the successive changes to which they have been exposed. They may be reduced to the three great classes of clays, including the different kinds of clay slates, shales, etc., sandstone and limestone.

It must be added also, that many of them contain carbonaceous matters produced by the decomposition of early races of plants and animals, and that mixtures of two or more of the different classes are frequent. The purest clays are produced by the decomposition of felspar, but almost all the crystalline rocks may produce them by the removal of their alkalies, iron, lime, etc.

Where circumstances have been favourable, the whole of these substances are removed, and the clay which remains consists almost entirely of silica and alumina, and yields a soil which is almost barren, not merely on account of the deficiency of many of the necessary elements of plants, but because it is so stiff and impenetrable that the roots find their way into it with difficulty.

The sandstones are derived from the siliceous particles of granite and other rocks, and consist in many cases of nearly pure silica, in which case their disintegration produces a barren sand, but they more frequently contain an admixture of clay and micaceous scales, which sometimes form a by no means inconsiderable portion of them.

Such sandstones yield soils of better quality, but they are always light and poor. Where they occur interstratified with clays, still better soils are produced, the mutual admixture of the disintegrated rocks affording a substance of intermediate properties, in which the heaviness of the clay is tempered by the lightness of the sandstone. Limestone is one of the most widely distributed of the stratified rocks, and in different localities occurs of very different composition.

Limestones are divided into two classes, common and magnesian; the former a nearly pure carbonate of lime, the latter a mixture of that substance with carbonate of magnesia. But while these are the principal constituents, it is not uncommon to find small quantities of phosphate and sulphate of lime, which, however trifling their proportions, are not unimportant in an agricultural point of view.

These limestones are hard and possess to a greater or less extent a crystalline texture. They are replaced in later geological periods by others which are much softer, and often purer, of which the oolitic limestones, so called from their resemblance to the roe of a fish, and chalk are the most important. Other limestones are also known which contain an admixture of clay.

The soils produced by the disintegration of limestone and chalk are generally light and porous, but when mixed with clay, possess a very high degree of fertility, and this is particularly the case with chalk, which yields some of the most valuable of all soils. But it is true only of the common limestones, for experience has shown that those which contain magnesia in large quantity are often prejudicial to vegetation, and sometimes yield barren or inferior soils.

Such are the general characters of the three great classes of stratified rocks; any attempt to particularise the numerous varieties of each would lead us far beyond the limits of the present work. It is necessary, however, to remark, that in many instances one variety passes into the other, or, more correctly speaking, sedimentary rocks occur, which are mixtures of two or more of the three great classes.

In fact, the name given to each really expresses only the preponderating ingredient, and many sandstones contain much clay, shales and clay slates abound in lime, and limestones in sand or clay, so that it may sometimes be a matter of some difficulty to decide to which class they belong. Such mixtures usually produce better soils than either of their constituents separately, and accordingly, in those geological formations in which they occur, the soils are generally of excellent quality.

The same effect is produced where numerous thin beds of members of the different classes are interstratified, the disintegrated portions being gradually intermixed, and valuable soils formed. The fertility of the soils formed from the stratified rocks is also increased by the presence of organic remains which afford a supply of phosphoric acid, and which are sometimes so abundant as to form a by no means unimportant part of their mass.

They do not occur in the oldest sedimentary rocks, but as we ascend to the more recent geological epochs, they increase in abundance, until, in the greensands and other recent formations, whole beds of coprolites and other organic remains are met with. Great differences are observed in the quality of the soils yielded by different rocks.

In general, those formed by the disintegration of clay slates are cold, heavy, and very difficult and expensive to work; those of sandstone light and poor, and of limestone often poor and thin. These statements must, however, be considered as very general; for individual cases occur in which some of these substances may produce good soils, remarkable exceptions being offered by the lower chalk and some of the shales of the coal formation.

Little is at present known regarding the peculiar nature of many of these rocks, or their composition; and the cause of the differences in the fertility of the soil produced from them is a subject worthy of minute investigation.

Chemical Composition of the Soil

Reference has been already made to the division of the constituents of the soil into the two great classes of organic and inorganic. And when treating of the sources of the organic constituents of plants, we entered with some degree of minuteness into the composition and relations of the different members of the former class, and expressed the opinion that they did not admit of being directly absorbed by the plant.

But though the parts then stated lead to the inference that, as a direct source of these substances, humus is unimportant, it has other functions to perform which render it an essential constituent of all fertile soils. These functions are dependent partly on the power which it has of absorbing and entering into chemical composition with ammonia, and with certain of the soluble inorganic substances, and partly on the effect which the carbonic acid produced by its decomposition exerts on the mineral matters of the soil.

In the former way, its effects are strikingly seen in the manner in which ammonia is absorbed by peat; for it suffices merely to pour upon some dried peat a small quantity of a dilute solution of ammonia to find its smell immediately disappear. This peculiar absorptive power extends also to the fixed alkalies, potash and soda, as well as to lime and magnesia, and has an important effect in preventing these substances being washed out of the soil—a property which, as we shall afterwards see, is possessed also by the clay contained in greater or less quantity in most soils.

On the other hand, the air and moisture which penetrate the soil cause its decomposition, and the carbonic acid so produced attacks the undecomposed minerals existing in it, and liberate the valuable substances they contain.

In considering the composition of a soil, it is important to bear in mind that it is a substance of great complexity, not merely because it contains a large number of chemical elements, but also because it is made up of a mixture of several minerals in a more or less decomposed state. The most cursory examination shows that it almost invariably contains sand and scales of mica, and other substances can often be detected in it.

Now it has been already observed that the minerals of which soils are composed, differ to a remarkable extent in the facility with which they undergo decomposition, and the bearing of this fact on its fertility is a matter of the highest importance, for it has been found that the mere presence of an abundant supply of all the essential constituents of plants is not always sufficient to constitute a fertile soil. Two soils, for instance, may be found on analysis to have exactly the same composition, although in practice one proves barren and the other fertile.

The cause of this difference lies in the particular state of combination in which the elements are contained in them, and unless this be such that the plant is capable of absorbing them, it is immaterial in what quantity they are present, for they are thus locked up from use, and condemn the soil to hopeless infertility.

It is admitted that unless the substances be present in a state in which they can be dissolved, the plant is incapable of absorbing them; but it is a matter of doubt whether it is necessary that they be actually dissolved in the water which permeates the soil, or whether the plant is capable of exercising a directly solvent action.

The latter view is the most probable, but at the same time it cannot be doubted, that if they are presented to the plant in solution, they will be absorbed in that state in preference to any other. Hence it has been considered important in the analysis of a soil, not to rest content with the determination of the quantity of each element it contains, but to obtain some indication of the state of combination in which it exists, so as to have some idea of the ease or difficulty with which they may be absorbed.

For this purpose it is necessary to determine :

- *1st,* The substances soluble in water;
- *2d,* Those insoluble in water, but soluble in acids;
- *3d,* Those insoluble both in water and acids; and if to these the organic constituents be added, there are four separate heads under which the components of a soil ought to be classified.

This classification is accordingly adopted in the most careful and minute analyses; but the difficulty and labour attending them has hitherto precluded the possibility of making them except in a few instances; and, generally speaking, chemists have been contented with treating the soil with an acid, and determining in the solution all that is dissolved.

Such analyses are often useful for practical purposes, as for example, when they show the absence of lime, or any other individual substance, by the addition of which we may rectify the deficiency of the soil; but they are of comparatively little scientific value, and throw but little light on the true constitution of the soil, and the sources of its fertility.

Nor is it likely that much satisfactory information will be obtained until the number of minute analyses is so far extended as to establish the fundamental principles on which the various properties of the soil depends.

The separation of the constituents of a soil into the four great groups already mentioned, is effected in the following manner : — A given quantity of the soil is boiled with three or four successive quantities of water, which dissolves out all the soluble matters. These generally amount to about one-half per cent of the whole soil, and consist of nearly equal proportions of organic and inorganic substances.

In very light and sandy soils, it occasionally happens that not more than one or two-tenths per cent dissolve in water, and in peaty soils, on the other hand, the proportion is sometimes considerably increased, principally owing to the abundance of soluble organic matters. When the residue of this operation is treated with dilute hydrochloric acid, the matters soluble in acids are obtained in the fluid.

The proportion of these substances is liable to very great variations, and in some soils of excellent quality, and well adapted to the growth of wheat, it does not exceed 3 per cent; while in calcareous soils, such as those of the chalk formation, it may reach as much as 50 or 60 per cent. In general, however, it amounts to about 10 per cent.

The organic constituents are also very variable in amount; ordinary soils of good quality containing from 2 to 10 per cent, while in peat soils they not unfrequently reach 30 or even 50 per cent. But these cannot be considered *fertile* soils. The insoluble constituents are likewise subject to great variations, but, in the ordinary clay and sandy soils of this country, they generally form from 70 to 85 per cent of the whole.

The distribution of the constituents under these different heads will be best illustrated by a few analyses of soils of good quality, and for this purpose we shall select two, noted for the excellent crops of wheat they produce, and for their general fertility. The analyses were made from the upper 10 inches, and a quantity of the 10 inches immediately subjacent was analysed as sub-soil.

The first is the ordinary wheat soil of the county of Mid-Lothian, the other the alluvial soil of the Carse of Gowrie in Perthshire, so celebrated for the abundance and luxuriance of the crops it produces.

In examining these analyses, it is particularly worthy of notice that by far the larger proportion of the substances soluble in water consists of organic matter, lime, and sulphuric acid, the two last being in combination as sulphate of lime, while some of those substances which are usually considered to be the most important mineral constituents of plants are present in very small quantity — potash, for instance, forming not more than 1-25,000th of the whole soil, and phosphoric acid being entirely absent.

On the other hand, this portion contains the whole of the chlorine which exists in the soil, and this might be anticipated from the ready solubility in water of the compounds of that substance. The portion soluble in acids consists of alumina and oxide of iron, both of which are comparatively unimportant to the plant, but very important, as we shall afterwards see, in relation to the physical properties of the soil.

The remainder of the substances soluble in acids, amounting to from 1 and 2 per cent, is composed of some of the most essential constituents of plants. Lime, magnesia, potash, and soda, appear again in larger quantity than in the soluble part, and along with them we have the phosphoric acid to the amount of from 0·2 to 0·4 per cent of the whole soil, and sulphuric acid in much smaller quantity.

The insoluble matters differ remarkably in the two soils, that from the Carse of Gowrie being characterised by a large quantity of potash and soda, indicating an important difference in the materials from which they have been formed. In the Perthshire soil it is obvious that the felspathic element has been abundant, and that its decomposition has been arrested at a time, when it still contained a large quantity of alkalies.

And this difference is of great practical importance, because those soils, which contain a large quantity of potash in their insoluble portion, have within them a source of permanent fertility, the alkali being gradually liberated by the decomposition which is constantly in progress, owing to the air and moisture permeating the soil.

As regards the special distribution of the inorganic matters, it is to be observed that some of them occur in each of the three heads under which they are arranged, while others are confined to one or two. Silica and the alkalies occur generally, though not invariably, in all three. Chlorine is met with only in the part soluble in water, phosphoric acid only in that soluble in acids, while sulphuric acid occurs in both the last-named divisions. The greater part of the organic matters are insoluble both in water and acids. At least it is generally believed that any portion dissolved by strong acids, in the course of analysis, has been entirely decomposed, and is in a completely different state from that in which it existed actually in the soil.

As an example of a calcareous soil, forming a striking contrast to those given above, we select one from the island of Antigua, from which very large crops of sugar-cane are obtained. The soil is of great depth, and analyses of the sub-soil at the depth of 18 inches and 5 feet are given. These last analyses are not so minute as that of the soil itself, the soluble matters not having been separately determined, but included in that soluble in acids.

In this soil there is a general resemblance in the composition of the portion soluble in water to those of the wheat soils. But the part soluble in acids is distinguished by the great abundance of carbonate of lime. The sub-soil contains also a large quantity of protoxide of iron, a substance frequently found in sub-soils containing much organic matter, and to which the air has imperfect access.

Under these circumstances peroxide of iron is reduced to protoxide; and when present abundantly in the soil in that form, iron has been found to exercise a very injurious influence on vegetation; and it has frequently happened that when sub-soils containing it have been brought up to the surface, they have in the first instance caused a manifest deterioration of the soil, although after some time, when it had become peroxidised by the action of the air, it ceased to be injurious.

It is unnecessary to multiply analyses of fertile soils, those now given being sufficient to show their general composition. They are all characterised by the presence, in considerable quantity, of all the essential constituents of plants, in a state in which they may be readily absorbed.

The absence of one or more of these substances immediately diminishes or altogether destroys the fertility of the soil; and the extent to which this occurs is illustrated by the following analysis of a soil from Pumpherston, Mid-Lothian, forming a small patch in the lower part of a field, and on which nothing would grow. Being naturally wet, it had been drained and sowed with oats, which died out about six weeks after sowing, and left a bare soil on which weeds did not show the slightest disposition to grow.

Soluble in Acids

In this instance the barrenness of the soil is distinctly traceable to the deficiency of phosphoric acid, sulphuric acid, and chlorine. There is also a remarkably large quantity of oxide of iron, which, when acted on by the humic acid, is well known to be highly prejudicial to vegetation, and that this took place was shown by the fact that the drains, a couple of months after being laid, were almost stopped up by humate of iron.

They must, however, be considered as in a great measure exceptional cases, as it is but rarely that so large a number of constituents is absent, and it is much more frequent to find the deficiency restricted to one or two substances. They are illustrations of barrenness dependent on different circumstances. The first shows the unimportance of the organic matters of the soil, which are here unusually abundant, without in any way counteracting the infertility dependent on the absence of the other constituents.

The second is that of a nearly pure sand; and the third, though it contains a greater number of the essential ingredients of the ash, is still rendered unfruitful by the deficiency of alkalies, sulphuric acid, and chlorine. An examination of the foregoing analyses indicates pretty clearly some of the conditions of fertility of the soil, which must obviously contain all the constituents of the plants destined to grow upon it. But it by no means exhausts the subject, for numerous instances are known of soils containing all the essential elements of plants in abundance, but on which they nevertheless refuse to grow.

In these instances the defect is due either to the presence of some substance injurious to the plant, or to the state of combination of those it requires being such as to prevent their absorption. Reference has been already made to the bad effects of protoxide of iron, and it would appear that organic matter is sometimes injurious.

Even water, by excluding air, and so preventing those decompositions which play so important a part in liberating the essential elements from their more permanent compounds, although it cannot render a soil absolutely barren, not unfrequently materially diminishes its fertility.

The state of combination of the soil constituents unquestionably exercise a most important influence on its fertility. That this must be the case is an inference which may be easily drawn from the statements already made regarding the different minerals from which it is directly or indirectly produced.

If, for instance, a soil consist to a large extent of mica, it would be found on analysis to contain abundance of potash and some other matters, and yet our knowledge of the difficulty with which that mineral is decomposed, would enable us to pronounce unfavourably of the soil; and practical experience here fully confirms the scientific inference. The forms of combination most favourable to fertility is a subject on which our information is at present comparatively limited. It was at one time believed that solubility in water was an indispensable requisite, but recent investigations appear to lead to a directly contrary conclusion. The analyses of soils already given, show that the part directly soluble in water embraces only a certain number of the constituents of the plant, and of those dissolved the quantity is very small.

This becomes still more apparent if we estimate from the analyses the actual quantities of those substances contained in an acre of soil. It is generally assumed that the soil on an imperial acre of land 10 inches deep weighs in round numbers about 1000 tons; and calculating from this, we find that the quantity of potash soluble in water in the Mid-Lothian wheat soil, amounts to no more than 70 lb. per acre.

But a crop of hay carries off from the soil about 38 lb. of potash, and one of turnips, including tops, not less than 200 lb., so that if only the matters soluble in water could be taken up by the plant, such soils could not possess the amount of fertility which they are actually found to have.

It is to be remembered, also, that in these analyses the experiment is made under the most favourable circumstances for ascertaining the whole quantity of matters which are capable of dissolving in water; that practically dissolved is very different. The average rainfall in Kent, where the waters he examined were obtained, is 25 inches. Now, it appears that about two-fifths of all the rain which falls escapes through the drains, and the rest is got rid of by evaporation.

An inch of rain falling on an imperial acre weighs rather more than a hundred tons; hence, in the course of a year, there must pass off by the drains about 1000 tons of drainage water, carrying with it, out of the reach of the plants, such substances as it has dissolved, and 1500 tons must remain to give to the plant all that it holds in solution. These 1500 tons of water must, if they have the same composition as that which escapes, contain only two and a half pounds of potash, and less than a pound of ammonia.

It may be alleged that the water which remains, lying longer in contact with the soil, may contain a larger quantity of matters in solution; but even admitting this to be the case, it cannot for a moment be supposed that they can ever amount to more than a very small fraction of what is required for a single crop. It may therefore be stated with certainty that solubility in water is not essential to the absorption of substances by the plant, which must possess the power of itself directly attacking, acting chemically on, and dissolving them.

The mode in which it does this is entirely unknown, but it in all probability depends on very feeble chemical actions, and hence the importance of having the soil constituents, not in solution, but in such a state that they may be readily made soluble by the plants. Many of the minerals from which fertile soils are formed are probably not attackable by plants when in their natural condition, and even after disintegration the quantity of the essential elements of their food, which are present in an easily assimilable state, is at no one time very large.

But this is of comparatively little importance, for the soil is not an inert unchangeable substance; it is the theatre of an important series of chemical changes effected by the action of air and moisture, and producing a continued liberation of its constituents. This decomposition is effected partly by the carbonic acid of the atmosphere, but to a much larger extent by its oxygen acting upon the organic matters of the soil, and causing a constant though slow evolution of that acid, which in its turn attacks the mineral matters.

From these analyses it appears that the air contained in the pores of the soil is much richer in carbonic acid than the atmosphere, the poorest soil containing about 25 times, and a recently manured soil 250 times as much. This carbonic acid, which is obviously produced by the decomposition of the vegetable matters and manure, acting partly as gas and partly dissolved in the soil water, exerts a solvent action on its constituents.

And, though a very feeble acid, its continuous action produces in the course of time a large effect; while, during the interval, the constituents of the soil are safely stored up, and liberated only as the plant requires them, by which bountiful provision of nature they are exposed to fewer risks of loss than if they had been all along in a state in which they could be absorbed.

Carbonic acid not only assists in effecting the decomposition of the minerals of the soil, but its aqueous solution acts as a solvent of many substances, which are quite insoluble in pure water.

It is in this way that much of the lime contained in natural waters is held in solution, and it has been ascertained that magnesia, iron, and even phosphate of lime, may also be dissolved by it. It is probable that when these substances are dissolved, the plants will take them from solution in place of themselves attacking the insoluble matters; but of the extent to which this may occur nothing is yet known — the action of solvents on the soil being a subject which is as yet scarcely examined.

Carbonic acid is, however, a most important agent in producing the chemical changes in the soil, and the particular value of humus lies in its affording a supply of that substance exactly when it is wanted; but the carbonic acid of the atmosphere also takes part in these changes, although with different degrees of rapidity according to the character of the soil, acting rapidly in light, and slowly in stiff, clay soils.

The solvent action of the carbonic acid is, no doubt, principally exerted on the substances soluble in acids, but not entirely, for it is known that the insoluble part is gradually being disintegrated and made soluble; and hence it is that the composition of that part of the soil which resists the action of acids, and which at first sight might appear of no moment, is really important. It is obvious that this circumstance must at once confer on the soil of the Carse of Gowrie a great superiority over those of Mid-Lothian and most other districts; for it contains in its insoluble part a quantity of alkalies which must necessarily form a source of continued fertility.

Accordingly, experience has all along shown the great superiority of that soil, and of alluvial soils generally, which are all more or less similar to it. The facility with which these matters are attackable by carbonic acid is also an important element of the fertility of a soil, and it is to the existence of compounds which are readily decomposed by it that we attribute the high fertility of the trap soils.

By a further examination of the analyses of fertile soils, it is at once apparent that the most essential constituents of plants are by no means very abundant in them. In fact, phosphoric and sulphuric acids, lime, magnesia, and the alkalies, which in most instances make up nine-tenths of the ash of plants, form but a small portion of even the most fertile soils; while silica, which, except in the grasses, occurs in small quantity, oxide of iron which is a limited, and alumina a rare, constituent of the ash, constitute by far their larger part.

Thus the total amount of potash, soda, lime, magnesia, phosphoric and sulphuric acids and chlorine, contained in the Mid-Lothian wheat soil amounts only to 3·5888 per cent, and in the Perthshire to 6·4385, the entire remainder being substances which enter into the plant for the most part in much smaller quantity.

And, as these small quantities of the more important substances are capable of supplying the wants of the plant, it must be obvious that a very small fraction of the silica, oxide of iron, and alumina, which the soils contain, would afford to it the whole quantity of these substances it requires, and that the remainder must have some other functions to perform.

The soil must be considered not merely as the source of the inorganic food of plants, for it has to act also as a support for them while growing, and to retain a sufficient quantity of moisture to support their life; and unless it possess the properties which fit it for this purpose, it may contain all the elements of the food of plants, and yet be nearly or altogether barren.

The adaptation of the soil to this function is dependent to a great extent on its mechanical texture, and on this considerable light is frequently thrown by a kind of mechanical analysis. If a soil be shaken up with water and allowed to stand for a few minutes, it rapidly deposits a quantity of grains which are at once recognised as common sand; and if the water be then poured off into another vessel and allowed to stand for a longer time, a fine soft powder, having the properties and composition of common clay, is deposited, while the clear fluid retains the soluble matters.

By a more careful treatment it is possible to distinguish and separate humus, and in soils lying on chalk or limestone, calcareous matter or carbonate of lime. In this way the components can be classified into four groups, a mixture of two or more of which in variable proportions is found in all soils. The relative proportions in which these substances exist in soils are, as we shall afterwards see, the foundation of their classification into the light, heavy, calcareous, and other sub-divisions.

But they are also intimately connected with certain chemical and mechanical peculiarities which have an important bearing on its fertility. It is a familiar fact, that particular soils are specially adapted to the growth of certain crops; and we talk of a wheat or a turnip soil as readily distinguishable. It is to be observed, however, that in many such instances the mere analysis may show no difference, or, at least, none sufficient to account for the peculiarity.

In this instance such difference as exists is rather in favour of the soil on which clover fails, but it is exceedingly trifling; and it is necessary to seek an explanation in the special properties of its mechanical constituents. These properties are partly mechanical and partly chemical, and in both ways exercise an important influence on the fertility of the soil. Sand and clay, the most important of the mechanical constituents, confer on the soil diametrically opposite properties; the former, when present in large quantity, producing what are designated as light, the latter stiff or heavy soils.

The hard indestructible siliceous grains, of which sand is composed, form a soil of an open texture, through which water readily permeates; while clay, from its fine state of division, and peculiar adhesiveness or plasticity, gives it a close-textured and retentive character, and their proper intermixture produces a light fertile loam, each tempering the peculiar properties of the other. Indeed, their mixture is manifestly essential, for sand alone contains little or none of the essential ingredients of plants; and if present in large quantity, the openness of the soil is excessive, water flows through it with rapidity, manures are rapidly wasted, and on the accession of drought, the plants growing upon it soon languish and die.

Clay, on the other hand, is by itself equally objectionable; the closeness of its texture prevents the spreading of the roots of plants, and the access of carbonic acid, which, as we have already seen, is so important an agent in the changes occurring in the soil. In fact, a pure clay, that is to say, a clay unmixed with sand, even though it may contain all the essential constituents of the plant, is for this reason unfertile.

Practically, of course, these extreme cases rarely occur; the heaviest clay soils being mixtures of true clay with sand, and the most sandy containing their proportion of clay; but frequently the preponderance of the one over the other is so great, as to produce soils greatly inferior to those in which the mixture is more uniform.

It is easy to understand how the proportions in which sand and clay are mixed must affect the suitability of soils to particular crops, and that an open soil must be favourable to the turnip, and a heavy clay, owing to the resistance it offers to the expansion of the bulbs, unfavourable. But these substances also exercise an important chemical action on the soluble constituents of the food of plants, combining with them, and converting them into an insoluble, or nearly insoluble state, so as to prevent their being washed away by the rain or other water which percolates through the soil.

It has long been known to chemists that clay has a tendency to absorb a small proportion of ammonia, and even when brought up from a great depth frequently contains that substance. It is to Mr. Thompson of Moat Hall, however, that we owe the important observation, that arable soils rapidly remove ammonia from solution, and Way, who pursued this investigation, showed that not only ammonia, but potash, and several of the other important elements of the food of plants, are thus absorbed.

The removal of these substances from solution is easily illustrated by a simple experiment. It suffices to take a tall cylindrical vessel open at both ends, and filled with the soil to be operated upon, which is retained by a piece of rag tied over its lower end. A quantity of a dilute solution of ammonia being then poured upon the surface of the soil, and allowed to percolate, the first quantity which flows away is found to have entirely lost its peculiar smell and taste; and in a similar manner the removal of potash may be illustrated.

This action is by no means confined to those substances when in the free state, but is equally marked when they are combined with acids in the form of salts, and in the latter case the absorption is attended with a true chemical decomposition, the base only being retained, and the acid escaping most commonly in combination with lime.

Thus, if sulphate of ammonia be employed, the water which flows from the soil contains sulphate of lime, and if muriate of ammonia be used, it is muriate of lime which escapes. This absorbent action is most remarkably manifested in the case of ammonia and potash, but it takes place also with magnesia and soda.

With the latter, however, it is incomplete, only a half or a fourth of the soda being removed from solution, the difference depending to some extent on the acid with which it is in combination. The extent to which absorption takes place varies also with the nature of the soil, and the state of combination of the substance used.

It appears also, as far as absorption goes, to be immaterial whether the ammonia is free or combined. But it is different with potash, which is absorbed from the nitrate to the extent of about 0.6 per cent, and from a caustic solution of potash to double that amount. The circumstances under which absorption takes place modify, in a manner which cannot well be explained, the amount absorbed by the same soil.

It is found generally to be most complete with very dilute solutions, and if a soil be agitated with a quantity of ammonia larger than it can take up, it will absorb only a certain amount of that substance, but by a further increase of the amount of ammonia a still larger quantity will be absorbed.

It is important to observe that when a salt is used, the base only is absorbed, and the acid escapes in combination with lime; even nitric acid, notwithstanding its importance as a food of plants, being in this predicament. From this it may be gathered that lime is not readily absorbed from solutions of its salts; indeed, it would appear that the only salt of that substance liable to absorption is the bicarbonate, from which it is taken to the extent of 1·4 per cent by the soil.

The absorption of lime from this salt, and that of phosphoric acid, which takes place to a considerable extent, probably occurs, however, quite independently of the clay present in the soil, and is occasioned by its *lime*, which forms an insoluble compound with phosphoric acid, and by removing half the carbonic acid of the bicarbonate of lime converts it also into an insoluble state.

In addition to these mineral substances, organic matters are also removed from solution. This is conspicuously seen in the case of putrid urine, which not only loses its ammonia, but also its smell and colour, when allowed to percolate through soil; and an equally marked result was obtained with flax water, from which the organic matter was entirely abstracted.

The cause of this absorptive power is still very imperfectly known. Mr. Way having observed that sand has no such property, while clay, even when obtained from a considerable depth, always possesses it, supposed that the absorption was entirely due to that substance. A difficulty, however, presents itself in explaining how it should happen that while a pure clay absorbs only 0·2847 of ammonia, a loamy soil, of which one-half probably is sand, should absorb a larger quantity.

The inference is, that the effect cannot be due to the clay as a whole, and Mr. Way has sought to explain it by supposing that there exist in the soil particular double silicates of alumina and lime. He has shown that felspar and the other minerals from which the soil is produced have no absorbent power, but that artificial compounds can be formed which act upon solutions of ammonia and potash in a manner very similar to the soil; but there is not the slightest evidence that these compounds exist in the soil, and in the year 1853[I] I pointed out the probability that clay is not the only agent at work, but that the organic matters take part in the process.

So powerful indeed is the affinity of these substances for ammonia, that chemists are at one as to the difficulty of obtaining humic and other similar acids pure, owing to the obstinacy with which they retain it; and there cannot be a doubt that in many soils these substances are in this point of view of much importance.

This is particularly the case in peat soils, which, though naturally barren, may be made to produce good crops by the application of sand or gravel; and as neither of these can cause any absorption of the valuable matters, we must attribute this effect to the organic matter. Referring to an earlier series of experiments made in 1850, I showed that, if a quantity of dry peat be taken and ammonia poured on it, its smell disappears; and this may be continued until upwards of 1.5 per cent of dry ammonia has been absorbed, and this quantity is *retained* by the peat.

In this case pure ammonia was used, but Way's experiments having shown that this alkali is not absorbed from its salts by organic matters, I expressed the opinion that humate of lime (which certainly exists in most soils) ought on chemical grounds to decompose the salts of ammonia and cause the retention of their base. The recent researches of Brustlein have shown that lime does cause the organic matters to absorb ammonia from its salts.

He confirms the fact that pure ammonia is absorbed by peat, and shows that decayed wood has the same effect, although both are without action on solutions of its salts. A stiff clay, on the other hand, containing organic matters and much carbonate of lime, readily absorbed ammonia, both when pure and combined; but after extracting the lime by means of a dilute acid, it lost the power of taking it from its salts, although it retained the free alkali as completely as before.

On the addition of a small quantity of lime, it again acquired the power of withdrawing ammonia from its compounds. These experiments may be explained, either on the supposition of the presence of humate of lime, or by supposing that the carbonate of lime first decomposed the salts of ammonia, and that the liberated alkali combined with the organic matter. It must be admitted, however, that it is very doubtful whether the ammonia and other substances are fixed in the soil by a true chemical combination.

They are certainly retained by a very feeble attraction, for it appears from Brustlein's experiments that ammonia may be, to a considerable extent, removed by washing with abundance of water, and that if the soil which has absorbed ammonia be allowed to become dry in the air, it loses half its ammonia, and after four times moistening and drying, three-fourths have disappeared.

These facts are certainly not incompatible with the presence of a true chemical compound, for the humate of ammonia is not absolutely insoluble, and many cases occur of actions taking place in the presence of water, which are entirely reversed when that fluid is removed; and it is quite possible that when humate of ammonia is dried in contact with carbonate of lime, it may be decomposed, and carbonate of ammonia escape.

There are other circumstances, however, which render it, on the whole, most probable that the combination is not wholly chemical, but rather of a physical character, among which may be more especially mentioned the fact, that the quantity of the substances retained by the soil is dependent on the degree of dilution of the fluid from which they are taken; and that the quantity absorbed never exceeds a very small fraction of the weight of the soil.

The practical inferences to be drawn from these facts regarding the value of soils are of the highest importance. It is obvious that two soils having exactly the same chemical composition may differ widely in absorptive power, and that which possesses it most largely must have the highest agricultural value.

The examination of different soils, in this point of view, is a subject of much importance, and deserves the best attention of both farmers and chemists, although little has as yet been done in regard to it, and the results which have been obtained are not of a very satisfactory character. Liebig states, that in his experiments, all the arable soils examined possessed the same absorptive power, whether they contained a large or a small proportion of lime or alumina.

It can scarcely be expected, however, that this should be true in all cases, and there are many facts which seem to indicate that differences must exist. It is well known that there are some soils in which the manure is very rapidly exhausted, and it is more than probable that this effect is due to deficient absorptive power, which leaves the soluble matters at the mercy of the weather, and liable at any moment to be washed out by a heavy fall of rain.

The more strictly mechanical properties of the soil, such as its relations to heat and moisture, are not less important than its chemical composition. It is known that soils differ so greatly in these respects as sometimes materially to affect their productive capacity. Thus, for instance, two soils may be identical in composition, but one may be highly hygrometric, that is, may absorb moisture readily from the air, while the other may be very deficient in that property.

Under ordinary circumstances no difference will be apparent in their produce, but in a dry season the crop upon the former may be in a flourishing condition, while that on the latter is languishing and enfeebled, merely from its inability to absorb from the air, and supply to the plant the quantity of water required for its growth.

In the same way, a soil which absorbs much heat from the sun's rays surpasses another which has not that property; and though in many cases this effect is comparatively unimportant, in others it may make the difference between successful and unsuccessful cultivation in soils which lie in an unfavourable climate or exposure.

The investigation of the physical characters of soils has attracted little attention, and we owe all our present knowledge of the subject to a very elaborate series of researches on this subject, published by Schübler, nearly thirty years ago. He determined :

- *1st*, The specific gravity of the soils;
- *2d*, The quantity of water which they are capable of imbibing;
- *3d*, The rapidity with which they give off by evaporation the water they have imbibed; that is, their tendency to become dry;
- *4th*, The extent to which they shrink in drying;
- *5th*, Their hygrometric power;
- *6th*, The extent to which they are heated by the sun's rays;

- *7th*, The rapidity with which a heated soil cools down, which indicates its power of *retaining* heat;

- *8th*, Their tenacity, or the resistance they offer to the passage of agricultural implements;

- *9th*, Their power of absorbing oxygen from the air.

Each of these experiments was performed on several different soils, and on their mechanical constituents. Schübler's experiments are undoubtedly important, and though the methods employed are some of them not altogether beyond cavil, they have apparently been performed with great care.

It is nevertheless desirable that they should be repeated, for such facts ought not to rest on the authority of one experimenter, however skilful and conscientious, nor on a single series of soils, which may not give a fair representation of their general physical properties. In fact, Schübler appears to imagine that having once determined the extent to which the sand, clay, and other mechanical constituents of the soil possess these properties, we are in a condition to predicate the effect of their mixture in variable proportions, although this is by no means probable.

In examining these properties, Schübler selected for experiment, pure siliceous sand, calcareous sand (carbonate of lime in coarse grains), finely powdered carbonate of lime, pure clay, humus, and powdered gypsum.

He used also a heavy clay consisting of 11 per cent of sand and 89 of pure clay, a somewhat stiff clay containing 24 per cent of sand and 76 of clay, a light clay with 40 per cent of sand and 60 of pure clay, a garden soil consisting of 52·4 per cent of clay, 36·5 of siliceous sand, 1·8 of calcareous sand, 2 per cent of finely divided carbonate of lime, and 7·2 of humus, and two arable soils, one from Hoffwyl, and one from a valley in the Jura, the former a somewhat stiff, the latter a light soil.

The superiority of a retentive over an open soil is sufficiently familiar in practice, and though this is no doubt partly due to the former absorbing and retaining more completely the ammonia and other valuable constituents of the manures applied to it, it is also dependent to an equal if not greater extent upon the power it possesses of retaining moisture.

This may be considered as representing the quantity of water retained by these different soils when thoroughly saturated by long continued rains. The column immediately succeeding gives the quantity of that water which escapes by evaporation from the same soil after exposure for four hours to dry air at the temperature of 66°. The fifth, sixth, seventh, and eighth columns indicate the quantity of moisture absorbed, when the soil, previously artificially dried, is exposed to moist air for different periods. These characters are dependent principally, though not entirely, on the porosity of the soil.

The last may also be in some measure due to the presence of particular salts, such as common salt, which has a great affinity for moisture, but is chiefly occasioned by their peculiar structure. It is to be remarked that clay and humus are two of the most highly hygrometric substances known, and it is peculiarly interesting to observe, that by a beneficent provision of nature, they also form a principal part of all fertile soils.

The quantity of water imbibed by the soil is important to its fertility, in so far as it prevents it becoming rapidly dry after having been moistened by the rains. It is valuable also in another point of view, because if the soil be incapable of absorbing much water, it becomes saturated by a moderate fall of rain, and when a larger quantity falls, the excess of necessity percolates through the soil, and carries off with it a certain quantity of the soluble salts. Important as this property is, however, it must not be possessed in too high a degree, but must permit the *evaporation* of the water retained with a certain degree of rapidity.

Soils which do not admit of this taking place are the cause of much inconvenience and injury in practice. By becoming thoroughly saturated with moisture during winter, they remain for a long time in a wet and unworkable condition, in consequence of which they cannot be prepared and sown until late in the season, and though chemically unexceptionable, they are always disadvantageous, and in some seasons greatly disappoint the hopes of the farmer.

The extent to which the imbibition and evaporation of water takes place is very variable, but they are obviously related to one another, the soils which absorb it least abundantly parting with it again with the greatest, facility; for it appears that siliceous sand absorbs only one-fourth of its weight of water, and again gives off in the course of four hours four-fifths of that it had taken up, while humus, which imbibes nearly twice its weight, retains nine-tenths of that quantity after four hours' exposure.

Long-continued and slow evaporation of the water absorbed by a soil is injurious in another way, for it makes the soil "cold" — a term of practical origin, but which very correctly expresses the peculiarity in question. It is due to the fact, that when water evaporates it absorbs a very large quantity of heat, which prevents the soil acquiring a sufficiently high temperature from the sun's rays.

The soils which have absorbed a large quantity of moisture shrink more or less in the process of drying, and form cracks, which often break the delicate fibres of the roots of the plants, and cause considerable injury : the extent of this shrinking is given in the fourth column. The relation of the soils to heat divides itself into two considerations : the amount of heat absorbed by the soil, and the degree in which it is retained.

The former is dependent on so many special considerations, that the results cannot be tabulated in a satisfactory manner. It is independent of the chemical nature of the soil, but varies to a great extent according to its colour, the angle of incidence of the sun's rays, and its state of moisture.

It is, however, an important character, and has been found by Girardin to exercise a considerable influence on the rapidity with which the crop ripens. He found in a particular year that, on the 25th of August, 26 varieties of potatoes were ripe on a very dark-coloured sandy vegetable mould, 20 on an ordinary sandy soil, 19 on a loamy soil, and only 16 on a nearly white calcareous soil.

The tenacity of the soil is very variable, and indicates the great differences in the amount of power which must be expended in working them. According to Schübler, a soil whose tenacity does not exceed 10, is easily tilled, but when it reaches 40 it becomes very difficult and heavy to work.

Its mechanical properties are concerned, humus is a substance of the very highest importance, for it confers on the soil, in a high degree, the power of absorbing and retaining water, diminishes its tenacity and permits its being more easily worked, adds to its hygrometric power and property of absorbing oxygen from the air, and finally, from its dark colour, causes the more rapid absorption of heat from the sun's rays.

It will be thus understood, that though it does not directly supply food to the plant, it ministers indirectly in a most important manner to its well-being, and that to so great an extent that it must be considered an indispensable constituent of a fertile soil. But it is important to observe that it must not be present in too large a quantity, for an excess does away with all the good effects of a smaller supply, and produces soils notorious for their infertility.

Such are the important physical properties of the soil, and it is greatly to be desired that they should be more extensively examined. The great labour which this involves has, however, hitherto prevented its being done, and will, in all probability, render it impossible except in a limited number of cases. Some of these characters are, however, of minor importance, and for ordinary purposes it might be sufficient to determine the specific gravity of the soil in the dry and moist state, the power of imbibing and retaining water, its hygrometric power, its tenacity, and its colour.

With these data we should be in a condition to draw probable conclusions regarding the others; for the higher the specific gravity in the dry state, the greater is the power of the soil to retain heat, and the darker its colour the more readily does it absorb it. The greater its tenacity the more difficult is it to work, and the greater difficulty will the roots of the young plant find in pushing their way through it.

The greater the power of imbibing water, the more it shrinks in drying; and the more slowly the water evaporates, the colder is the soil produced. The hygrometric power is so important a character that Davy and other chemists have even believed it possible to make it the measure of the fertility of a soil; but though this may be true within certain limits, it must not be too broadly assumed, the results of recent experiments by no means confirming the opinion in its integrity, but indicating only some relation between the two.

The Sub-soil

The term soil is strictly confined to that portion of the surface turned over by the plough working at ordinary depth; which, as a general rule, may be taken at 10 inches. The portion immediately subjacent is called the sub-soil, and it has considerable agricultural importance, and requires a short notice. In many instances, soil and sub-soil are separated by a purely imaginary line, and no striking difference can be observed either in their chemical or physical characters.

In such cases it has been the practice with some persons not to limit the term soil to the upper portion, but to apply it to the whole depth, however great it may be, which agrees in characters with the upper part, and only to call that sub-soil which manifestly differs from it. This principle is perhaps theoretically the more correct, but great practical advantages are derived from limiting the name of soil to the depth actually worked in common agricultural operations.

The sub-soil is always analogous in its general characters to a soil, but it may be either identical with that which overlies it or not. Of the former, striking illustrations are seen in the wheat sub-soils, the analyses of which have been already given. In the latter case great differences may exist, and a heavy clay is often found lying on an open and porous sand, or on peat, and *vice versa*.

Even where the characters of the sub-soil appear the same as those of the soil, appreciable chemical differences are generally observed, especially in the quantity of organic matter, which is increased in the soil by the decay of plants growing upon it and by the manure added. In general, then, all that we have said regarding the characters of soils both chemically and physically, will apply to the sub-soils, except that, owing to the difficulty with which the air reaches the latter, some minor peculiarities are observed.

The most important is the effect of the decay of vegetable matter, without access of air, which is attended by the reduction of the peroxide of iron to the state of protoxide, and not unfrequently by the production of sulphuret of iron, compounds which are extremely prejudicial to vegetation, and occasionally give rise to some difficulties when the sub-soil is brought to the surface, as we shall afterwards have to notice.

The physical characters of the sub-soil are often of much importance to the soil itself. As, for instance, where a light soil lies on a clay sub-soil, in which case its value is much higher than if it reposed on an open or sandy sub-soil. And in many similar modes an important influence is exerted; but these belong more strictly to the practical department of agriculture, and need not be mentioned here.

CLASSIFICATION OF SOILS

Numerous attempts have been made to form a classification of soils according to their characters and value, but they have not hitherto proved very successful; and the result of more recent chemical investigations has not been such as to encourage a farther attempt. We have not at present data sufficient for the purpose, nor, if we had, would it be possible to arrange any soil in its class except after an elaborate chemical examination.

The only classification at present possible must be founded on the general physical characters of the soil; and the ordinary mode followed in practice of dividing them into clays, loams, etc. etc., which we need not here particularise, fulfils all that can be done until we have more minute information regarding a large number of soils. Those of our readers who desire more full information on this point are referred to the works of Thaer, Schübler, and others, where the subject is minutely discussed.

Chapter 3

SOIL BIODIVERSITY

Soil Biodiversity is an intriguing, largely unappreciated facet of global biodiversity. There are many phyla, even "domains," within soils, which are largely unseen, making use of the uniquely diverse physicochemical complexity of soils, which is an inter-section of mineral, organic, aquatic, and aerial habitats. Organisms have evolved in soils literally since pre-Cambrian times (more than 600 million years ago). They are still largely un-described, and this is particularly true for the prokaryotes, which have awaited die development of new techniques to characterise diem. By linking several organismal groups to major processes in global biogeochemistry, it is proving possible to appreciate die wide array and diverse nature of soil organism functions in the biosphere.

SOILS AS COMPONENTS OF ECOSYSTEMS

Soil-Forming Factors

Soils are an intriguing, relatively diin (often <1 m depth) zone of physical-chemical and biological weadiering of the earth's land surface. Soils are formed by an array of factors, namely climate, organisms, parent material, the extent of slope, and aspect (relief) operating over time. These factors affect major ecosystem processes, such as primary production, decomposition, and nutrient cycling, which lead to the development of ecosystem properties unique to that soil type, as a result of its previous history. For example, a deep loess soil in Iowa, with a very fertile and deep surface or "A" horizon, containing considerable amounts of organic matter, will be very different from an "A" horizon developed in the Nebraska sandhills, with much greater porosity and lower water retention due to the nature of the sandy surface material. As noted in the soil-forming factors diagram, the array of biota — namely microbiota, vegetation, and consumers (herbivores, carnivores, detritivores) — is influenced by soil processes and in turn has an impact on the soil system.

Poly-Phasic Nature of Soils, Influence on the Biota

Soils are perhaps the ultimate in interface media, located at the inter-section of four principal entities : the atmosphere, biosphere, lithosphere, and pedosphere. Soils provide a wide range and variety of microhabitats, thus accommodating a very diverse biota.

The microbes (bacteria and fungi) are found in numerous microsites, well-aerated or not; bacteria may thus respire either aerobically or anaerobically. There is an enormous amount of surface area (hundreds of nr per gram of soil) on the soil particles, which range in size classes from clays (0.1-2 μm in diameter), to silts (2-50 µm in diameter), and sands (0.05-2 mm diameter). Numerous microbes and micro- and meso-fauna (protozoa and nematodes) exist in water films on these particles and in or on the surfaces of micro-aggregates formed from the primary particles.

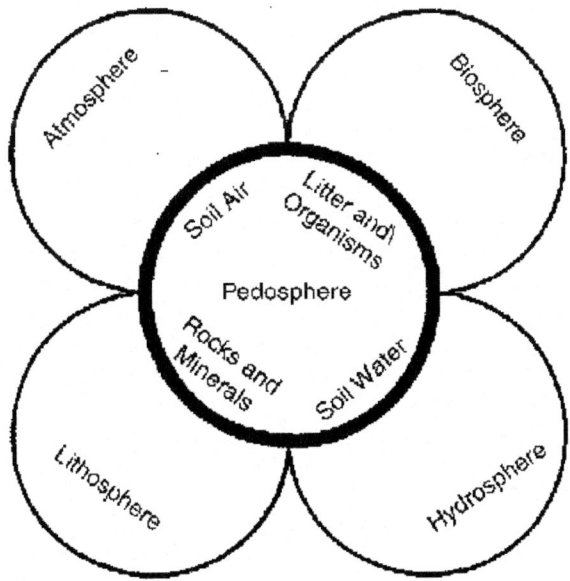

Fig. The Pedosphere Showing Interactions of Abiotic and Biotic Entities in the Soil Matrix.

In turn, the more mobile fauna, from collembola and mites (larger meso-fauna) to the macrofauna (earthworms, millipedes, ants, termites, and fossorial or earth-dwelling vertebrates), move through macro- and micro-pores in the soil. The macrofauna play a role in moving parts of the soil profile around and form many sorts of burrows and pores; they are often termed "ecological engineers."

SOILS AS ORGANISING CENTERS IN ECOSYSTEMS

Soils may be viewed as the organising centers for terrestrial ecosystems. Major functions such as ecosystem production, respiration, and nutrient recycling are controlled by the rates at which nutrients are released by decomposition in the soil and litter horizons and transported to the photo-synthetic layers of the ecosystem. This is particularly true for less heavily managed, near natural ecosystems, many of which occur on soils of relatively poor nutrient status. In these systems, mycorrhizas are often obligate partners in the obtaining of adequate nutrients for the growing plants. Mycorrhizas are known to be efficient at extracting nutrients from both mineral and organic sources, enabling plants to thrive in habitats that are considered poor in nutrients. We need to be aware of these and other mutualistic associations between micro-organisms and roots, such as rhizobia and actinorrhizas, various root-associated symbiotic bacteria that facilitate the nitrogen fixation process on which the entire ecosystem often depends. These associations have arisen in soils over evolutionary time and are key to an understanding of ecosystem function.

MAJOR SOIL PROCESSES

Decomposition : Immobilisation and Mineralisation

A very large proportion (greater than 90 per cent) of the terrestrial net primary production is returned to the soil as dead organic litter. This litter, consisting of leaves, roots, and wood from trees and organic residues from agricultural fields, is decomposed on or in the soil, and the nutrients contained within it recycled for further use. The decomposition process drives complex food webs in the soil, with numerous interactions between the initial agents of decomposition, the bacteria and fungi, and the fauna which hi turn feed on them. Decomposition is the catabolism of organic compounds in plant litter and other organic detritus. Decomposition is principally the result of microbial activities; few soil animals have cellulases in their guts, which allows them to hydrolyse the celluloses in plant residues. The decomposition of organic residues involves the activities of a variety of soil biota, including both microbes and fauna, which interact conjointly in the process. For example, the initial breaking up of plant litter usually is conducted by the chewing and macerating action of both large and small animals. This initial breaking into smaller pieces, or "comminution," is a process that benefits the fauna, which derive nutritional benefit from the litter or microbes initially colonising the plant material. The increased surface area and further inoculation of the smaller pieces enhances the microbial access to, and breakdown of, these tissues.

Nitrogen Cycle : Major Processes

Nitrogen enters the ecosystem via nitrogen fixation, in which the dinitrogen molecule (N_2) is separated into two nitrogen atoms, with considerable expenditure of energy and the assistance of the nitrogenase enzyme, to break the triple covalent bond. The atoms are ammonified and then used in the production of amino acids and proteins in the plant. Another avenue for nitrogen entry into soils is by lightning fixation, in which the extensive high-voltage energy in the lightning charge ruptures the dinitrogen molecule, hydrogens are attached, and then the ammonium is brought in by rainfall. As shown in figure, nitrogen is lost from the system via harvest and erosion of organic forms of N, it can be ammonified in decomposition, and then undergoes nitrification to nitrate (NO^{3-}), whereupon it can be taken up by biota, eidier plant roots or into microbial tissues. If there is adequate energy and low amounts of oxygen present, there can be denitrification, in which the nitrogen is lost as either nitrogen gas (N_2) or N_2O, nitrous oxide. For further details, consult text-books on ecology or ecosystem studies. The nitrogen cycle is of critical importance to biodiversity considerations, because key points in the cycle are dependent on relatively species-poor assemblages of microbes, including die nitrogen fixation and nitrification steps. There are only a few species of nitrogen fixing rhizobia, in the genera *Rhizobium* and *Bradyrhizobium*. The other principal nitrogen-fixing symbiont, the bacterium *Frankia* (Actinomycetaceae) forming the actinorrhiza (literally actinomycete-root), contains only a few species in the genus. However, approximately 194 plant species in eight families and four different subclasses of flowering plants have been identified as hosts. These plants share die general tendency to grow in marginal soils and play an important role as pioneer species in early successional habitats. In the nitrification steps, noted earlier, there are only a few genera and species of nitrifiers. Most of them are autotrophic and quite sensitive to changes in soil pH. This means that these organisms may be unusually prone to being diminished or eliminated in regions where there is considerable acid rain.

BIODIVERSITY IN SOILS

Evolutionary History

Soils, as we know them, with well-differentiated profiles, probably developed concurrently with the origin of a land flora in die early Devonian era, about 425 million years ago. The micro-organisms that inhabit die soils, particularly die prokaryotic microbes such as die cyanobacteria, originated perhaps 3 billion years ago.

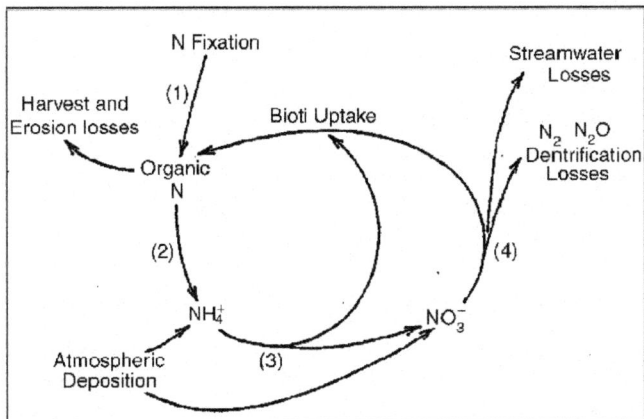

Fig. Inputs and outputs of N, which make up the IUntersystem Transfers in an Ecosystem. Numbers Indicate groups of Organisms active at a given stage of the N cycle.

1. Rhizobia, *Frankia,*

2. Ammonifiers,

3. Nitrifying bacteria,

4. Denitrifying bacteria.

Denitrifying bacteria must compete with biotic uptake of NO_3 by other microbes and plants; thus the rate of nitrification sets an upper limit on denitrification losses. Stream water losses represent the excess of available N over that taken up by biotic processes.

Diversity of Biota

Biodiversity is an inclusive concept, including a wide range of functional attributes in ecosystems in addition to being concerned with numbers of species present in the system. This differentiates it from the concept of species diversity, which is concerned with the identity and distribution of species in a given habitat or region. Soil biodiversity is best considered by focusing on the groups of soil organisms that play key roles in ecosystem functioning. Spheres of influence (SOI) of soil biota are recognised, such as the root biota, the shredders of organic matter, and the soil bioturbators. These organisms influence or control ecosystem processes and have further influence via their interactions with key soil biota (*e.g.,* plants). What is the extent of redundancy within functional groups within these SOI? Some soil organisms, such as the fungus and litter-consuming micro-arthropods, are very speciose. For example, there are up to 170 species in one Order of mites, Oribatida (members of the Arachnida, eight-legged arthropods), in the forest floor of one watershed in western North Carolina. The soil biota considered at present to be most at risk are some of the species poor functional groups, such as specialised bacteria, that is, nitrifiers and nitrogen fixers. Others include fungi forming mycorrhiza (literally fungus-root), a symbiotic association that benefits both plant and fungus, with the plant supplying high-quality carbon to the fungus, and the fungal hyphae exploring a greater volume of the soil, obtaining scarce mineral nutrients, particularly phosphorus. Other species-poor functional groups include macrofaunal shredders of organic matter (*e.g.,* millipedes) and bioturbators of soils, which includes various types of earthworms and termites.

Three Great "Domains" of Organisms on Earth

All of life exists in three great "urkingdoms," or domains. These domains are:

- The *Bacteria* (eubacteria), which are the bacteria as generally considered;
- *Archaea* (archaebacteria), which include the methanogens (methane-producers), most extreme halophiles (ones living in hypersaline environments), and hyperthermophiles (ones living in volcanic hot springs, and in mid-sea ocean hot-water vents);
- *Eucarya* (eukaryotes).

The first two domains are prokaryotes, which are unicellular organisms, lacking a unit membrane-bound nucleus and other organelles, usually having their DNA in a single circular molecule. Eukaryotes, in comparison, consist of all of the organisms that have a unit membrane-bound nucleus and other organelles, such as mitochondria. Eukaryotic organisms are often multicellular. This scheme is based on an increasing body of evidence from ribosomal RNA (rRNA) phylogenies, that the archaebacteria are worthy of the same taxonomic status as eukaryotes and bacteria. As shown in Figure, the universal rRNA tree develops from a postulated "cenancestor," leading to the relative positions of the three great domains.

Number of Species of Prokaryotes

Recent estimates of the number of prokaryotic species range from 100,000 to 10 million. Interestingly, the number of described species of bacteria in soil amount only to about 4000. This discrepancy is due largely to the fact that only a small proportion, usually less than 1 per cent, of the bacteria present in soil or any other medium are amenable to culturing and subsequent microscopic observation. It should be noted that, on the basis of the accepted criterion for separating taxa in microbial studies, which is a greater than 70 per cent DNA homology, a mouse and a human would be considered as being in the same species. This leads to complications, as we shall see, in discussing the total amount of genetic diversity of all organisms, including the as-yet largely unknown diversity of Archaea and Eubacteria. The latter now are estimated to have an array of 36 kingdoms, which are genetically as diverse as the Kingdoms Animalia, Plantae, and Fungi in older classification systems.

Biomass and Numbers of Bacterial Species on Earth

This figure is vastly underestimated. We are just now delineating the overall genetic makeup of isolates taken from soils, which are determined by the use of molecular probes. The total numbers of bacteria on earth in all habitats is truly staggering : 4-6 × 10^{30} cells, or 350 to 550 petagrams of Carbon. One petagram is 10^{15} g, or one billion metric tonnes. The amount of the total that is calculated to exist in soils is approximately 2.6 × 10^{29} cells, or about 5 per cent of the total on earth. A majority of bacteria exist in oceanic and terrestrial subsurfaces, especially in the deep mantle regions, extending several kilometers below the earth's surface.

Viruses as Quasi-Organisms

Viruses are quasi-organisms, not included in the three domains. Viruses are RNA or DNA molecules contained within protein envelopes. Viral particles are metabolically inert, carrying out neither biosynthetic nor respiratory functions. They multiply only within host cells, by inducing a living host cell to produce the necessary viral components. Once assembled, the replicated viruses escape from the cells. Viruses infect all sorts of animals, plants, and microbes. Viruses parasitising bacterial cells are commonly called bacteriophages, or simply phages.

Although little is known about the ecology of viruses, they can persist in soils for many years and decades. Some research on viruses in deserts showed that they were inactivated in soils at acid pH levels between 4.5 and 6. There is little information on the overall species diversity of viruses in soils. Current estimates are 5000 species known and perhaps 130,000 in existence.

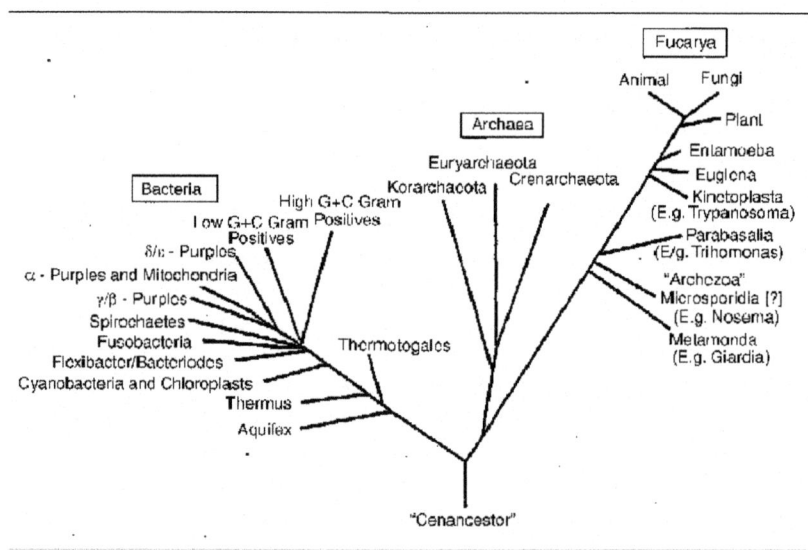

Fig. Schematic Drawing of a Universal rRNA tree Showing the Relative Positions of Evolutionary Pivotal groups in the Domains Bacteria, *Archaeo,* and *Eucarya.*

The location of the root (the cenancestor) corresponds to that proposed by reciprocally rooted gene phytogenies. The question mark beside the Archezoa group Microsporidia denotes recent suggestions that it might branch higher in the eukaryotic portion of the tree.

Numbers and Biodiversity of Eukaryotes

Fungal Diversity

Fungi are multicellular eukaryotes that are found in many habitats worldwide. They have long, ramifying strands (hyphae), which can grow into and explore many micro-habitats, and are used for obtaining water and nutrients. The hyphae secrete a considerable array of enzymes, such as cellulases, and even lignases in some specialised forms, decomposing substrates *in situ*, imbibing the decomposed sub-units and translocating them back through the hyphal network. Fungi are very abundant, particularly hi undisturbed forest floors in which literally thousands of kilometers of hyphal filaments will occur per gram of leaf litter. Fungi are still little-described, with possibly less than 5 per cent of them known to Science (69,000 described; perhaps 1,500,000 in existence). This is largely because of the fact that so many fungi are associated with tropical plants and animals, and these in rum have not been described.

Table. Comparison of the Numbers of Know and Estimated Total Species Globally of Sellected Groups or Organisms.

Group	Known Species	Estimated Total Species	Percentage Known
Vascular Plants	220,000	270,000	81
Bryophytes	17,000	25,000	68
Algae	40,000	60,000	67
Fungi	69, 000	1,500,000	5
Bacteria	3,000	30,000	10
Viruses	5,000	130,000	

As noted earlier, the roles of mycorrhizas in soil systems are being increasingly viewed as central to much of terrestrial ecosystem function. The total number of mycorrhizal species may be just 1000 or 2000, but they are essential to the growth and reproduction of numerous families of plants. Recent experimental studies have noted that species richness, namely with large versus small numbers of species of Arbuscular mycorrhiza, has a positive impact on plant primary production in macro-cosms of North American old fields (fields undergoing succession and not intensively managed).

MICRO-FAUNA

The unicellular eukaryotes, or Protoctista, include a wide range of organisms, which are more often called protozoans. These include the flagellates, naked amoebae, testacea, and ciliates. These organisms range in size from a few cubic micrometers in volume to larger ciliates, which may be up to 500 micrometers in length and 20 to 30 micrometers in width. Protozoa are quite numerous, reaching densities of from 100,000 to 200,000 per gram of soil. Bacteria, their principal prey, often exist in numbers up to 1 billion per gram of soil. All of these organisms are true water-film dwellers and become dormant or inactive during episodes of drying in the soil. They can exist in inactive or resting stages for literally decades at a time hi very exeric environments. About 40,000 extant protozoan species have been described, but many more undoubtedly are awaiting scientific discovery. Foissner notes that about 360 protozoan species per year are being discovered. In an extensive survey of soils from Africa, Australia, and Antarctica, in some cases nearly half of the total species described were new to science. This was particularly true in Africa, where of 507 species identified, 240 of diem, or 47 per cent , were previously un-described. Even in a more extensively investigated region, Australia, 43 per cent of the total of 361 species were new to science. In Antarctica, 95 species were described, with only 14, or 15 per cent , being unknown. Because many habitats have been uninvestigated yet, and the isolation procedures are still imperfect, from 70 to 80 per cent of all soil ciliates may yet be unknown. This high proportion may hold true for the other protozoan groups as well.

MESOFAUNA

Nematodes

Nematodes feed on a wide range of foods. A general trophic grouping is bacterial feeders, fungal feeders, plant feeders, and predators and omnivores. Anterior (stomal or mouth) structures can be used to differentiate general feeding or trophic groups. The feeding categories are a good introduction, but feeding habits of many genera are complex or poorly known. For example, some genera in immature phases will feed on bacteria and then become predators on other fauna once they have matured.

Because of the wide range of feeding types and the fact that nematodes seem to reflect ages of the systems in which they occur (*e.g.*, annual *versus* perennial crops, or old fields and pastures and more mature forests), they have been used as indicators of overall ecosystem condition. This is a growing area of research in soil ecology, and one in which the inter-section of community analysis and eco- system function could prove very fruitful. Current species described total some 5000, and upward of 20,000 may exist.

Collembola

Collembolans, or "springtails," are primitive Apterygote (wingless) insects. They are called "springtails" because many of diem have a spring-like lever, or furcula, which enables them to move many body lengths away' from predators by use of it, in a springing fashion. Collembolans are ubiquitous members of die soil fauna, often reaching abundances on 100,000 or more per square meter. They occur throughout the soil profile, where their major diet is decaying vegetation and associated microbes (usually fungi). However, like many members of die soil fauna, collembolans defy placement in exact trophic groups. Many collembolan species will eat nematodes when those are abundant. Some feed on live plants or their roots. One family (Onychiuridae) may feed in die rhizosphere and ingest mycorrhizae or even plant padiogenic fungi. Eight families of Collembolans occur in soils. Many Collembolans are opportunistic species, capable of rapid popula- tion growth under suitable conditions. Eggs are laid in groups. Collembolans become sexually mature with die fifth or sixth-instar, but they continue to molt throughout life. Although many species are bisexual, some of die common spe- cies are parthenogenic, consisting of females only. Collembolan "blooms" are a phenomenon of late winter or early spring, when some species may appear in large numbers on the surface of snow banks, on die surface ice of pond water, or on lichen-covered granite outcrops. There are some 6500 described species and possibly more than 10,000 in existence.

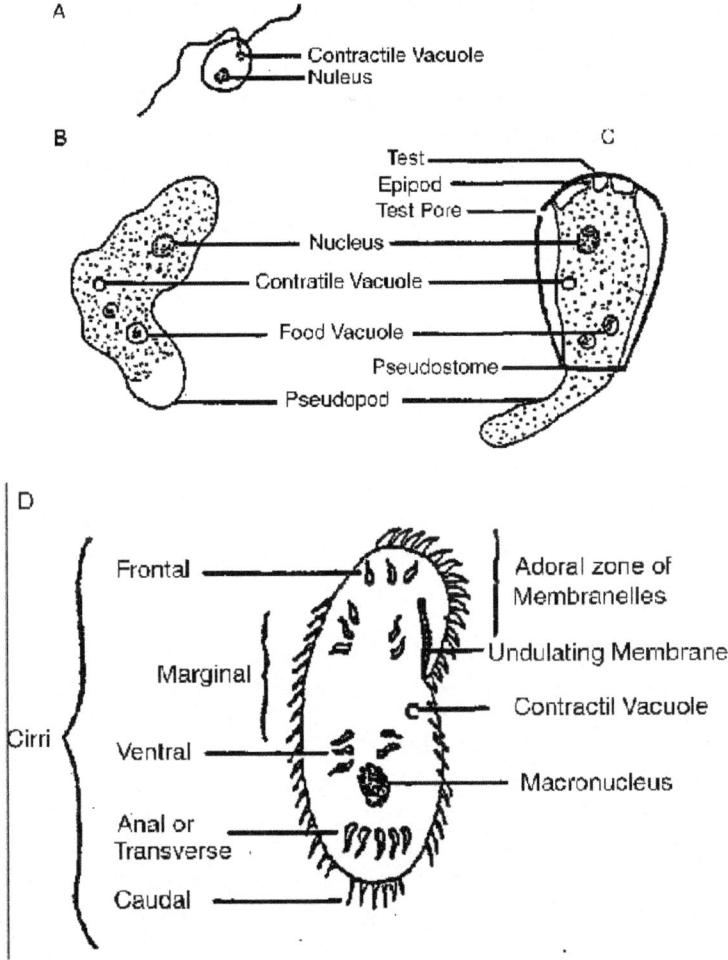

Fig. Morphology of four types of soil Protozoa : (a) flagellate *(Bodo)*, (b) naked amoeba *(Naegleria)*, (c) testacean *(Hyalosphenia)*, and (d) dilate (Oxytridia).

Mites (Acari)

The soil mites, Acari, are chelicerate arthropods related to the spiders. They are often the most abundant micro-arthropods in many types of soils. A 100-g sample may contain as many as 500 mites representing nearly 100 genera. This diverse array includes participants in three or more trophic levels, with varied strategies for feeding, reproduction, and dispersal. Four suborders of mites occur frequently in soils : the Oribatei, Prostigmata, Mesostigmata, and Astigmata. Occasionally, mites from other habitats are extracted from soil samples. These include, for example, plant mites (also called spider mites), predaceous mites, normally found on green vegetation, and parasites of vertebrates or invertebrates. The most numerous ones are the true soil mites. The oribatid mites (Oribatei) are the characteristic mites of the soil and are usually fungivorous or detritivorous. Mesostigmatid mites are nearly all predators on other small fauna, although a few species are fungivores and may become numerous at times. Astigmatid mites are associated with rich, decomposing nitrogen sources and are rare except in agricultural soils. The Prostigmata contains a broad diversity of mites with several feeding habits. Very little is known of the niches or ecological requirements of most soil mite species, but some interesting information is emerging. For further details on the life-history characteristics of these interesting animals, refer to Coleman and Crossley. About 20,000 species have been described and possibly in excess of 80,000 exist.

Macro-fauna

Termites

Termites (Isoptera) are one of the major ecosystem "engineers" particularly in tropical regions. Termites are social insects with a well-developed caste system. By their ability to digest wood, they have become economic pests of major importance in some regions of the world. Termites are arranged in five different families. The termites in a more primitive family, the Kalotermitidae, possess a gut flora of protozoans, which enables them to digest cellulose. Their normal food is wood that has come into contact with soil. Many species of termites construct runways of soil, or along root channels, and some are builders of large, spectacular mounds. Members of the phylogenetically advanced family Termitidae possess a formidable array of microbial symbionts (bacteria and fungi, but not protozoa), which enable them to process and digest the humified organic matter in tropical soils and to grow and thrive on such a diet. Although termites are mainly tropical in distribution, they occur in temperate zones and deserts as well Termites are often considered the tropical analogs of earthworms since they reach large abundances in the tropics and process large amounts of litter. Termites parallel earthworms in ingestive and soil turnover functions. The principal difference is that earthworms egest much of what they ingest in altered form (that enriches microbial action), whereas termites can transfer large amounts of soil/organic material into building nests and mounds (carbon sinks). More than 2000 species of termites have been described, and probably up to 10,000 exist.

Earthworms

The earthworm fauna of North America is surprisingly poorly known, given the importance of these animals to soil processes and soil structure. Much of the evidence for earthworm effects on soil processes comes from agro-ecosystems and involves a small group of European lumbricids (family Lumbricidae in the order Oligochaeta). In North America, south of the southern limit of the Wisconsinan glaciation, several native genera exist. However, exotic (often peregrine European lumbricids) earthworm species have been introduced into much of this area following human population changes and colonisations. Impacts of exotic earthworms on native species are not well understood, although there is evidence that when native habitat is destroyed and native earthworm species extirpated, exotic earthworms colonise the newly empty habitat. As more extensive studies are carried out, it is becoming clear that earthworms are present in a wide variety of tropical as well as temperate ecosystems. Earthworms have important roles in the fragmentation, breakdown, and incorporation of soil organic matter (SOM). This affects the distribution of SOM and also its chemical and physical characteristics. Changes in any of these soil parameters may have significant effects on other soil biota, by changing their resource base (*e.g.*, distribution and quality of SOM, microbes, or micro-arthropods) or by changing the physical structure of the soil. Recent evidence indicates that earthworm activities impact the communities of other soil biota through their effects on the chemical and physical characteristics of SOM, causing changes in oribatid species richness and micro-arthropod abundances. It is probable that earthworm-induced changes in the microbial and micro-arthropod communities will also have impacts both higher and lower in the soil food web. Some 3650 species of earthworms have been described and possibly as many as 8000 exist.

Chapter 4

SOIL-HABITAT

SOIL BIOTA AND ECOSYSTEM FUNCTIONING

Biological activity in soils is mainly concentrated in the top soils. The biological component occupy a small fraction (<0.5 per cent) of the total soil volume and make less than 10 per cent of total organic matter in soil. This living component of the soil organic matter consists of 5-15 per cent of plant roots, 85-90 per cent soil organisms. The variation in soil profile, resource availability, micro-climate, chemical and physical structure influence the soil biota size, composition and distribution. The profile of an undisturbed mature soil is divisible into a number of strata, which also includes surface organic layers (hemiedaphon) and the underlying mineral soils (enedaphon). The vegetation layer above the soil surface is termed as epigeon. The Hemiedaphon may be further divided into hydrophile, mesophile and xerophile based on the moisture content. In an attempt to better conceptualise soil systems into biologically relevant regions on the basis of their spatial and temporal heterogeneity; the activity of soil organisms in different layers of soil profile has been conceptualises as "sphere of influence". The soil profile may be divided into five major layers *viz.*,

1. *Detritosphere* : Above soil surface, comprise of litter fermentation, humification, which have considerable root mycorrhizal and saprophytic activity and grasing by the soil fauna.
2. *The drilosphere* : The soil layer influenced by the activities of earthworm and other insects;
3. *The porosphere* : Which is the region of water films occupied by bacteria, protozoa, nematodes and of channels between aggregates occupied by micro-arthropods and the aerial hyphae of fungi;
4. *The aggregatosphere* : the region where the activity of microbes and fauna is concentrated in the voids between micro-aggregates and macro-aggregates.
5. *The rhizosphere* : The zone of soil influenced by roots associated by mychorrhizal hyphae and other products.

The spores are formed and maintained by biological influence that operates at different spatial and temporal scales. Moreover, each sphere has distinct properties that regulate inter-actions among organisms and biological properties that they mediate.

Diversity

The life on planet earth has originated about 4 billion years ago, just after the earth surface became cool enough to have water in liquid state. These organisms were like today's prokaryotes – bacteria. The modern studies have shown the living organisms can be broadly classified into two radically different kinds; prokaryotes (less complex cell structure) and eukaryotes (organisms with true nucleus), which further divided into many other forms. Over the times the living things have occupied different ecological niche. These niches are determined by competition between species determined by the availability of nutrients, water availability and soil environment such as temperature, pH, and salt concentration. The micro-organisms employed different strategies to survive and prosper in different environments. That may be combative strategies (c–selected) which maximises occupation and exploitation of resources under non-stressed conditions; stress strategies (s–selected) which have involved the development of adoption which allow survival and endurance of continuous stress environment and ruderal strategies (r–selected) characterised by a short span with a high reproductive potential which often enables success in severely disturbed but favourable for a short period. These three strategies can merge to give secondary strategies (C-R, S-R, C-S and CSR) which form part of a continuum with transition zone between them. The free-living components of soil biota are bacteria, fungi, algae and the fauna. They may be grouped either on the basis of body width *viz.*, micro, meso and macro-organisms or on the basis of functional groups *viz.*, mycophagous/ herbivores, omnivores and predators, period of soil inhabitance, habitat preference or biological activity.

Table. Groups of Soil Biota Based on Body Size.

Organism	Size(mm)	Examples
Microflora	<1	Bacteria, Algae, Fungi, Actionomycetes
Microfauna	<2	Protozoa, Nematode
Mesofauna	2–10	Collembola, Acarina
Macrofauna	>10	Earthworms, Termites, Snails, Arachnids

Feeding and locomotion are the other two main activities that divide the organisms in different groups. Based on locomotion the soil animals can be distinguished as burrowing ones from the others that move on the soil surface or through the pore spaces/channels/ cavities in soils. In terms of feeding activity the soil animals can be classified in five major groups.

Table. Classification of Soil Fauna based on their Activity.

Organism	Activity
Carnivores	Predator
	Animal Parasites
Phytophagus	Above ground green Plant Material
	Root Systems
	Wood Material
Saproghagus	Caprophagus
	Xylophagus
	Necrophagus
	Detrivores
Symbionts	VAM and Endophytes
Microphytic Feeder	Fungal hyphae, spores, algae, lichens, bacteria feeder
Misellaneous Feeders	Omnivores (food varies from site to site).

The temporary and permanent members of soil fauna can be distinguished based on their period / stages of life cycle spent in soil. The permanent inhabitants of soil are termed as 'geobionts'. The temporary category belongs to the insects that enter in soil as adults and escape during unfavourable conditions or insects that undergo part of their development, as eggs or larvae, in the soil, termed as 'geophiles' *e.g.*, Coleoptera, Thysanoptera, Heteroptera, Diptera etc. The geophils are further distinguished as inactive and active geophiles. Inactive geophiles include adult insects which seeks the shelter afforded by loose or decaying leaf litter/wood and in surface soil. The temporary geophiles have little or no contri- bution in soil structure. The active geophiles pass different stages of development in soil and are closely associated with soil. The inactive geophiles are also termed as transients and the active geophiles as periodic and temporary based on their period of presence in soil.

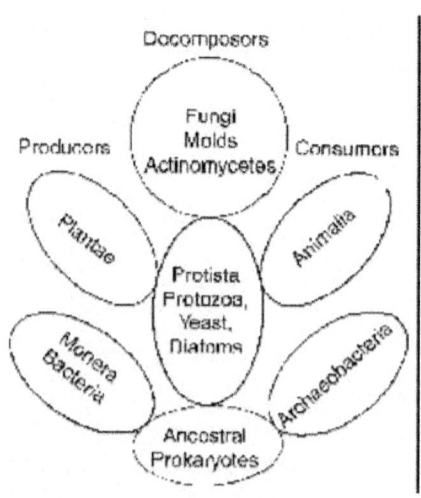

Fig. Schematic diagram showing five kingdoms of living organisms.

Clasification of Soil Biota

All living things can be classified into one of the five fundamental kingdoms of life namely, Monera, Protista, Fungi, Plantae, Animalia and are well represented in soil ecosystem.

1. *Kingdom Monera* : includes prokaryotes-single cell organisms that do not possess nucleus *e.g.* bacteria, actinomycetes and the blue green algae

2. *Kingdom Protista* : includes single cell organisms that do possess nucleus *e.g.* nucleated algae and slime moulds

3. *Kingdom Fungi* : These non- motile eukaryotes lack flagella and developed spores like yeast, moulds and mushrooms

4. *Kingdom Plantae* : these eukaryotes develops from embryos and use chlorophyll like mosses and vascular plants

5. *Kingdom Animalia* : the multicellular eukaryotes develops from a blastula (a halo ball of cells).

Ecosystem Processes

Soil organisms are an integral part of agricultural and forestry ecosystems. They play a significant role in maintaining soil health, ecosystem functions and production. Each organism has a specific role in the complex web of life in the soil. The sustained use of the agriculture and water resources is dependent upon maintaining the health of the living biota and their key role in the primary productivity of the ecosystem they inhabit.

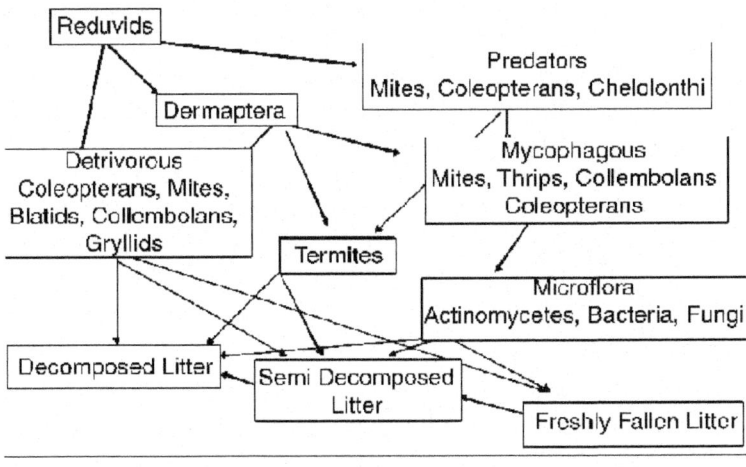

Fig. A typical food chain of forest soil ecosystem.

Their diverse role can be categories into :

(a) Facilitating nutrient acquisition by the vegetation through the mycorrhiza and N- fixing organisms,

(b) Regulating the flow of nutrients through decomposition, mineralisation and immobilisation,

(c) Mediating and breakdown of organic matter,

(d) Modification of soil structure which influence water availability to the plants,

(e) Modifying the health of the plant by parasitism and pathogenesis.

Soil micro-flora is the primary consumer in the detritus food web. They act on the organic wastes and convert them either into useful or into innocuous and less harmful substance. In nature fungi, bacteria and invertebrates interact and influence decomposition and functioning of rhizosphere. Soil fauna governs the distribution, abundance and activity of soil fungi and bacteria. In saprophytic succession, six mechanisms of interaction are important.

Two control fungal distribution and abundance :

1. Selective grasing of fungi,

2. Dispersal of fungal inoculums.

Four mechanisms stimulate microbial activity :

1. Direct supply of mineral nutrients in urine and faces,

2. Stimulation of bacterial activity by faunal activity,

3. Compensatory fungal growth due to periodic grasing,

4. Release of fungi from competitive stasis.

In rhizosphere, the mechanisms of interactions are dispersal and selective grasing of flora. Micro and meso fauna carry fungal propogules including root pathogens, to root surface. They also graze on root surface, and they selectively graze saprophytic fungi. It has been shown that dispersal of pathogens to the rhizosphere is less important than preferential grasing on pathogens. Soil organisms can also be used to reduce or eliminate environmental hazards resulting from accumulations of toxic chemicals or other hazardous wastes. This action is known as bioremediation. A number of soil organisms can be detrimental to plant growth, for example, the build up of nematodes under certain cropping systems. However, they can also protect crops from pest and disease outbreaks through biological control and reduced susceptibility.

Soil Health Indicator

Soil health defined as 'continued capacity of soil to function as a vital living system, within ecosystem and land use boundaries, to sustain biological productivity, promote the quality of air and water environments, and maintain plant. Soil health animal and human health' is a term that is used synonymously with soil quality. Soil organisms contribute to maintenance of soil health by regulating decomposition, mineralisation, formation and maintenance of soil structure. These functions and their regulatory role make them potential bio-indicators of soil health. The organisms respond to the soil management in time scale (months/

years) and are highly sensitive to any change, whereas the chemical and physical parameters of soil are largely fixed by geographic constrains and changes can be perceived or measured after the damage was already done to the system. Soil properties which have desirable properties to develop or used as bio-ndicators are listed by Doran and Safley.

Table. Soil Properties of Relevance as Indicators of Soil Quality and Soil Health.

Indicators	Relevance to	
	Soil Quality	Soil Health
Physical Indicators		
Mineral Composition	+	–
Texture	+	–
Depth	+	–
Bulk density	+	+
Water holding Capacity	+	+
Potosity	+	+
Chemical Indicators		
pH	+	+
Electrical Conductivity	+	+
Cation Exchange capacity	+	+
Organic Matter	+	+
Major Elements	+	+
Heavy Metals	+	+
Biological Indicators		
Microgical biomass	+	+
Soil Respiration	+	+
Mineralisable N	+	+
Enzyme Activity	+	+
Abundance of Microflora	+	+
Abundance of soil fauna	+	+
Root disease	+	+
Soil Bio-diversity	–	+
Food web structure	–	+
Plant growth	+	+
Plant Biodiversity	–	+

WHAT ARE SOIL BIOTA

Soil biota, the biologically active power house of soil, include an incredible diversity of organisms. Tons of soil biota, including micro-organisms (bacteria, fungi, and algae) and soil "animals" (protozoa, nematodes, mites, springtails, spiders, insects, and earthworms), can live in an acre of soil and are more diverse than the community of plants and animals above ground. Soil biota are concentrated in plant litter, the upper few inches of soil, and along roots. Soil organisms interact with one another, with plant roots, and with their environment, forming the soil food web.

WHAT DO SOIL BIOTA DO?

As soil organisms consume organic matter and each other, nutrients and energy are exchanged through the food web and are made available to plants. Each soil organism plays a role in the decomposition of plant residue, dead roots, and animal remains. The larger soil organisms, such as millipedes and earthworms, shred dead leaves and residue, mix them with the soil, and make organic material more accessible to immobile bacteria. Earthworms can completely mix the top 6 inches of a humid grassland soil in 10 to 20 years. Ants and termites mix and tunnel through soils in areas of arid and semiarid rangeland. Predators in the soil food web include scorpions, centipedes, spiders, mites, some ants, insects, and beetles. They control the population of soil biota. The smaller organisms, including mites, springtails, nematodes, and one-celled protozoa, graze on bacteria and fungi. Other organisms feed on dead roots, shredded residue, and the fecal by-products of the larger organisms. The smallest soil organisms, microscopic bacteria and fungi, make up the bulk of the biota in the soil. They finish the process of decomposition by breaking down the remaining material and storing its energy and nutrients in their cells. Algae and fungi are the first organisms to colonise rock and form "new soil" by releasing substances that disintegrate rock.

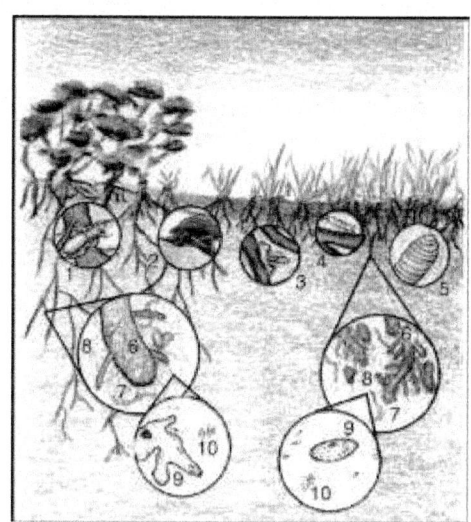

IMPORTANCE OF SOIL BIOTA

Through their inter-actions in the soil food web, the activities of soil biota link soil with the plants and animals above ground. Soil organisms perform essential functions that allow soil to resist degradation and provide benefits to all living things.

Residue Decomposition

Without the soil food web, the remains of dead plants and animals would accumulate on the earth's surface, making nutrients unavailable to plants. Soil biota decompose these organic residues and some forms of organic matter in the soil. They convert these materials into new forms of organic matter and release carbon dioxide into the air. Many of the biota can break down pesticides and pollutants.

Nutrient Storage and Release

Most of the annual nutrient needs of rangeland plants are supplied through decomposition of organic matter in the soil. As soil organisms consume organic materials, they retain (immobilise) nutrients in their cells. This process prevents the loss of nutrients, such as nitrogen, from the root zone. When fungi and bacteria die or are eaten by other organisms, nutrients are mineralised, that is, slowly released to the soil in plant-available forms. Nutrient immobilisation and mineralisation occur continuously throughout the year. Some bacteria and fungi provide nutrients to plants in exchange for carbon. Special types of bacteria, called nitrogen fixers, infect the roots of clover and other legumes, forming visible nodules. The bacteria convert nitrogen from the air in the soil into a form that the plant host can use. When the leaves and roots die and decompose, nitrogen levels increase in the surrounding soil, improving the growth of other plants. Fungi produce hyphae that frequently look like fine white entangled threads in the soil. Some fungal hyphae (mycorrhizal fungi) attach to plant roots and act like an extended root system, providing nutrients and water to the plant.

Water Storage, Infiltration, and Resistance to Erosion

Soil biota form water-stable aggregates that store water and are more resistant to water erosion and wind erosion than individual soil particles. Threads of fungal hyphae bind soil particles together. Bacteria and algae excrete material that "glues" soil into aggregates. As they tunnel through the soil, the larger soil biota form channels and large pores between aggregates, increasing the water infiltration rate and reducing the run-off rate.

WHAT AFFECTS SOIL BIOTA?

Soil biota multiply rapidly when organic material, roots, and plant litter, their food source, are available and the soil is moist and warm. Seasonal patterns of biological activity coincide with plant growth stages, litter fall, and root die-off. To be active, bacteria require films of water in soil pores, whereas fungi can function in drier conditions. When the soil is too dry, bacteria and fungi become less active or temporarily shut down, protozoa form dormant cysts, and the number of most other organisms declines. When the soil is saturated and anaerobic, the number of denitrifying bacteria increases. Organisms affect each other through predation and competition for food and space. Small soil pores can restrict the movement of large soil organisms. Different types of vegetation produce different types of litter and plant residue and thus provide different food sources for soil biota. Changes in the vegetation or the pattern of plant distribution affect the soil organisms.

Management Considerations

Grasing

Proper management of the plant community is the best strategy for maintaining the benefits of the soil food web. Plant production and the supply of organic matter can be maintained or enhanced by timely grasing, the proper frequency of grasing, and control of the amount of vegetation removed. If the plant community is overgrazed, a reduction in the amount of surface plant material and roots will result in less food for soil organisms. As biological activity decreases, a downward spiral of the important functions of soil organisms results in a lower content of organic matter and impedes nutrient cycling, water infiltration, and water storage. Heavy grasing also can reduce the abundance of nitrogen-fixing plants, causing a decrease in the supply of nitrogen for the entire plant community.

Erosion

Erosion removes or redistributes the surface layer of the soil, the layer with the greatest concentration of soil organisms, organic matter, and plant nutrients. Run-off and wind erosion redistribute litter from one area of rangeland to a surrounding area. The loss of organic matter reduces the activity of soil biota in the areas from which the litter has been removed.

Compaction by Grasing Animals and Vehicles

Soil compaction reduces the larger pores and pathways, thus reducing the amount of habitat for nematodes and the larger soil organisms. Compaction can also cause the soil to become anaerobic, increasing losses of nitrogen to the atmosphere.

Fire and Pest Control

Fire can kill some soil organisms and reduce their food source while also increasing the availability of some nutrients. Pesticides that kill above-ground insects can also kill beneficial soil insects. Herbicides and foliar insecticides applied at recommended rates have a smaller impact on soil organisms. Fungicides and fumigants have a much greater impact on the soil organisms.

Chapter 5

PRINCIPALS OF SOIL ORGANISMS

Soil is a complex mix of organic and inorganic matter that includes thousands of different species, the vast majority of which are still undescribed. Some of the organisms are pests which cause significant crop losses while others perform 'environmental services' such as biological control of pests, aeration, drainage, and nutrient and water cycling. As a dynamic living resource, soil is the basis of sustainable agriculture, as well as the physical support for most other human activities. In many areas of the world, soil fertility is declining and erosion is getting worse. Marginal lands are especially susceptible. Worldwide, an estimated 2 billion hectares of land are considered degraded, that is, less productive due to deterioration of essential soil processes. The deterioration usually results from interactions among three types of processes : physical, such as erosion, crusting, and sealing; chemical, such as nutrient depletion, acidification, and pollution; and biological, such as organic matter depletion and loss of soil flora and fauna. Asia, Africa, and Latin America together account for an estimated 75 per cent of the global area of degraded land, with 750, 490 and 240 million hectares, respectively. North America, Europe, and Australia each have an estimated 100–200 million hectares of degraded land. Loss of soil structure and fertility, together with the increasing incidence of pests, weeds, and diseases, are often responsible for the migration of small-scale farmers practicing shifting or semishifting cultivation, as in the forest and savanna areas of Africa, South America, and Asia. Attempts to reverse this global trend by means of more sustainable agricultural practices depend on a thorough understanding of soil structure and function. However, while the physical and chemical characteristics of soils have been extensively studied, the great diversity of soil organisms and the complex interactions among them remain poorly documented and understood. It is only comparatively recently that the importance of the soil biota in maintaining soil quality, plant health, and soil resilience (the ability to recover from natural or anthropogenic disturbance) has been recognised. Despite growing interest in developing more sustainable agricultural systems, soil biota management remains a largely neglected area of agricultural research-for-development.

SOIL HEALTH AND PLANT HEALTH

Soil ecosystems are among the most complex of all terrestrial communities, and the role of the soil biota in maintaining plant health is not fully understood. The composition of the soil biota is strongly influenced not only by the nature of the underlying organic matter and mineral components, but also by environmental variables such as temperature, pH, and moisture. Numbers and types of soil organisms thus vary widely both in time and space, and are greatly influenced by agricultural activities such as tillage and cropping practices. There is a strong relationship between soil fertility and plant health, in the sense of the plant's ability to resist pests and diseases. Poor land management and declining soil fertility often result in a negative feedback cycle characterised in part by an increase in soil-borne pests and diseases. Agricultural practices such as adding lime, inorganic fertilizers, and pesticides can change the physical and chemical nature of the soil environment, thereby altering the number of organisms and the ratio of different groups of organisms. Since plant health is intimately linked to soil health, managing the soil in ways that conserve and enhance the soil biota can improve crop yields and quality. A diverse soil community will not only help prevent losses due to soil-borne pests and diseases but also speed up decomposition of organic matter and toxic compounds, and improve nutrient cycling and soil structure. Colonisation of roots by mycorrhizae or other endophytic fungi, for example, can confer resistance to root-feeding insects. Conversely, a plant's ability to compensate for root damage can be compromised by nutrient deficiency if soil fertility is low.

Elusive Targets : Soil-borne Pests and Diseases

From a crop management perspective, soil pests and diseases pose special problems. They are hidden from view and hard to detect until the sudden appearance of damage creates an urgent need for rapid, curative treatments. Even then, damage due to soil pests and pathogens may be mis-diagnosed and ascribed to other causes, such as nutrient deficiencies. In general, farmers are less aware of soil-borne pests and diseases (and less able to recognise them) than those attacking the above-ground parts of the plant. In the case of some insect pests, for example, farmers may fail to make the connection between the damaging, soil-based larval stage of the insect and its above-ground adult stage during which little or no damage occurs. At the research level, the complexity and diversity of subterranean ecosystems pose unique challenges to those seeking to quantify the effects of individual taxa or species assemblages. This work is made even more difficult by the complex nature of many tropical cropping systems. Population densities of soil organisms can vary widely within a few meters, with their distribution being greatly influenced by the physical, chemical, and biological characteristics of the soil. Furthermore, different types of soil organisms require different methods for their extraction, identification, and quantification, and for the estimation of crop losses caused. The diversity of available techniques may make it difficult to compare results from different research groups.

SOIL ORGANISMS

Scientists have devised various schemes for characterising and classifying soil organisms in order to be able to cope with their great diversity. Apart from the conventional taxonomic approach, there are classification schemes based on body size, function (decomposition, etc.), and the role of organisms from an anthropocentric point of view. The last-mentioned of these systems recognises 'productive' biota (such as crop plants), 'destructive' biota (pests, pathogens, and weeds) and 'resource' biota (species that contribute to soil processes such as decomposition but that do not produce a harvestable product). Under body size, there are three main groups : the microbiota (<100 μm diameter), the mesobiota (100 μm to 2 mm diameter), and the macrobiota (>2 mm diameter). Micro-organisms are the most abundant members of the soil biota. They include species responsible for nutrient mineralisation and cycling, antagonists (biological control agents against plant pests and diseases), species that produce substances capable of modifying plant growth, and species that form mutually beneficial (symbiotic) relationships with plant roots. This last group includes mycorrhizal fungi, various actinomycetes, and some bacteria. Within the soil biota, the most important groups of both destructive and resource organisms are the bacteria, fungi, nematodes, arthropods (such as mites and insects), earthworms (mostly beneficial), and weeds.

Bacteria

The most abundant members of the vast community of soil organisms are bacteria. Per gram of soil, they can reach densities of one bllion (one thousand million) individuals — and an estimated 20 000 to 40 000 species. The magnitude of bacterial biodiversity has only recently been revealed through molecular techniques that can differentiate between hard-to-culture taxa. Bacteria play important roles in many soil processes, including the cycling of nitrogen, carbon, and phosphorus, and the degradation of pesticides and other potential pollutants. They can multiply rapidly under favourable conditions, although very high growth rates are generally limited by the availability of nutrients. Bacterial populations are typically greater and more diverse close to plant roots. It has been estimated that 5 to 10 per cent of root surfaces may be occupied by bacteria. Various endophytic bacteria (*e.g.*, species of *Azotobacter*, *Acetobacter*, and *Azospirillium*) not only fix nitrogen but also stimulate the production of root hairs. This increases the host plant's capacity to take up water and nutrients and compensate to some extent for grasing by root-feeding pests. Various bacteria that colonise seed coats and plant roots produce compounds capable of affecting plant growth. These growth-promoting rhizobacteria (including species of *Pseudomonas*, *Bacillus*, *Serratia* and *Arthrobacter*) can also increase plant resistance to pests and diseases through a variety of mechanisms, including the production of structural materials that strengthen root tissues and antibiotic metabolites that directly affect other elements of the soil biota. In some cases, these beneficial bacteria displace potential pathogens by competing for nutrients. Or, they may stimulate increased synthesis of defensive compounds by the plant itself, resulting in so-called induced resistance or systemic acquired resistance. *Rhizobium etli*, for example, induces systemic resistance in potato roots to the potato cyst nematode (*Globodera pallida*). Some rhizobacteria are able to suppress weed growth while leaving crops unaffected. The relationship between bacterial endophytes and their host plants is a subject of increasing interest to those seeking better ways to manage the soil biota. Plant roots themselves produce a number of compounds that affect microbial populations in their vicinity (the so-called rhizosphere). These compounds may act as attractants, repellents, biocides, or biostats (compounds that inhibit microbial growth), and as such offer new possibilities for the manipulation of both pathogens and beneficial species.

Fungi

As with the bacteria, the great diversity of fungi remains poorly documented. It has been estimated that only about 5 per cent of fungi have so far been described. Fungi play as important a role in soil processes as do bacteria, but tend to be more abundant in slightly acid soils. They vary widely in size, preferred habitat, and mode of life. Fungal plant pathogens (*e.g.*, some species of *Fusarium* and *Verticillium*) can cause diseases such as root rots and vascular wilts that are significant problems in many parts of the world. In the early 1990s, for example, the increasing prevalence of fungal root rots severely restricted the viability of bean crops in parts of Kenya. Many soil-borne fungal pathogens (*e.g.*, *Rhizoctonia solani* and *Pythium* spp.) are capable of infecting a range of plant genera. Furthermore, plants infected with fungal diseases may be more vulnerable to attack by soil-dwelling insects. In Malawi, for example, it was found that groundnuts were more vulnerable to attack by termites if they were infected by fungal pathogens such as *F. solani*. However, some fungi form symbiotic relationships with plant roots, enhancing the plant's ability to take up nutrients. There is some evidence that such relationships can help less competitive plants to become established in pasture systems. Many species of soil fungi are saprophytic (*i.e.*, grow on dead organic matter), while others are parasitic on animals or plants. Some are important antagonists or biological control agents of soil-borne pests or diseases (*e.g.*, species of *Dactylaria* and *Arthobotrys* for nematodes, *Beauveria* and *Metarhizium* for insect pests, and *Trichoderma* and *Coniothyrium* for plantpathogenic fungi).

Nematodes

Nematodes (roundworms or eelworms) are the most abundant micro-fauna in the soil and are particularly numerous in the top 5 cm. In the top 2 cm of soil, their numbers may exceed two billion per hectare. The number of nematode species worldwide has been estimated at 80 000 to 100 000 species. The majority of soil-dwelling nematodes feed on bacteria and fungi, but many prey on other nematodes, protozoa and rotifers, and some species are parasitic on insect pests. More than 2000 species are parasites of higher plants. These include cyst and root-knot nematodes, such as certain species of *Heterodera*, *Globodera*, *Cactodera*, and *Meloidogyne*), and 'migratory' species such as *Pratylenchus* spp., *Ditylenchus destructor*, and *Scutellonema bradys*. Their feeding lowers crop yields by disrupting water and nutrient uptake or by decreasing fruit or tuber quality or size; it can also allow fungal and bacterial pathogens to gain access to damaged roots, causing secondary infections. The presence of root-knot ematodes, for example, is known to increase the incidence of root rots and fusarium wilts in a wide range of crops. Plant parasitic nematodes tend to be more damaging pests in the tropics than in temperate zones. This is because their population growth rate is favoured by warm, humid climates and the longer growing seasons in these regions, which allows for more reproductive cycles per year. Nematode pests are associated with almost all crops grown in the tropics and cause losses of millions of dollars each year. It has been estimated that, worldwide, plant-parasitic nematodes annually reduce agricultural production by about 12 per cent. Losses are particularly severe in developing countries due to insufficient expertise for species identification, inadequate quantification of the pest problem, and a limited range of management options. Worldwide, root-knot and cyst nematodes are especially important pests and infestations by them continue to undermine national efforts to improve food security and alleviate poverty. Losses to individual crops often go unnoticed or are attributed to other causes. This is because most symptoms of nematode damage such as chlorosis (yellowing), patchy growth stunting, and wilting in hot weather, are easily confused with nutrient deficiencies or sometimes with bacterial or fungal diseases. Compared with the plant-parasitic nematodes, much less is known of the biology and ecology of other nematodes. This is particularly true of the bacteria-feeding nematodes, which are generally considered to be beneficial or harmless. These species, often concentrated in the root zone of higher plants, play an important role in soil nutrient cycling and can help distribute bacteria that promote plant growth. Although nematodes are not capable of widespread dispersal on their own, in agricultural systems they may be spread via contaminated machinery, land levelling, irrigation, and soil erosion.

Mites

Although these are among the most common soil-dwelling mesofauna, they tend to be restricted to the leaf litter and surface layers. They are more diverse than any other single group of soil arthropods (including the insects), and this diversity is reflected in their feeding habits and life history. In general, they are more important in their role as resource biota than as destructive biota. Some species feed on fungal spores, while numerous predatory species attack nematodes, other mites, insect eggs, and larvae. Other species feed on plant debris, dung, or carrion and are important members of the decomposer community. Although some mite species feed on living plants and others are parasitic on livestock, in general these potential pests are only a minor component of the mite fauna.

Insects

The number of taxa of soil-dwelling insects is relatively small compared with the diversity that marks other major groups of soil organisms. With many species, only part of the life cycle (egg, larva and/or pupa) is spent in the soil, although some taxa are associated with specific soil types. Root-feeding species can reduce crop yields by reducing a plant's ability to absorb water and nutrients and may cause further losses by facilitating the entry of soil-borne pathogens. In contrast, predatory and parasitic insects can contribute to the biological control of both invertebrate pests and plant pathogens. Some *Collembola*, for example, show a marked preference for feeding on the spores of fungal pathogens, including *Rhizoctonia solani* and *Fusarium oxysporum*. Species that feed on organic matter (detritivores) may be important members of the decomposer community. Some insects have ambivalent roles. For example, although termites are usually viewed as pests, in arid tropical soils with low earthworm populations the burrowing activities of termites can also help decompose organic matter and improve soil structure and porosity. Termites and ants are usually the dominant components of the soil insect biomass. As such, they are probably more important than all other insects in their effects on soil structure. They can also be significant pests. In parts of Malawi, for example, it has been estimated that the two most destructive termites (*Pseudacathotermes militaris* and *Macro-termes michaelseni*) damage crops on 72 per cent and 49 per cent of all small holdings, respectively. In Africa as a whole, the sub-family *Macrotermitinae* is considered the most damaging group, affecting a wide range of crops, including tree and pasture species. Their success as pests has been attributed to their ability to survive cultivation, feed on both living and dead plant material, and cultivate saprophytic fungi that serve as food in the dry season. Other important root-feeding insects include the 'white grub' larvae of various scarab beetles (*Scarabeidae*), weevils (*Curculionidae*), wireworms (*Elateridae*), and false wireworms (*Tenebrionidae*). Some *Hemiptera*, *Orthoptera*, and larval *Lepidoptera* (*e.g.*, cutworms) are likewise important in parts of the world. The significance of different species as pests is related to their host range. Those capable of feeding on a wide variety of plants, so called polyphagous species, are generally the most difficult to control. The distribution and damage potential of many soil-dwelling insects is also strongly influenced by soil moisture and rainfall patterns, with the more successful species being able to survive periodic drought. As noted earlier, damage by root-feeding insects can result in secondary infections by plant pathogens. Root crops are particularly susceptible to this type of damage, since the harvested part of the plant is directly affected. Potatoes, for example, soon become infected with bacteria and fungi if damaged by larvae of the potato tuber moth (*Phthorimaea operculella*), resulting in the rapid rotting of the tuber. The same is true of sweet potato tubers damaged by larvae of the sweet potato weevil (*Cylas formicarius*). In groundnuts, the level of aflatoxin (caused by the fungus *Aspergillus flavus*) can increase if the pods are damaged by termites, since the latter can transmit fungal spores.

Earthworms

The Greek philosopher Aristotle referred to earthworms as the "intestines of the earth". The description is apt not only because these invertebrates physically resemble digestive organs but also because they perform something of a digestive function when they break down plant residues in the soil and recycle nutrients. Experiments with maise, for example, have shown that plant residues can degrade 30 per cent faster in surface soil containing earthworms than in soil without them. Earthworms perform a variety of other ecosystem services as well. For instance, they enhance soil porosity thereby reducing rainwater run-off and allowing for infiltration of moisture to plant roots. They also play a vital role in water purification, agro-chemical detoxification, and maintenance of soil stability. When it comes to sheer biomass, earthworms usually account for the bulk of invertebrates that make their home in the soil. Along with ants and termites, they have a special distinction as 'ecological engineers'. This is due to their ability to excavate remarkably large amounts of soil and to create organomineral structures through their casts (excrement). To date about 3700 species of earthworms have been scientifically described. However, the actual number is estimated to be at least twice that, with the overall knowledge gap being significantly greater for tropical earthworm species than species in temperate zones. Although many aspects of earthworms' soil engineering role have been documented, interactions with other members of the soil biota, and how those affect plant growth and health, positively or negatively, are much less well understood.

Weeds

In the tropics, crop losses due to competition with weeds are on the order of 25 per cent , with weed removal often the most labour intensive task in small holder systems. In most areas, the weed population centers on about 20 particularly troublesome species. The individual taxa may be indigenous or introduced, but they usually share a few key traits, especially rapid vegetative growth, high fecundity, and persistence in the soil seed bank. Weed species that resemble the crop in their early stages (*e.g.*, grass weeds in cereal systems) are particularly difficult to deal with when hand weeding is the main control technique. Besides competing with the crop for water and nutrients, weeds can also act as alternative hosts for pest nematodes and plant pathogens. And by boosting the humidity around the base of crop plants, weeds also increase the likelihood of infections by pathogens such as *R. solani* and *Sclerotium rolfsii*. Some weed species are parasitic on certain staple food crops, in some cases causing total crop failure. The role of weeds in relation to soil-dwelling insect pests has not received a lot of scientific attention, although in some cases weeds are known to act as alternative hosts, maintaining the insect pest between crop cycles. Wild *Ipomoea* species, for example, can support the sweet potato weevil (*Cylas formicarius*) between crops, and in Zimbabwe, the groundnut plant hopper (*Hilda patruelis*) can survive the dry season on a variety of different weeds, allowing it to invade groundnuts as soon as they emerge. But weeds may also play a constructive role by supporting higher numbers of soil-dwelling predators that serve as pest control agents.

MANAGING THE SOIL BIOTA

In natural ecosystems, out-breaks of pests and diseases, including those mediated by soil, are comparatively rare. In this respect, it is generally accepted that biological diversity contributes to the relative stability of natural systems. This is probably as true for subterranean communities as it is for those above ground. Agroecosystems, in contrast, are typically much less diverse than natural ecosystems, and are subject to frequent anthropogenic disturbances. They have much simpler food webs and more open nutrient cycles, which makes the maintenance of soil biodiversity — and ecosystem stability — much more difficult. However, there are many direct and indirect means of influencing the soil biota to improve plant health in the short term and to increase soil quality and productivity in the long term.

Direct Methods

These fall into two broad categories : temporary reduction of pest populations and longer-term enhancement of beneficial species.

Reduction : Bio-fumigation and Solarisation

Synthetic soil fumigants like methyl bromide have sometimes been used to control soil-borne plant pathogens and nematodes, particularly for high-value cash crops. However, environmental concerns have stimulated considerable interest in the search for less hazardous alternatives. Bio-fumigation is one such option. It involves the use of plants (or other organic materials) which, as they decompose, produce toxins in concentrations high enough to suppress soil pests and diseases. The decomposition of various brassica species, for example, results in the release of toxic levels of isothiocyanates. A considerable amount of research has been conducted on how best to exploit this effect. Appropriate brassica species are generally selected on the basis of their suitability to local conditions and the type and quantity of isothiocyanates they produce. The plants are usually incorporated into the soil just before flowering, when levels of glucosinolates (isothiocyanate precursors) are generally at their peak. Another approach, for high-value crops, is solarisation. Moist soil is covered with a plastic film so that temperatures beneath the film reach levels lethal to soil-borne pathogens. This technique has been used in Malawi for control of club-root disease of brassicas, and has reduced field populations of plantparasitic nematodes (including species of *Meloidogyne*, *Pratylenchus*, and *Tylenchus*) by up to 97 per cent at depths of 20 centimeters. It has also been used successfully for managing weeds. In some cases, a higher degree of control has been obtained by combining bio-fumigation with solarisation. Combining poultry manure applications with solarisation, for example, was found to be more effective at controlling plant pathogens than either approach alone. In effect, higher temperatures associated with solarisation enhanced the release of toxic levels of ammonia from the manure. However, while techniques such as these may be useful in the short term, they inevitably lead to negative effects on the beneficial components of the soil biota. Collection and destruction of soil-borne pests by hand also constitute a direct control method. In India, farmers have reduced adult populations of *Holotrichia* spp. (*Scarabeidae*) in this way. When adult beetles congregate in trees before mating and laying eggs, farmers shake the trees and kill the fallen beetles on the ground. Fires and lamps have also been used to lure and kill adults at night. However, collective action across a large area is needed to achieve a significant reduction in damage by larvae. Overall, high labour requirements limit the usefulness of manual collection methods.

Enhancing Beneficial Organisms

Over the years many efforts have gone into culturing and releasing beneficial members of the soil biota such as mycorrhizal fungi, plant growth-promoting rhizobacteria, and biological control agents. Various biological control agents for all groups of soil pests occur naturally, and attempts have been made to isolate these natural enemies (mainly fungi, bacteria, nematodes, and insects), increase their numbers artificially, and then release them back into the field at relatively high densities. Large-scale use of this approach depends on low-cost production techniques, delivery at the right time and in the right quantities, and, in the case of commercial production, meeting local registration requirements. Such an approach is perhaps better suited to growers of high-value crops in developed countries than to small-scale farmers in developing countries. Nevertheless, recent research has generated promising results, particularly in the area of seed dressings. Mixtures of the mycorrhizal fungus *Glomus mosseae*, the nitrogen-fixing bacterium *Bradyrhizobium japonicum*, and the antagonistic fungus *Trichoderma pseudokoningii* have been shown to improve soybean growth and reduce damage due to nematodes (*Meloidogyne* spp.). Treating seeds with species of *Streptomyces* and *Bacillus* has provided some control of dry root rot (caused by *Macrophomina phaseolina*) in chickpea, and of fusarium wilt in pigeonpea. Similarly, cultures of *Rhizobia* have been used to inoculate legumes, and mycorrhizal preparations enhance the establishment of newly planted trees. In each case, the use of organisms as seed dressings rather than as more diffuse soil treatments makes maximum use of microbial resources. One highly sophisticated approach to the manipulation of soil fungal communities involves efforts to replace toxigenic strains of *Aspergillus flavus* with more competitive atoxigenic strains of the fungus, thereby reducing the incidence of aflatoxin contamination of crops such as maise.

Indirect Methods

Indirect methods involve manipulating the factors that influence biotic activity (such as micro-habitat structure, micro-climate, and nutrient availability) rather than the organisms themselves. A number of techniques have the potential to increase soil biodiversity, in turn reducing the incidence of soilborne pests and pathogens. Such methods involve the adoption or modification of agronomic practices to conserve and enhance naturally occurring populations of beneficial taxa, or reduce populations of injurious species. Opportunities for modifying existing cropping systems with these aims in mind occur at all points of the crop cycle. They include adding organic matter to the soil (through mulches, green manures, or animal waste), reducing tillage, and adopting more diverse cropping systems. In some cases, only minor modifications of existing practices are required, making such methods more readily applicable to the needs of poor farmers than are the more resource-intensive direct methods. However, applying such techniques in the most effective manner requires a thorough understanding of below-ground food webs. Many of these techniques are currently under-utilised in tropical small-scale agriculture because they have yet to be critically evaluated in such situations.

Supplying Organic Matter

Organic matter can help modify soil structure and is of fundamental importance to many soil functions, including carbon cycling and sequestration and nutrient storage. Incorporation of rich and varied sources of organic matter not only supplies plant nutrients, but also helps to increase below-ground biodiversity by providing an array of substrates capable of supporting diverse soil organisms. Increased biodiversity in turn contributes to the ability of the soil to suppress plant pests and diseases. In Tanzania, for example, application of manure to bean fields resulted in a reduction in root damage by larvae of bean leaf beetles (*Ootheca* spp.), either because of higher levels of predators, parasites, or pathogens in such fields, or because the plants' own defences were stimulated in some way. The interactions responsible for the observed beneficial effects of organic amendments on plant health can be both subtle and complex. For example, it has been shown that briefly exposing the eggs of two pest nematodes (*Meloidogyne incognita* and *Heterodera schachtii*) to aqueous extracts of various organic amendments (including poultry manure and composted tree bark) significantly increased their susceptibility to the fungal pathogen *Verticillium chlamydosporium*, presumably by degrading the egg membrane in some way. There is also evidence to suggest that soils with a high level of organic matter tend to have higher populations of beneficial nematodes and lower populations of plant-parasitic species, as well as a greater proportion of bacterial isolates capable of suppressing the growth of certain weeds. Thus, growth of the weed *Chenopodium album* was reduced by 75 per cent in soils receiving annual inputs of green manure, composted plant residues, and cattle manure, compared with populations in non-amended soils. Suitable sources of organic matter include animal wastes, green manures, crop residues, and composted vegetation. It is important to note, however, that the effects of organic amendments can vary not only with the nature of the material added, but also with soil pH. Thus, at pH 5 microsclerotia (resistant dormant stages) of the pathogenic fungus *Verticillium dahliae* were killed in direct proportion to the concentration of pig slurry incorporated into the soil, but no such effect was observed in soils of pH 6.5. Similarly, research on bean root rots caused by *Fusarium solani* has shown that while a green manure of *Calliandra* sp. reduced populations of the pathogen (and subsequent disease severity), additions of farmyard manure increased the severity of the disease. In the case of *V. dahliae*, the toxicity effects of organic amendments have been attributed to various volatile fatty acids (including acetic, propionic, butyric, isobutyric, and valeric acids); manures that contained all of these compounds killed the microsclerotia, whereas those that contained only acetic acid and small amounts of a few other acids did not. Thus, the effects of organic amendments can depend not only on the characteristics of the organic matter (particularly the ratio of carbon to nitrogen), but also on the nature of the existing soil microbiota, and how individual taxa react to the addition of different organic amendments. For example, while *V. dahliae* populations were reduced by the application of pig slurry, populations of antagonistic *Trichoderma* spp. increased, further helping to suppress populations of the pathogen. In so-called 'disease-suppressive' soils, the microflora is usually dominated by species that produce large amounts of antibiotics (*e.g.*, various species within the fungal genera *Trichoderma*, *Penicillium*, and *Aspergillus*, and the actinomycete genus *Streptomyces*). Such soils typically have good physical structure, and even if very fresh, high-nitrogen organic matter is applied, the production of putrescent (poorly oxidised) products is very low. In contrast, most of the world's agricultural soils have a much higher proportion of pathogenic micro-organisms, and the application of fresh organic material can result in the production of relatively high concentrations of poorly oxidised and malodorous compounds that can be toxic to crops. Addition of organic amendments to such soils stimulates microbial activity in general and increases the baseline populations of beneficial soil fungi, although different species may be more prevalent in some soils than others. The increase in disease suppression observed following regular additions of organic matter may be due to a variety of mechanisms. For example, the plant's own defences may be stimulated by the activities of micro-organisms, or pathogenic species may be suppressed by more competitive species, or inhibited by species that produce toxic secondary metabolites or extra-cellular enzymes. Developing disease-suppressive soils through the use of organic amend ments takes time. Over the years, though, benefits are enjoyed in the form of better plant health and soil structure.

Increasing Plant Diversity

As noted previously, the observation that 'diversity promotes stability' has been made in relation to both natural and managed ecosystems. In an agricultural context, intensive cropping— particularly in monoculture—tends to be associated with 'instability' in the form of short-term increases in the populations of soil-borne pathogens, invertebrate pests, and weeds, occasionally resulting in 'outbreak' situations.

Increasing plant diversity can help reduce the frequency of such events and can be tackled at various levels :

- Taxonomic or species diversity (mixed cropping of various kinds)
- Genetic diversity within species (planting different varieties of the same species)
- Vertical diversity (planting crops of different heights)
- Horizontal diversity (arrangement of plants in space)
- Temporal diversity (*e.g.*, seasonal differences in plant cover and diversity of growth stages achieved through staggered planting or sowing).

Different crops exploit soil resources in different ways. Maximising the diversity of cropping systems in both time and space (by rotations, inter-cropping, and so on) creates a mosaic of soil resources and niches which in turn enhances belowground biodiversity and improves the resilience of the system as a whole. Certain cropping sequences, for example, favour the build-up of various beneficial bacteria that promote plant growth, while the availability of the host crop is known to be the biggest single factor influencing the number and diversity of plant parasitic nematodes in the soil. Differences in root morphology and biomass, and in patterns of root exudation and carbon allocation, can all influence the population density and activity of other members of the soil biota. Furthermore, maintaining some kind of continuous plant cover through the use of living crops or mulches moderates fluctuations in soil temperature and moisture, and further enhances stability. Cropping patterns thus have a profound effect on soil pests and beneficial species. The cropping sequence over time, for example, can greatly affect populations of soil organisms that are relatively host-specific (both pests and beneficial organisms). Thus, natural populations of arbuscular mycorrhizae can be increased by growing a suitable host prior to planting the main crop, while populations of host-specific pests or pathogens can be decreased by planting non-host plants as part of the rotation. However, while the benefits of crop rotation are often well understood by farmers, their land holdings may be too small to allow them to practice it, and continuous cultivation may then lead to a gradual increase in soilborne pests and diseases. Mixed cropping (in various forms) is widely practiced by small-scale farmers in many parts of the world, with inter-cropping being the most widespread cropping system in the tropics. These systems have many advantages over mono-cultures. The presence of non-host plants, for example, can interfere with the host-finding and establishment of mobile pests, and there is also the possibility of yield compensation by the inter-crop if one species is damaged. Inter-cropping also helps to maximise food production from small areas of land and to minimise the risk of catastrophic crop failures. Considerable research has been conducted on the reduction of above-ground pests and diseases through inter-cropping and attention is now turning to its effects on below-ground problems. However, the results obtained so far have been rather variable. In Malawi, for example, inter-cropped maise plots suffered less damage from termites than did mono-culture plots; but, in India, inter-cropping groundnut with sunn hemp (*Crotolaria juncea*) had no effect on either termite abundance or crop damage. In Peru, infestations of potato tuber moth larvae (*P. operculella*) were reduced by inter-cropping potato with beans, onions, tomatoes, or maise, and in Colombia, inter-cropping cassava with *Crotolaria* sp. has been found to significantly reduce damage caused by the subterranean burrowing bug (*Cyrtomenus bergi*). Inter-cropping can also reduce disease and weed problems. In Burundi, inter-cropping potato with maise reduced the incidence of bacterial potato wilt caused by *Pseudomonas solanacearum*, while *Striga* spp. can be suppressed by inter-cropping maise with legumes such as *Desmodium uncinatum*. Inter-planting young plantations of tree crops such as cocoa, coffee, and oil palm with annual food crops is a useful weed suppression strategy for the first few years, not only because of the shading effect of the latter, but also because, on small holder farms, the food staples are more likely to be weeded, with incidental benefit to the tree crop. It is difficult to generalise about the effects of inter-cropping. The outcome is likely to vary not only with the pest and crop species concerned, but also with agronomic factors such as planting date and the frequency and timing of cultivation. The choice of crops will also affect the quality and quantity of plant residues left on or in the soil, which in turn can influence the growth, survival, and reproduction of both pest species and their antagonists.

Mulching

A mulch has been defined as any form of covering applied to the soil surface. By this broad definition, it includes crop residues, weeds, green manures, and other plant material cut and carried in from elsewhere, as well as artificial materials such as paper and plastic. The organic mulches, which are more relevant to resource — poor farmers in developing countries, are quite common in the traditional farming systems of the humid tropics. Besides reducing soil erosion and improving nutrient cycling, mulching can also help suppress weeds, pests, and diseases. Herbicide use or time spent weeding by hand may be significantly reduced by mulching, and notable successes have been achieved by using mulches to suppress soil-borne plant pathogens. In Kenya, for example, black rot of cabbage caused by the bacterium *Xanthamonas campestris* was controlled by a grass mulch applied immediately after transplanting. In such cases, it is thought that the effect of the mulch is due to a combination of its role as a physical barrier (reducing rain splash of the pathogen onto the crop), together with its ability to change the micro-climate at the soil surface and enhance the activity of beneficial soil micro-organisms capable of suppressing pathogens. Mulching has also been used to divert termites from crops, and in various parts of Africa, mulching with the weed *Tithonia diversifolia* has been shown to reduce nematode damage and improve crop growth. In Uganda, mulching of banana plantations appeared to reduce numbers of the nematode *Radopholus similis*, possibly because the mulch reduced soil temperatures, thereby slowing nematode feeding and reproduction. Conversely, the presence of crop residues on the soil surface may enhance the biological control of insect pests by entomo-pathogenic nematodes. It has been shown, for example, that such residues increase the persistence of *Steinernema carpocapsae*, probably by protecting it from desiccation or ultra-violet light. Mulching nevertheless requires careful management : if crop residues are only partially decomposed by the time the next crop is planted, the incidence of some seedling diseases (such as those caused by *Rhizoctonia solani*, *Sclerotium rolfsii*, and *Macrophomina phaseolina*) can increase. While burial of crop residues increases the rate of decomposition and can help to improve soil structure, these benefits must be weighed against the generally adverse effects of tillage on the soil biota.

Host Plant Resistance

Genetically based host plant resistance to pests or pathogens may be constitutive (continually expressed) or induced (expressed only in response to damage). This so-called intrinsic resistance can be one of the most economical and sustainable options for the integrated control of many plant pathogens and some invertebrate pests, particularly nematodes. Recent research at the International Center for Agricultural Research in the Dry Areas (ICARDA) has resulted, for example, in the identification of several lentil lines resistant to vascular wilt caused by *Fusarium oxysporum*, while new varieties of beans resistant to root rots have been widely adopted by farmers in western Kenya. However, there are a number of obstacles to the development and use of resistant cultivars. First, in situations where different strains or pathotypes of a pathogen or pest occur, the resistance may be of variable value. Second, selection for resistance-breaking biotypes can be rapid if resistant varieties are not deployed with care. Third, developing resistant germplasm is often a slow and costly process and the resulting seeds of improved varieties may either be too expensive for the majority of small holders, or supplies may be inadequate to meet demand. Where this is the case, it is sometimes possible to obtain an acceptable reduction in soil-borne pathogens or nematodes by Inter-planting resistant varieties with local landraces. In some instances, new varieties may not be adopted because of differences in taste or other characteristics that render them unacceptable to farmers or consumers. Such difficulties can be avoided by involving both farmers and consumers in breeding programmes. When new cultivars resistant to pests and diseases are being developed, their effects on weeds should also be considered. This is particularly true for varieties targeted on small holder systems, in which weeding may account for up to 60 per cent of all pre-harvest labour. Failure to take into account the effects of new varieties on weeds can create additional problems. For example, attempts to replace traditional tall varieties of rice with high yielding semid-warf cultivars with erect (rather than drooping) leaves resulted in increased weed problems in some areas. On the other hand, cassava cultivars that spread and establish a complete ground cover in about 12 weeks require less weeding than do non-branching forms. Some success has also been achieved in breeding crop varieties resistant to the parasitic witchweeds (*Alectra* and *Striga* species).

Reduced Tillage

Conventional tillage immediately changes the structure of the soil microbial community, even if total microbial biomass is little affected. Under conventional tillage regimes, bacteria-based food webs predominate, and flushes of mineralisation related to cultivation can lead to increased losses of nutrients and organic matter from the soil. In this way, tillage can increase the potential both for nitrate leaching and the emission of greenhouse gases such as carbon dioxide and nitrous oxide. In the long term, it can have deleterious effects on soil structure and biodiversity. Conventional tillage practices are generally unfavourable to soil mesofauna, macrofauna, and various fungi. Indeed, many of these organisms are extremely sensitive to perturbations of the soil. If tillage is minimised and crop residues are left on the surface, natural populations of many species in these groups can be enhanced and the spatial and temporal diversity of the subterranean food web increased. In zero-till or minimal tillage systems, fungus-based soil food webs predominate and nutrient leaching is reduced because nitrate levels tend to be lower. The choice of tillage system also has the potential to alter the level of fungal pathogens and the population density and diversity of nematode species. Conventional tillage, for example, can lead to soil compaction and a consequent increase in plant-parasitic nematodes at the expense of beneficial bacteria-feeding species. Reduced tillage also has the notable benefit of cutting labour costs. Nevertheless, conversion of a conventional system to reduced or zero tillage may result in short-term increases in plant pathogens, invertebrate pests, and/or weeds. Thus, careful management is required in the first few years until the balance is tipped in favour of beneficial species.

Sanitation

Destruction of crop residues after harvest (to reduce pathogen inoculum or harbourage areas for invertebrate pests) is a relatively simple technique for growers to adopt. However, in some cases, unexpected results have been obtained. For example, in one study, mechanical or chemical destruction of melon roots after harvest resulted in increased reproduction of the root-infecting fungus *Monosporascus* sp. compared with untreated controls. Again, however, such results may be the result of a relatively impoverished soil biota, with few antagonistic species.

Ridging and Terracing

Growing crops on ridges or in raised beds can improve plants' ability to resist diseases exacerbated by compacted soils, such as fusarium root rot of beans. Such practices also reduce the incidence of diseases associated with wet soils, such as the seedling diseases and root rots caused by species of *Rhizoctonia* and *Pythium*. In the highlands of Africa, terracing has been shown to enhance the populations of some mycorrhizal fungi such as *Glomus callosum* and *G. occultum*, with consequent beneficial effects on crop growth.

Irrigation

Where irrigation is possible, it can help reduce damage by soil pests, both because some pests will be killed out-right, and because well watered crops are better able to resist (or compensate for) damage to their roots.

Effects of Other Agricultural Practices

Development of a truly integrated approach to the management of soil biota will depend not only on adopting techniques for enhancing beneficial species, but also on techniques that mitigate or minimise the adverse effects of other agricultural practices.

Seed Saving

Subsistence farmers generally retain their own seed for planting, a practice that has the advantage of preserving local cultivars adapted to the prevailing environmental conditions. While farmer-saved seed is often of good quality, problems can arise if it is contaminated with pathogens. If the seed germinates, it may produce seedlings of low vigour which are then predisposed to other soil-borne infections. Farmers should therefore be encouraged to save their best seed for planting, and perhaps to improve seed quality by careful management of the mother plants.

Pesticide Use

Synthetic pesticides are rarely used by small holder farmers, since they are normally prohibitively expensive except when subsidised by the State. In some cases, too, shortage of water is a constraint on their use. Where they are used, herbicides and foliar insecticides rarely reach the soil in sufficient concentration to directly affect most of the soil biota, although nitrifying bacteria are occasionally affected. It has been reported that the decomposition of plant material treated with certain herbicides can trigger short-term increases in some plant pathogens. Fungicides and soil fumigants are the most damaging compounds, but again are rarely used except for some high-value crops. Seed dressings are the most economical means of deploying pesticides against soil pests. However, with the withdrawal of the persistent organochlorines (*e.g.*, DDT and aldrin) in the 1980s, even seed dressings have become too expensive for most smallholders. Thus, interest in non-chemical methods of control has been revived. In recent years this interest has been further stimulated by the impending ban on methyl bromide. Under the Montreal Protocol, use of this pesticide, which is also a powerful ozone depleter, is to be eliminated in industrial countries by 2005, with some exemptions recently negotiated, and by 2015 in developing countries.

Application of Inorganic Fertilizers

Most fertilizers can inhibit local microbial activity, especially when they are applied in high concentrations. Some nitrogenous fertilizers can produce biocidal levels of ammonia. Furthermore, high levels of inorganic fertilizer, particularly in tropical soils, tend to reduce populations of mycorrhizal fungi. Some species may even disappear under such circumstances.

Clear Cutting

Clear cutting forests or scrub can affect the soil biota for many years, with the severity of the impact being determined by soil characteristics, climate, and the nature of the subsequent plant cover. In general, the diversity of microbial species is little affected, but their relative abundance alters. Increases in the populations of some members of the microbiota may persist for 5 to 10 years after clear cutting, often accompanied by siseable losses of soil nitrogen and other nutrients. In termite-infected areas, clearing of woodland removes food and nesting sites. The species whose numbers increase after clearing are normally those with deeper subterranean nests (*e.g.*, *Microtermes* spp.) or those that construct large mounds (*e.g.*, *Macrotermes* spp.). *Microtermes* species, for example, have been known to increase from a mean density of 500 individuals per m² in woodland to 4000 in fields cultivated for eight years or more.

Burning

In contrast to the situation following clear cutting, the number of soil micro-organisms generally declines after burning, mainly because soil organic matter is destroyed. However, although there is a perception that soil-borne pests and diseases are reduced by burning, this is seldom the case for plant pathogens. Temperatures are not usually high enough to kill all resistant (dormant) stages in the upper layers of the soil. The microbiota usually recovers rapidly following plant growth and litter accumulation, with the microflora probably re-establishing itself from the deeper layers of the soil and the micro-fauna from small patches of litter that escape burning. Some soil-borne plant diseases (*e.g.*, vascular wilts caused by *Verticillium dahliae*) can thus occur even in newly cleared land, while root rots caused by fungi such as *Armillaria mellea* and *Rosellinia necatrix* can be a problem immediately after forest clearance but may quickly decline after a few cropping seasons. Furthermore, it has recently been found that viable bacteria and fungal spores can be transported considerable distances in the smoke from burning plant material, with the concomitant risk of spreading diseases to distant fields. The weed flora also changes after bush clearance, with broad-leaved species predominating in the first few years and grass weeds gradually becoming more common thereafter.

CROP LOSSES FROM SOIL ORGANISMS

In the tropics, soil-borne pests and diseases are recognised as major causes of reduced crop yields and therefore of economic losses to farmers. Plants grown on marginal land with low soil fertility often lack vigour and are particularly susceptible to the ill effects of these organisms. Soil-borne pests, for example, can undermine the development of healthy root systems. The damage caused by soil organisms appears to be on the rise in the tropics, in part due to the pressures on the land from continually expanding human populations and associated changes in cropping patterns. Nevertheless, these agricultural losses and their economic significance are often poorly quantified. Here are three examples of the significant damage caused by soil organisms.

- In Central and South America, white grubs (*Holotrichia* spp.) cause major losses in maise, sorghum, bean, potato, and rice production, at times exceeding 50 per cent of the total yield. Similarly, white grubs have been reported to seriously damage bean and groundnut crops in Africa.

- Sweet potato weevils of the genus *Cylas* are considered the world's most destructive pest of sweet potato, with crop losses of up to 73 per cent reported from Uganda. In the highlands of Peru, farmers ranked the Andean potato weevil (*Premnotrypes* spp.) as the most important pest of that tuber crop, with losses in single fields of up to 75 per cent.

- In recent years, burrower bugs (*Cyrtomenus bergi* and other soil-borne hemipterans) have gravely damaged a range of crops in Central and South America. When burrower bugs feed on cassava roots, they can introduce soil-borne pathogens such as *Aspergillus, Diplodia, Fusarium, Phytophthora,* and *Pythium.* The resulting lesions can reduce root starch content by more than 50 per cent , and may wipe out a crop's entire commercial value.

Mycorrhizal Fungi

A wide range of soil-borne fungi can invade the roots of higher plants to form mutually beneficial relationships called mycorrhizae. The best known of these associations are arbuscular mycorrhizae which are especially important in acid soils, where phosphorus is often the limiting nutrient. The hyphae (threads of mycelium) enhance the uptake and translocation of nutrients (particularly phosphorus) into the host plant, and the fungus in turn receives carbohydrates from the plant. Mycorrhizal fungi can also increase the drought tolerance of their plant partners and enhance their resistance to plant pathogens, nematodes, and toxicity caused by heavy metals. Mycorrhizal fungi can be found in the roots of grasses, some trees and shrubs, and most agricultural crops. In mature roots, the proportion of root weight attributable to a mycorrhizal fungus can vary from about 3 per cent in sorghum to about 16 per cent in soybean. The mycelium of a mycorrhizal fungus, which can extend several centimeters around the plant root, enhances the formation of soil aggregates, a particularly valuable trait in coarse, sandy soils. In addition, the mycelium is an important resource for fungus-grasing insects (*e.g.,* some *Collembola*) and conducts root exudates further out into the soil, enhancing the carbon supply for other members of the soil biota. Dense populations of bacteria have been found in this part of the rhizosphere, and synergistic interactions between arbuscular mycorrhizal fungi and other beneficial members of the soil microbial community (such as *Trichoderma* spp.) have been observed. Agricultural practices can have major impacts on mycorrhizal fungi, both positive and negative. On the one hand, conventional tillage disrupts the networks of fungal hyphae in the soil, delaying colonisation of crop plants and reducing their phosphorus uptake. As a result, crop yields are lower than those obtained in zero-tillage systems. The absence of suitable host plants (over winter, for example), long-term fallowing, or continuous mono-culture can all cause severe declines in soil fungal populations. On the other hand, well planned crop rotations, mixed cropping, and the use of cover crops can all conserve or enhance this important resource. Some crops are more dependent on mycorrhizal relationships than others. Faba beans and maise, for example, depend more heavily on mycorrhizae than do wheat and potatoes, while brassicas and beets do not support mycorrhizal relationships at all. Including the latter in crop rotations can therefore reduce or delay root colonisation in the crops that follow.

Sustainable Management of Banana Weevil

The banana weevil (*Cosmopolites sordidus*) is a major pan-tropical pest of banana and plantain (*Musa* spp.). Adults lay their eggs in the plant's pseudostem close to ground and the larvae bore into the rhizome, causing tissue at the edge of their tunnels to turn brown and rot. The plant is weakened and its ability to take up nutrients is reduced. If the infestation is heavy, the plant is small, or the variety particularly susceptible, the plant can die. Yield loss can thus be due to the death of the plant, reduced bunch size, failure to produce new suckers, and/or shortened plantation life. At present, there is no standard protocol for sampling the weevil, and assessment of larval populations involves destruction of the plant. Nevertheless, a number of management techniques have been developed and adopted, and new approaches are being considered. Several resistant varieties of banana have been produced and released, providing a sound basis for an integrated management strategy that also includes cultural and biological controls. Banana weevils can reproduce in crop debris (including cut stumps); indeed, in some cultivars, weevil survival seems to be higher in crop residues than in growing plants. Hence, sanitation (removing and destroying crop debris immediately after harvest) is an obvious approach to cultural control. Research in Uganda has shown that such sanitation practices can significantly reduce populations of weevils and lead to yield increases of up to 70 per cent . Since both banana weevils and nematodes can be transferred to new fields through infested planting material, simple cleaning strategies such as corm paring prior to planting have also been introduced and adopted by farmers in parts of Kenya and Tanzania. Biological control agents are also under investigation. Research conducted at the International Institute of Tropical Agriculture (IITA) in Nigeria, for example, has resulted in the identification of a promising hymenopteran egg parasitoid and some dipteran larval parasitoids from Sumatra that might form the basis of a classical biological control programme. There is also interest in exploiting the potential of native ants.

Ants are generalist predators, and in Cuba some species have been reported to be effective predators of the banana weevil. Surveys in Tanzania and Uganda have recorded a diverse array of ants in banana plantations, with the number of species on individual farms ranging from 19 to 34, generally with *Pheidole* species predominating. Nests of these species occur in the soil, in crop debris, and in standing plants, and if their populations can be manipulated, they may be a useful addition to an integrated control strategy. Microbial control with the entomopathogenic fungus *Beauvaria bassiana* is another option currently under investigation. Other approaches include attempting to identify and utilise various species of endophytic fungi that are known to colonise banana tissues. Certain strains of these fungi—which penetrate the host plant and grow internally for at least part of their life cycle— can increase host plant growth and/or confer a degree of resistance to weevil damage. Research with Ugandan isolates of some of these species has shown that they can be successfully inoculated into tissue- cultured banana plants, which themselves can form part of an integrated control strategy by ensuring that new plantations are pest- and disease-free at establishment. Various other fungi associated with the banana rhizosphere are also being assessed for their ability to control *Radopholus similis*, the burrowing nematode.

Planting for Pest Management

The repellent or biocidal effects of particular plants can sometimes be exploited for the control of soil-borne pests and diseases. For example, the inclusion of some species of marigold (*Tagetes* spp.) in crop rotations has been shown to be a promising means of controlling some nematode species (*e.g.*, *Meloidogyne* spp. and *Pratylenchus penetrans*). In Uganda, a toxic plant is the basis of a novel means of managing East African mole rat or root rat (*Tachyoryctes splendens*). This mammal lives in subterranean burrows and causes considerable yield losses in many crops, especially cassava and sweet potato, by feeding on the roots and lower stems. Attempts to control the rats by traps, snares, and digging are only partially effective. A promising new approach involves planting *Tephrosia vogellii*, an indigenous leguminous shrub, around field margins and as scattered individuals throughout the field. The leaves and roots of *T. vogellii* contain rotenone, a natural insecticide that is also toxic to fish — and root rats. The shrub is easily established from seed. In fields heavily infested with root rats, it should be planted at 3 × 3 m spacings, with additional plantings at 1 m intervals around the field boundary. Other crops may be grown in the same field. After one year, the field should be clear of root rats, and the within-field *T. vogellii* plants can be removed. However, because *T. vogellii* is known to host some damaging species of root-knot nematodes, crops tolerant of these pests, such as maise, should be planted immediately afterwards. Weeds, too, can be affected by the plant species growing around them. Allelopathy, the adverse effect of one species on another, may be mediated by the production of germination inhibitors or other root exudates. Such interactions can be exploited for low-input weed control. For example, Inter-planting of Kenyan maise fields with *Desmodium uncinatum* or *Calliandra calothyrsus* inhibits germination and growth of *Striga* spp.

Development of a New Pest : The Subterranean Burrower Bug

In the early 1980s, a new pest, the subterranean burrower bug (*Cyrtomenus bergi* Froeschner), was found on cassava roots in parts of Colombia and Panama. The bulk of the population is found in the top 20 cm of the soil. Feeding by both adults and immature burrower bugs allows soil-borne pathogens to enter the root tissue, resulting in the development of brownblack lesions and a considerable reduction in yield. Since its discovery, the bug has spread to other parts of South and Central America. It attacks a wide range of crops, including potato, groundnut, sorghum, coffee, onion, asparagus, maise, and sugarcane. The rapid rise of *C. bergi* to pest status is thought to be related to the reduction in plant diversity and simplification of agro-ecosystems that resulted from the intensification of local cropping patterns in its center of origin. Current research on the management of the bug focuses on inter-cropping, identifying sources of resistance, and the use of pathogens and parasites of insects, including fungi and *Heterorhabditis* nematodes. Research in Costa Rica has also shown that populations of the bug are consistently lower in zero-till maise plots than in conventionally plowed fields, presumably because plowing makes it easier for the insect to move through the soil.

THE WAY AHEAD

Research Priorities and Approaches

In 2001, the Food and Agriculture Organisation of the United Nations (FAO) informally surveyed the global community of soil biodiversity experts. It concluded that while a considerable amount of relevant research was being conducted on both natural and agricultural systems in various parts of the world, sub-tropical and arid regions were under-represented. Furthermore, while inputs of organic matter were the focus of considerable research effort, much less attention was being given to other important areas such as the biology of fungal root pathogens and the effects on the soil biota of tillage, agricultural chemicals, pH adjustments, and so on. The ecological effects of agricultural bio-technology and the relationship between soil biodiversity and the biological control of pests were also identified as areas that would benefit from more concerted interdisciplinary work. Some concern was also expressed over the general lack of expertise in natural resource management and development issues among those currently doing soil biota work. These concerns should be addressed as a matter of urgency, since maintaining or enhancing soil biodiversity produces significant benefits both at the farm level, by improving returns on labour and other inputs, and at the national level, by preventing land degradation and by improving water quality. A rich and varied soil biota reduces the need for synthetic pesticides through the suppressive effects of naturally occurring biological control agents, helps prevent pollution by rapid detoxification of agricultural chemicals, and stems land degradation by improving soil structure. Furthermore, by improving nutrient cycling and reducing losses through leaching, a healthy and diverse soil biota can reduce fertilizer requirements. To fully capitalise on the services provided by the soil biota, scientists and farmers in various ecological zones need to collaborate on the development of robust diagnostic tools and key principles for managing soil organisms. Standard protocols are needed for sampling each of the major groups of soil organisms, and also for determining their spatial and temporal variability. In the future, geographic information systems may contribute much to the documentation of different soil types and land-use histories. However, better understanding of the interactions between crops, soil fertility, and the soil biota will still be needed. The following five research themes have been identified as being particularly important in this regard :

- The influence of the physical and chemical characteristics of soils on the distribution and abundance of soil pests, pathogens, and their natural enemies.

- Techniques for reducing crop losses to pests through better management of soil fertility and plant health (including cultural practices and the effects of organic amendments).

- The effects of agro-chemicals (including inorganic fertilizers) on key members of the soil biota.

- Techniques for restoring soil fertility, with emphasis on methods suitable for resource-poor farmers in marginal areas.
- The value of fungi, bacteria, and other organisms as indicators of soil health and fertility.

The research methods employed are also important since they directly influence the relevance of solutions to farmers, the foremost stewards of agricultural soils. Innovations that minimise pest and disease problems while maximising the beneficial elements of the soil biota must take into account farmers' socio-economic environment as well as their current understanding of biophysical constraints on production. Even if farmers typically lack advanced training in biology and soil science, they are usually highly observant, and many have considerable knowledge of the pests, weeds, and diseases that occur on their holdings. Participatory approaches to research for development offer practical means of maximising farmer contributions to problem analysis and the design of solutions. This increases the chances of adoption, in part through an increased sense of local ownership of technology, and provides valuable feedback to scientists collaborating with farmers. Farmer organisations provide a practical venue for such joint learning.

The SP-IPM Response

The SP-IPM's Soil Biota Thematic Group is formulating a global project on soils that will promote sustainable and productive agriculture in Asia, Africa, and tropical America. Titled 'Better Lives from Healthy Soils : Integrated Pest Management in Soil Agroecoststems', the research project will enhance the food security and livelihoods of poor farmers through soil ecosystem management for soil and plant health. It involves 11 international agricultural research centers and the national agricultural research systems (NARS) of 40 to 50 countries in the developing world. The participants will integrate research across centers, disciplines, crops, and ecological regions through a holistic approach. The context of soil-borne pests is highly complex, with interactions among beneficial and harmful organisms, abiotic factors, cropping practices, and multiple crops within different geographic regions. This calls for a multi-disciplinary approach that can be applied to different ecologies. The global project will take such an approach, creating a large pool of scientists who together will be able to achieve what no individual organisation can. The involvement of a range of international centers and advanced institutions together with NARS and small-holder farmers of many countries will establish an inter-institutional network for the study of soil biota that will lead to the sustainable management of soil-borne crop pests. Field work will be done with the full participation of farmers and communities, thus ensuring a demand-driven process of innovation linked to strategic research. A wide range of methodologies is currently being used in research on soil organisms. This poses a serious obstacle to scientific progress in the management of soil-borne pests. One of the objectives of the global project will be to standardise methodologies, thus increasing the value and comparability of the data generated. The development of databases and networks will speed up knowledge sharing and help participating scientists to make more efficient use of the data generated.

Supporting Public Policy

In many countries, legislation designed to protect the soil has lagged far behind measures intended to protect other natural resources such as air and water. Laws and policies to maintain or improve soil quality are urgently needed in many parts of the world. Population growth, land scarcity, and inappropriate land management techniques have all been identified as factors contributing to land degradation and hence to the need for policy intervention. However, if higher priority is to be given to the conservation and management of the soil and its associated biota, then policy-makers need a better understanding of soil-based ecosystem services and of their monetary value. Developing a suitable policy framework for sustainable land management will require input from specialists in a variety of disciplines, as well as constructive dialogue between stakeholders at both local and national levels. Research institutions can contribute to the policy-making process by ensuring that adequate information is available, although the ultimate success of any legislative reforms will depend on the level of community and political support. Strengthening farmer knowledge and improving institutional linkages to promote information exchange should therefore be considered priorities. In most cases, policy reform will involve harmonising agricultural and environmental legislation to reconcile the apparently conflicting goals of high agricultural productivity and environmental protection, including biodiversity conservation. The overall aim should be to create a duty of care with regard to soil conservation and land use that has the broad support of the majority of stake-holders. Ideally, national policies should incorporate mechanisms for constant monitoring and modification, so that they can be adjusted to suit local conditions and so achieve truly sustainable development. Issues such as agricultural intensification must be handled with care in order to avoid potential conflicts. Although reduction in biodiversity is often seen as an automatic consequence of intensification, this need not be the case if the process is carefully managed. Indeed, landscapes composed of a mosaic of different levels of intensification can sustain very high levels of diversity, both above and below ground. Recent initiatives aimed at developing a global policy framework for land use and soil conservation could be of tremendous benefit and should be encouraged. The aim should be to develop a set of guidelines encompassing the essential principles of sustainable soil management that will help individual countries develop suitable national strategies. Such an instrument would be invaluable in promoting the protection of soil resources and in helping to identify gaps in existing international legislation. The fundamental importance of the soil and its associated biota must be recognised at all levels if this valuable but often neglected resource is to be preserved and managed in the most appropriate way.

Chapter 6

SOIL MICROBES

These are soil fungi of which mushrooms and toadstools are the visible forms. And also the single-celled soil bacteria. You need a good compound microscope with high magnification to see bacteria and the single filaments of soil fungi.

SOIL MICROBES OVERVIEW

Soil looks lifeless but a quarter of a teaspoon of soil (1 gram) contains many microbes - 100 million bacteria, 1 million actinomycetes and 5 metres of fungal filaments.

Soil is certainly living! In pastures grazed by sheep near Armidale, NSW, Australia, we have measured the weight of microbes in soil and calculated that there is the equivalent weight of soil microbes of 88 sheep/ha.

Above Ground

Below Ground

Since the pastures were grazed with only 20 sheep/ha above-ground, there was the equivalent weight of 4 sheep (as microbes) in the soil to every real sheep grasing above. Scientists in England and New Zealand have calculated that they have equivalent weights of 110 sheep/ha in microbes in pasture soil. These English and New Zealand soils would be moister and have more organic matter food for microbes than our Australian pasture soil which has periodic droughts and where organic soils are rare.

Ways of Feeding

Bacteria and fungi have similar ways of feeding. Their main food is organic detritus in leaf litter, soil, dung and carrion. In contrast to invertebrate animals which engulf food and digest it within their bodies, microbes digest their food outside themselves by secreting digestive enzymes over their food. Under the action of the enzymes, organic matter breaks down into simpler molecules external to the microbe, which then re-absorbs the simpler products of digestion through its cell wall. Microbes have a vast number of enzymes to do this job – 50 to 60 different ones have been discovered. Few invertebrates possess this many digestive enzymes.

FUNGI

Most of the time fungi hide away in the murky depths of the soil and are not visible. But their microscopic, mouldy strands are growing through bits of organic matter and soil, through leaf litter, dead wood and dung and into carcasses of dead animals. Here, the fungi quietly go about their job of decomposing detritus and recycling the nutrients back to the soil.

It is only when these microscopic strands merge together to form the fruiting bodies of mushrooms and toadstools that we become aware of this invisible network beneath our feet. The spores borne on these fruiting bodies disperse in the wind, or in or on, the bodies of animals feeding on them *e.g.* springtails, flies.

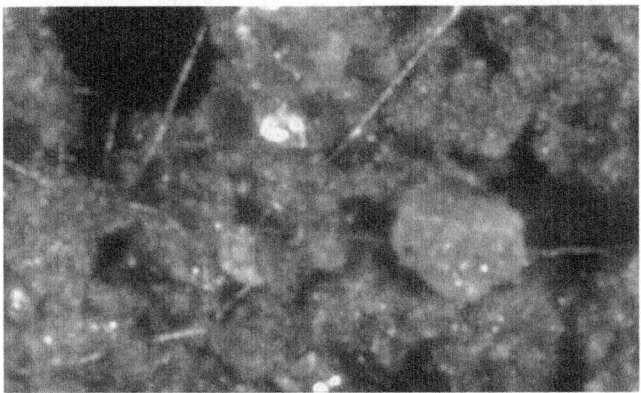

Fig. Microscopic Fungal Strands Growing Through Soil.

Fungal fruiting bodies are of many different and beautiful forms : mushrooms, toadstools, bracket fungi, puffballs, earthstars, coral fungi, large boletes and delicate crinoline fungi. Fungal fruiting bodies can also be rather gross. The starfish fungus smells revolting so as to attract blowflies which feed on the squishy mess in the centre of the fungus and transport its spores from place to place. Other soil fungi don't produce such conspicuous fruiting bodies, but form their spores or fruiting bodies within the soil *e.g.* the edible and delicious truffle which has to be sniffed out with the truffle-hunter's pigs or dogs.

Fungi grow by dividing and spreading out into a network of branches which grow through organic matter. Fungi can link different parts of the soil profile together. They can grow out from a piece of leaf litter, down into the soil, up into a dung pat and link all these together. It can absorb nutrients in one spot and transport them to another. This process is called translocation of nutrients and it is important in nutrient cycling. The networks of fungal filaments (mycelia) can explore their environment much more effectively than can soil bacteria which occur either as individual cells or as sedentary colonies.

The main food of fungi is organic matter in leaf litter, dung, soil and carrion. They (along with bacteria) are the main decomposer groups in soil because of their vast collective mass supported by the variety of digestive enzymes they can secrete onto organic matter to break it down.

Some fungi are edible and collecting wild fungi is a favourite pastime in Europe where, over the centuries, they have sorted out, by trial and error, those that are safe to eat – presumably the "errors" died! Australian fungi are a bit more problematical as we have not built up such a long-standing bed of knowledge.

There are no simple tests for edibility. Some old wives tales are : they are harmless if you can peel them, if they fail to turn dark on soaking in cold water or if a silver spoon cooked with the fungi, doesn't turn black. I wouldn't depend on these guidelines for edibility. To put these rules of thumb on a more scientific basis, don't eat mushrooms with yellow or greenish coloured gills, or have caps with yellow centres or caps which are thimble-shaped when young, but which smell of disinfectant (iodine) when stored in a plastic bag. Even then, it is best to stick to the fungi you can buy in the supermarket – there are many varieties to choose from these days.

Some soil fungi form a positive relationship with plant roots. They live in roots, but also extend their filament networks out into the soil where they absorb nutrients and water which they bring back to the plant to use. What the fungus gets out of this relationship is food in the form of sugars. This mutually beneficial relationship (or symbiosis) between root and fungus is called a mychorrizal association

The biggest organism in the world is invisible !

Not all soil fungi are goodies. The potato famine in Ireland in the mid-1800's was caused by a soil fungus. One in every nine people perished from starvation. Also, some soil-borne decomposer fungi can grow in grain and peanuts when these are stressed by drought. The fungus contaminates the grain and nuts with a toxin (aflatoxin) which causes liver damage in animals and humans.

Poisonous Fungi

Only a small proportion of Australian fungi are poisonous but since some fungi have fatal effects it's best to stick to the supermarket varieties. Recently, a prominent Australian newspaper in its life-style section had to quickly publish a correction when it got the captions to two fungi mixed up. The paper had incorrectly labelled the lethal fly agaric as "edible"!

Quirky Bit

Some microscopic fungi eat animals. Hard to imagine but the nematode-eating fungi grow little loops which soil nematode worms swim through. The loop tightens like a lassoo around the worm and the fungus oozes digestive enzymes onto it before absorbing the products of digestion into itself. Other nematode-eating fungi produce sticky knobs which stick to and trap nematodes.

BACTERIA

Bacteria are tiny one-celled micro-organisms, 0.001 mm in diameter, which need a light microscope with very high magnification to see them. They are best seen with an electron microscope.

Bacteria have a different way of growing to fungi. They multiply by cell division, one cell dividing to make 2 cells and so on. Sometimes the cells move apart and they live as separate individuals or they can stay together to form a colony (a few cubic mm in volume) as they do when cultured in the laboratory on an agar plate. Fungal cells are different. They divide and form branching networks of filaments (mycelia) in the soil and can explore the soil environment much more effectively than bacteria.

Bacteria form resistant spores when the climate turns harsh. They can survive droughts in a spore stage and when favourable conditions return, they can germinate and grow again.

The main food of bacteria is organic matter in leaf litter, dung, soil and carrion. They (along with fungi) are the main decomposer groups in soil because of the huge number of digestive enzymes they can secrete onto organic matter to break it down.

Some soil bacteria can fix atmospheric nitrogen, ultimately making it available to plants. Some of these live free in the soil *e.g.* Azotobacter, while others fix nitrogen while living in nodules on the roots of legumes *e.g.* Rhizobia. Here, these bacteria gain food in the form of sugars from the plants and the plant benefits through increased nitrogen fixed by the bacteria. This "win-win" relationship between bacteria and plant is called "symbiosis" – or a relationship where both organisms gain a mutual benefit from living together.

Diseases from Soil Bacteria

Not all soil bacteria are "goodies" as soil-borne diseases can attack plants and animals. Anthrax and tetanus, which infect domestic animals and humans, lurk in soil as does the bacterial blight of tomatoes.

OTHER MICROBES

Other microbes, apart from bacteria and fungi, also live in soil. These are actinomycetes, yeasts, viruses and algae. Not all of these other microbes are decomposer organisms.

- Actinomycetes are bacteria-like microbes, joined end to end in a filamentous strand to form a branched, fungus-like network. They seem to be half-way between a bacterium and a fungus. They are good decomposers and can break down particularly resistant forms of organic matter such as cellulose and chitin.

- Soil yeasts are unicellular fungi. They are decomposers.

- Soil viruses prey on soil bacteria but not a lot is known about these micro-organisms.
- Soil algae are also included in the microbes but these are different as they photo-synthesise like green plants do.

Quirky Bit

Oddly enough, soil actinomycetes are a good source of antibiotics. The soil actinomycete, Streptomyces, has produced many of our antibiotics and drugs *e.g.* streptomycin. The antiparasitic drug, avermectin, was first discovered in an actinomycete living in soil on a golf course in Japan ! It is used to protect domestic animals, pets and humans from both internal and external parasites such as roundworms, ticks and lice. In humans, it kills parasitic nematode worms which cause River Blindness in Africa and South America.

SOIL FOOD WEB

Soil biota feed on many different trophic (or feeding) levels. The energy base of the food chain in soil is organic matter such as dead leaves, twigs and logs, dead animals, animal excreta and degraded organic matter in soil. Sometimes it is called the "detrital" food chain as the primary food source is organic detritus. Food chains are formed as one organism feeds on another.

- Biota feeding directly on organic matter are called decomposers.
- Biota feeding on microbes are called microbivores.
- Biota preying on decomposers are predators.
- Sometimes predators feed on other predators.
- And if predators die, decomposers, in turn, feed on them.
- And so the food chains are inter-connected into food webs.

Food chains and food webs are depicted with arrows going from the organism that is eaten, to the organism that eats it. Eaten —> Eater.

A simple food chain in soil would be :

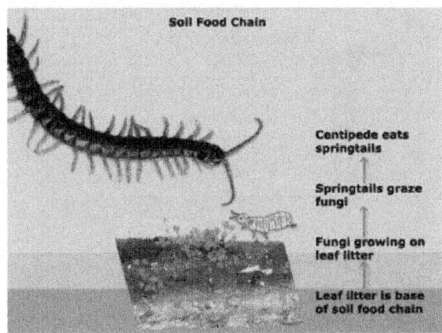

Fig. Soil Food Chain.

Pasture Food Web

In the two food webs which can be seen below, there are complex interactions between different trophic levels. In these food webs, we have concentrated on the smallest soil organisms. You need a microscope to see them. We focussed on the small organisms as it is the "little things that make the underworld work" and they are regarded as being the most important organisms in soil processes such as decomposition and nutrient cycling in soil. The smaller the size of a soil organism, the more numerous and the more metabolically active they are.

The two food webs still show only some of the possible inter-connections between soil organisms. The "top" predators in these food webs are the small predatory nematodes and mites. Linkages (arrows) could be drawn between each group and its decomposer groups. Remember, as each organism dies, it decays through the activities of the invertebrate and microbial decomposers. However, if we included all these inter-connecting arrows the diagram would look like an untidy bird's nest !

Unfertilised Native Pasture

Net Annual Web-flow of Mineralised N = 54 kg N/ha

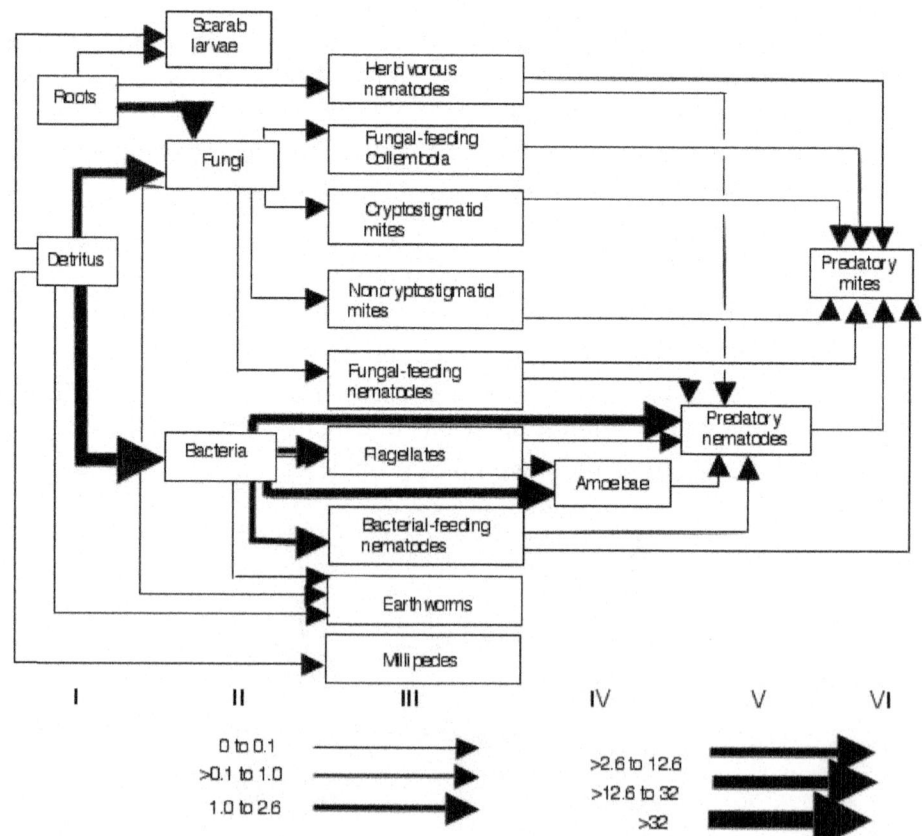

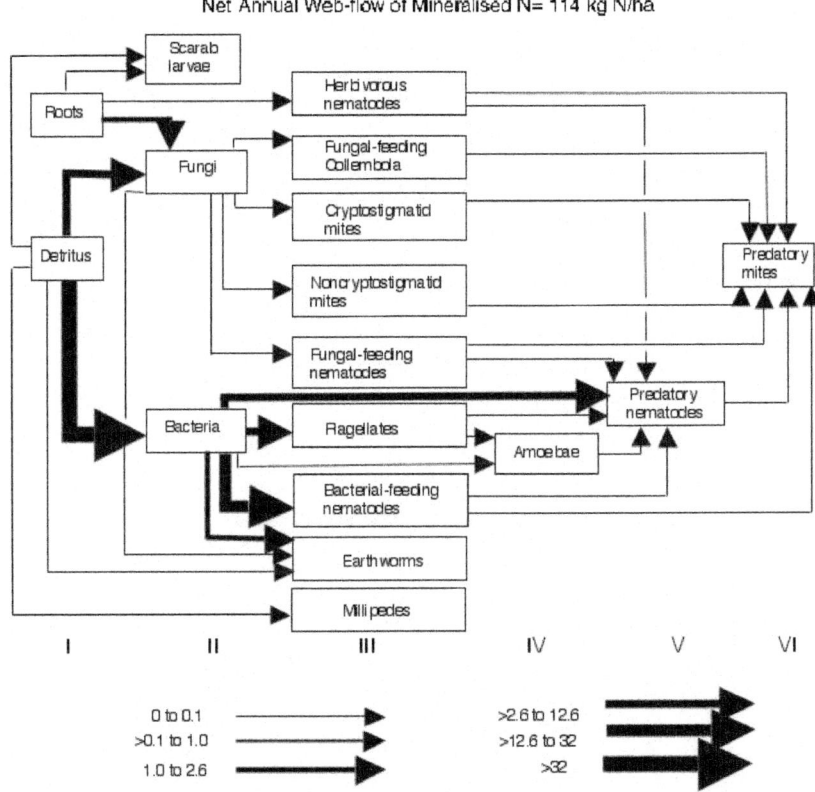

Fertilised Sown Pasture
Net Annual Web-flow of Mineralised N= 114 kg N/ha

Fig. Soil Food Webs of Small Organisms on Native, Unfertilized and Sown, Fertilized Pastures at Armidale NSW.

Food webs can be extended from a "who eats who" basis to include the amount of nutrients which flow through the various trophic (or feeding) levels. In the case of the two food webs below, different thicknesses of arrow denote the amount of nitrogen that is being mineralised or released by each different group from the organic matter base. The arrows represent net annual flows of nitrogen.

These two food webs were constructed from data obtained from two grasing systems at Armidale, NSW, Australia. One is a native, unfertilized pasture dominated by the native grasses, Themeda australis and Poa sieberiana. The other pasture is fertilized with superphosphate, and sown to the grass, Phalaris aquatica, and Trifolium repens (white clover). The clover in the introduced pasture fixes nitrogen so that both the soil and organic residues formed on the pasture are higher in nitrogen than that formed on the native pastures. Hence the thicker the arrows in the introduced pasture diagram.

ROLES OF SOIL ORGANISMS

Function of Soil Organisms

There are three main roles that soil organisms perform in soil :

1. Decompose organic residues
2. Re-cycle nutrients from organic residues
3. Enhance soil structure.

Decomposition and Nutrient Cycling

Decomposer invertebrate animals and microbes have quite distinct roles in breaking down organic detritus. Although invertebrates play only a small part in chemically degrading organic detritus, they help the more important microbes in many ways to do their job. During decomposition of organic matter, nutrients are released.

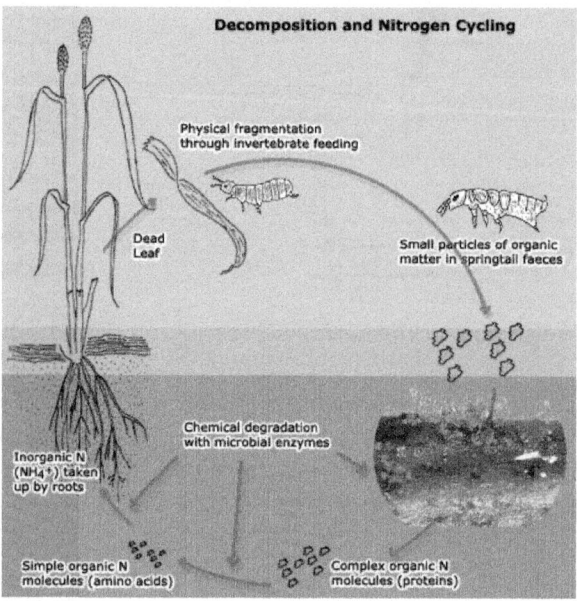

Fig. Nitrogen Cycling in a Pasture Soil. Springtails Fragment the Dead Grass while Microbes Chemically Degrade it.

Enhancement of Soil Structure

Good soil structure helps with physical fertility of soil. Both the large soil animals (*e.g.* earthworms) and the tiny microbes have roles in improving soil structure. Soil with good structure has many beneficial effects including enhanced water transmission into and through soil, lower bulk density and lower potential for soil erosion.

Large Soil Animals

Fig. Earthworm Tunnels - Horizontal and Vertical.

Large soil animals make tunnels through the soil. These are often called macropores as they form the larger pores in the soil. Some earthworms, dung beetles, spiders, ants and cicadas (as they emerge from the soil to change into adults), make vertical tunnels that open to the soil surface and down which water can infiltrate. Other tunnelers form macropores that don't open to the soil surface (earthworms, termites). The formation of holes (or macropores) in the soil, helps water transmission and soil hydrology. If water can easily enter the soil, less runs off to cause erosion.

Soil animals mix soil layers together and also mix organic matter that they eat with mineral soil layers. Ants, earthworms and termites bring organic matter into the soil from the surface and deposit it, thus increasing the organic matter content of soil. This helps water retention in soil.

Earthworm activity lowers soil bulk density and makes soil more friable–roots can penetrate this soil more easily.

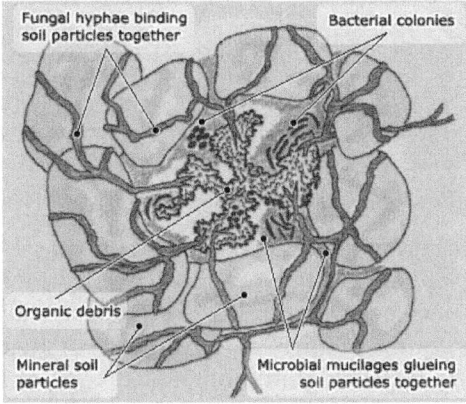

Fig. Diagram of a Soil Aggregate.

Microbes

Bacteria and fungi help in the formation of water-stable soil aggregates. Fungal hyphae grow around and between soil mineral and organic particles and physically bind them together. Both bacteria and fungi secrete polysaccharide mucilages which are sticky and glue the soil particles together into aggregates. These aggregates can be stable to the action of water for several months and help prevent slaking and dispersion of the soil.

Soil organisms clear away and degrade organic debris such as dead plants, animals and dung and use it as a source of food and nutrients, and in the process release chemical elements (nutrients) into the soil solution so that living plants may re-use them. This process sustains soil chemical fertility. But soil organisms are also responsible for soil physical fertility as well. They help the formation of good soil structure.

In addition, soil organisms degrade chemicals and pollutants that enter soil. Also, some soil organisms live in mutually beneficial relationships with plants, enhancing the plant's nutrition by increasing the nitrogen uptake from the atmosphere (Rhizobium) or phosphorus uptake from the soil by soil (mycorrhizae).

Role of Invertebrate Animals

Soil animals and soil microbes have different ways of decomposing organic detritus. Those soil animals which have mouthparts (*e.g.* arthropods) bite off bits of organic matter and fragment it into small pieces. Some earthworms select organic particles that are larger than their mouths, and can split them into smaller fragments. On passage through the invertebrate guts, further disintegration of the organic matter occurs.

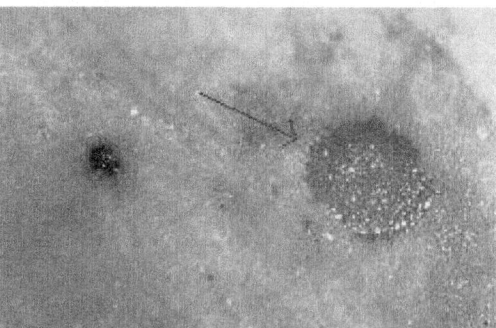

Fig. This Circular Patch on Dead Leaf Eaten by Springtail is Only 1mm Accross.

Watch as the springtail eats dead plant litter and passes a faecal pellet out the other end. The springtail is head down, tail up ! Springtails can deposit faecal pellets at the rate of once every 5 minutes. You can see how a large pile of dung has built up over the course of the 30 minutes as I watched it feed. The plant leaf is being fragmented as the springtail munches.

Role of Microbes

Microbes (bacteria and fungi) digest organic material externally by exuding a huge variety of digestive enzymes (>50 different ones) onto their food. These enzymes help chemically degrade organic detritus, breaking up large organic molecules (*e.g.* proteins) in which nutrient elements are bonded. Eventually, simpler molecules (*e.g.* amino acids) are split off and passed back through the microbial cell walls to be absorbed into their tissues and used for growth and maintenance. The simple organic molecules are eventually reduced to an inorganic form (*e.g.* NH_4^+) which plants can re-use.

MANAGEMENT EFFECTS ON SOIL BIOTA

What happens to the organisms that live in the soil when land is managed for agriculture, forestry and mining? The production of food, fibre, wood products and mineral ores sometimes involves profound disturbnces to the habitat of soil organisms. We clear the land of trees for grasing or cropping, we selectively log trees in forests or clear fell forests, we physically disturb the soil through ploughing or digging, we replace native plants with exotic ones, large herbivores graze the land, we use agro-chemicals for suppressing weeds, invertebrate and microbial pests of plants and the parasites of livestock. How do the activities listed below affect soil biota?

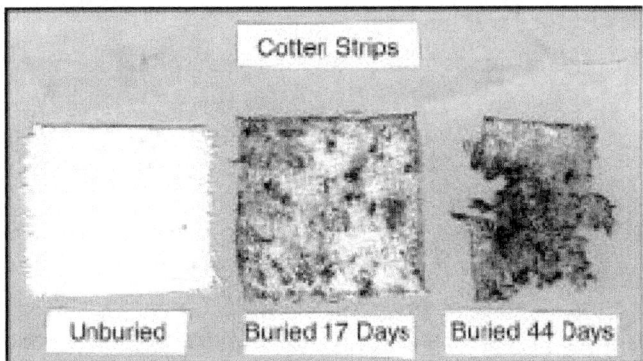

This photo shows cotton strips which have been buried for 17 and 44 days and a fresh, unburied strip. The microbes have been using the cellulose fibres in the cloth as a source of energy and, by 44 days in the soil, have started to rot the strip completely away in spots. You can also see how the colonies of bacteria and fungi colonising the strips have progressively stained the strips over time.

GRASING

Effects of Grasing on the Habitat, Micro-climate and Food Supply of Soil Biota

Fig. A Fenceline Photo with Heavily and Lightly Grazed Pastures Compared.

- Heavy grasing results in low plant litter levels. Litter provides both a living-space for soil animals as well as forming organic residues in the soil as food for soil biota.

- Litter protects the soil environment against climatic extremes. In pastures at Tamworth, in northern NSW, Australia, summer temperatures at 5 cm depth in overgrazed, bare pastures can reach 50 °C. However, the soil temperature can be <25 °C at the same depth in lightly grazed pastures with a good litter layer.

- Litter layers slow the evaporation of moisture from soil from 10 mm/day to <2 mm/day in summer in the same pastures.

- Trampling by high numbers of large herbivores such as sheep and cattle can compact the soil, squeesing the pores and making them smaller. It is then difficult for small soil animals to access the soil as they cannot form their own tunnels. In addition, compacted soil is hard for the tunneling animals such as earthworms to move through.

- Soil nutrients and organic matter (in the form of dung) are concentrated into stock camps on higher ground or in corners of paddocks, where stock congregate to rest. This aggregation occurs at the expense of nutrients and organic matter over the rest of the paddock.

Overgrasing affects the soil biota in following ways :

- Reduces the numbers and biomass of soil meso-fauna and macro-fauna
- Soil microbial activity is higher in the stock camps than over the general paddock
- Diversity of some species of soil biota declines (*e.g.* springtails).

CULTIVATION

Effects of Cultivation on the Habitat, Micro-climate and Food Supply of Soil Biota

Fig. Freshly ploughed soil.

- Burial of the protective layer of dead plants on the soil surface. This affects the living space for litter-dwellers and the soil micro-climate (increased soil moisture loss and extremes of temperature)
- Physical disruption of soil habitat *e.g.* pore channel continuity, pore size – this limits the mobility of those animals which do not tunnel
- Physical injury to animals
- Reduction of organic matter levels in soil
- Inversion of soil brings animals to surface where they desiccate and are eaten by birds
- Inversion of soil traps smaller soil biota.

Effects of cultivation on soil organisms :

- Numbers of large animals (earthworms, millipedes) are lowered and they can take up to two years to recover to pre-cultivation levels because they have long life cycles. Cropping soils that are repeatedly cultivated have fewer earthworms.

- Micro-arthropod numbers (springtails, mites) can also decline by 50 per cent after cultivation. They can recover their original numbers within 6 months as they have short life cycles.
- Soil microbial biomass declines after cultivation probably as a result of the reduced soil organic matter levels, their energy source. Microbial biomass can also be reduced, in part, by the physical disruption of extensive fungal hyphal networks.

FERTILIZERS

Fig. Native unfertilized Pasture.

Fig. Fertilized Pasture.

Fertilizers are used to increase agricultural productivity and to replace nutrients that are either removed in agricultural products (meat, wool, crops) or eroded by water and wind.

Agricultural fertilizers fall into several groups. Inorganic chemical fertilizers include phosphate, nitrogen and sulphur fertilizers while organic fertilizers include manures and compost. Lime is used mainly as a soil ameliorant to reduce acidity.

INORGANIC FERTILIZERS

Mostly, inorganic fertilizers increase soil biological activity because they increase plant production leaving organic residues of high quality as food for soil biota. There may be some short-term adverse effects, possibly due to pH change but, in the long-term, effects of inorganic fertilizers on overall biological activity are mostly stimulatory. However, species types and diversity of soil biota may change with fertilizer use.

Fig. Micro-arthropods (springtails and mites) from native unfertilized pasture (left) and fertilized pastures (right). Note larger numbers collected in same volume of fertilized pasture. Note greater diversity of types of animals in the native unfertilized pasture.

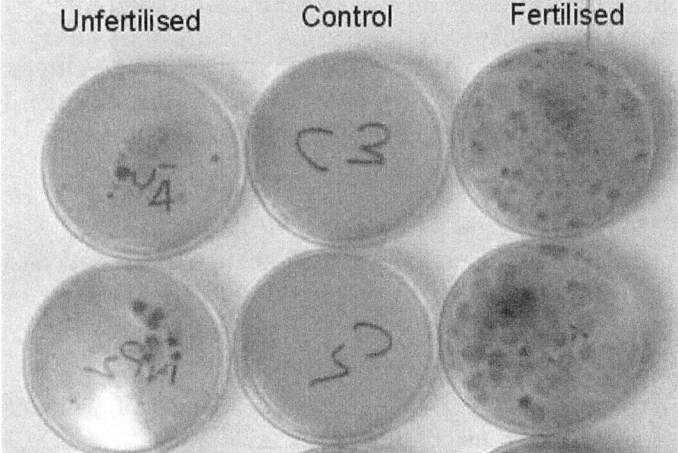

Fig. Agar Plates Growing Microbes from Native Unfertilized Pasture (left) and Fertilized Pasture (right). Note Larger Numbers of Microbes on Plates from Fertilized Pastures in same Volume of Soil.

P AND S FERTILIZERS

Superphosphate (a source of P and S) is used widely on both pastures and on cropping soils and is not considered to be toxic to soil biota in the long-term. The increased secondary production that follows from their use, along with the increased quality of organic residues that form on those farming systems, increases biological activity.

Effects of superphosphate on soil biota include :

• Increase in earthworm numbers by between 2 to 5 times.

• Increase in springtail, mite and nematode numbers by up to 3-4 times.

- Increase of microbial abundance in soil by 2 times.
- Species diversity of some groups decline (*e.g.* springtails).

Effects of elemental S fertilizers on soil biota include :

- Decrease in some types of Protozoa.
- Decrease in fungal biomass.
- Increase in abundance of sulphur -oxidising bacteria.

N FERTILIZERS

Nitrogen content of soil can be increased with the use of legumes in pastures or green manure rotations in cropping soils and this can increase soil biota abundance. Nitrogen can also be applied as inorganic fertilizers. There are different forms of nitrogen-containing fertilizer which are commonly used and they have variable effects on soil biota. As a general rule, if soil pH declines after application of the fertilizer, a reduction occurs in abundance of some soil biota. Urea is generally considered to be less harmful than the anhydrous ammonia as the pH change is less. These deleterious effects may be only short-term. Long-term effects of nitrogen fertilization can be stimulatory to soil biota with increased plant production providing better quality residues However, soil microbes decline if soil becomes acidic with prolonged use of anhydrous ammonia.

Effects of urea fertilizers on soil biota :

- Increase in free-living nematode abundance.
- Species diversity of nematodes can change : root and bacterial feeders can increase with N fertilizer use while fungal feeders and omnivores decrease.

Effects of anhydrous ammonia on soil biota :

- Decrease in soil microbes in immediate vicinity of fertilizer (short-term) or after prolonged use of this fertilizer.
- Decrease in earthworm abundance in immediate vicinity of fertilizer.

LIME

Effects of lime on soil biota include :

- Change in earthworm species from acid to alkaline tolerant. Overall numbers may remain the same.
- Increase in bacterial to fungal ratio. Fungi more acid tolerant than bacteria.

ORGANIC FERTILIZERS

Organic amendments in the form of composts or manures, generally increase biological activity in soil through an increased supply of energy (food for soil biota) and nutrients to the detrital food web.

Fig. Manure Spreading on Cropping Soil.

Effects of organic fertilizers on soil biota :

*   Increase in earthworm numbers.
*   Increase in nematode, springtail and mite abundance.
*   Increase in microbial biomass.

Pesticides

Chemicals used to control pests and diseases in plants and animals can have undesirable toxic side-effects on the non-target soil biota. Pesticides have variable effects on soil biota. The same pesticide may affect different species of soil biota in different ways. Also, effects of pesticides depend on the nature of the chemical, its dose, the method of application, temperature and moisture conditions in soil, crop residue management, the rate of decay of the chemical and the extent of leaching from the site. For example, some herbicides are more persistent if sprayed onto a mulch than if sprayed onto bare soil as they may be quickly inactivated by adsorption onto soil particles.

However, some generalisations of effects of pesticides on soil biota can be made. A general rule-of-thumb is that if the toxic effect of a pesticide on a non-target organism lasts longer than 60 days, then the chemical can be regarded as persistent and its toxic effects only slowly reversible.

Another general rule is that toxicity of pesticides from the least to the most toxic to soil biota, follows the order : herbicides < insecticides < fungicides. The following points can be made about the eco-toxic effects of pesticides.

Fig. Spraying Herbicides.

Herbicides

- Newer herbicides affect enzyme pathways in plants which are not found in soil biota.
- Older herbicides can have some toxic effects on soil biota *e.g.* springtails, mites.
- Many fungi are tolerant to herbicides.
- Earthworms can increase in numbers following herbicide application as their food supply, in the form of dead plants, is increased.

Insecticides

- Organochlorines are more active in soil than organophosphates or carbamates
- Earthworms are not very susceptible to insecticides except carbamates.

Fungicides

- Benomyl is toxic to earthworms.
- Some fungicides contain copper which is toxic to earthworms and microbes. Staining of the strips are areas where microbes have colonised.

Fig. Toxic Effect of a Copper-based Pesticide (+Cu Strip) on Soil Microbes Decomposing a Cotton Strip.

PLANT RESIDUE RETENTION

Fig. On Left a Cultivated Field with no mulch Retained Compared with a Field with Zero Tillage with a Thick Mulch Layer.

In zero tillage cropping systems, a layer of mulch is retained on the soil surface and is not cultivated into the soil. Under these systems soil biota generally increase in abundance.

Effects of retention of organic surface residues on soil biota include :

- Increase in earthworm numbers (x6 times).
- Increase in springtails and mites.
- Increase in microbial biomass.

Crop Rotations

Crop rotations occour when different crops are planted in succession, on the same area of land. Benefits of this practice are that they provide a disease break as diseases specific to one crop are denied a host for several years.

Crop rotations that include a legume phase (*e.g.* lucerne) increase the nitrogen content of organic residues and soil. The quality of the diet for decomposer organisms is increased.

Fig. Cropping soil with Lucerne Growing in Rotation.

Effects of legume in crop rotations on soil biota include :

- Increase in abundance of some small soil animals.
- Increase in earthworm abundance.
- Increase in soil microbial biomass.

Irrigation

The increased soil moisture which follows irrigation favours many soil biota. However, waterlogged soil conditions decreases the oxygen content of soil and there are few soil macro-fauna (*e.g.* earthworms) present under these conditions.

Effects of irrigation on soil biota :

- Irrigation allows earthworms to remain active in summer.
- Increase in springtail and mite abundance.
- Increase in Protozoa abundance.

- Increase in soil microbial biomass.

Fig. Irrigation Keeps Soil Moisture Levels High and
Prevents Death of Soil Biota in Drought.

FIRE

Burning of vegetation is used to remove stubble from cropping soils or in grasslands to improve the quality of rough pastures such as savannah grasslands where it eliminates woody plants and inedible dry vegetation. In addition, many wild-fires occour throughout Australia in the summer months which burn whole forests. The adverse effects of fire are generally restricted to litter dwellers and true soil species are little affected. However, repeated burning of crop residues or forests depletes the soil of organic matter and biological activity falls as the food supply to soil biota is reduced.

Fig. Regrowth after Fire in Semi-arid Australia.

Effects of fire on soil biota :

- Litter dwelling organisms die.
- True soil dwellers are little affected in uncultivated soils.

- Some large invertebrates can survive fires by sheltering under rocks.
- Soil microbes increase after fire due to "liming effects" of wood ash.

Fallow and Topsoil Storage

Topsoil is often scraped off and stored in heaps. Soil can be stored for months on building sites and then subsequently spread out again once building is finished. Soil is stored for longer periods (*e.g.* years) during surface mining and the soil re-spread and rehabilitated with plants once mining operations have finished. However, during storage, mycorrhizal fungal spores germinate and die in this stored soil as there are few plants for them to grow into and they are dependent on plants for survival. Vegetation planted into this topsoil once it has been spread, often fail to thrive. One explanation is that the soil is depleted of the spores of the mycorrhizal fungi which help in plant nutrition.

A similar thing happens during long fallow where the soil surface is bare of plants for many months.

TREE CLEARING

Fig. Forests Cleared to Make way for Cow Pastures.

Effects of tree clearing on soil biota have not been studied extensively.

Some effects that have been noted are :

- Micro-arthropod and microbial biomass abundance higher under eucalypt trees than in grassy interspaces between trees.
- Increase in earthworm numbers in temperate grassland after clearing deciduous European forests.
- Decline in earthworm numbers in tropical soils after clearing forests.

Where do Soil Biota Live?

Organisms are generally considered to be "soil" organisms if they live in soil and the layers of dead organic matter that overlie the soil. Thus, soil biota live in the soil, in the surface layers of dead plant leaves (or litter), in dead vertebrate and invertebrate animals (carrion), in dead trees and logs, dead roots and in dung pellets and pats.

Organisms are generally considered to be "soil" organisms if they live in soil and the layers of dead organic matter that overlie the soil. Thus, soil biota live in the soil, in the surface layers of dead plant leaves (or litter), in dead vertebrate and invertebrate animals (carrion), in dead trees and logs, dead roots and in dung pellets and pats.

There are some true soil dwelling biota that rarely see the light of day but many "soil" animals migrate up and down the soil/litter/dung/carrion profile as these layers are moistened and then dry out again. For example, springtails will migrate into the litter from the soil when the litter is moistened with rain or dew but retreat into the moister regions of the soil once the day warms up and dries the litter out again.

Earthworms also migrate from the soil below dung pats up into the pats and drag bits of dung down into their burrows. Migrations allow soil biota to exploit the rich food sources in the surface organic residues.

Even in the soil, biota are not uniformly distributed throughout. They aggregate at areas which are rich in organic matter such as around a small piece of dead leaf which has fallen into the soil, or in the faecal pellet of a springtail or around a dead root, or in the rhizosphere. The rhizosphere is that region which extends a few millimeters out from a root where root exudates and sloughed cells from roots provide an aggregation of organic materials which provide a rich food source for soil animals and microbes.

Ninety per cent of soil organisms live in the top 10 cm soil. This is also the zone where most plant roots live, where most organic matter in soil occurs, and most nutrients are also found. It's no wonder that soil organisms congregate in this region of highest nutrition.

BODY FORM AND LIFESTYLE

Coping with Living in Soil

Soil is a dense, dark habitat and soil biota cope by developing special adaptations of body form or by changing behaviour. However, there are advantages to living in the soil. There are generally lots of organic residues overlying the soil on which to feed; shelter from above-ground predators is provided and the temperature, relative humidity and the moisture content of the environment is more stable than in the above-ground environment. Here are some ways that soil biota use to adapt to life in the soil.

Adaptations of Body Form

- True soil animals live in the dark recesses of the soil and have little need for eyes. They often have small eyes or light sensitive areas of skin (*e.g.* earth worms) or are lacking eyes altogether (*e.g.* soil springtails).

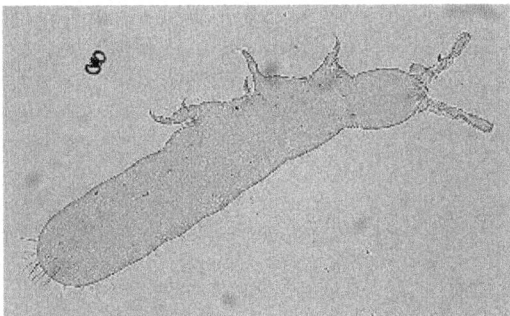

Fig. A Soil Dwelling Springtail with no Dark Eyespots.

- Many soil animals are wingless. Wings are no use as they can't be expanded and fluttered in close confinement in the soil. Springtails never developed any wings over evolutionary time and other animals lost theirs, all the better to live in soil and rotting logs (*e.g.*, wingless cockroaches).

Fig. A Flat Vative Cockroach Living in Rotting Wood.

- It is often the smaller species of any soil biota group which live in soil while their larger relatives inhabit the surface organic layers in plant litter and rotting logs (*e.g.* millipedes and centipedes).

Fig. A Large Dark litter-dwelling springtail. Compare its size with the white soil dwelling springtail on right.

- Some soil animals have small legs and other appendages (antennae, springing organ of springtails, legs of mites and springtails) as long appendages would only hamper movement in the cramped environment of the soil. However, predatory soil animals still tend to have long legs for chasing prey (*e.g.* predatory soil mites).

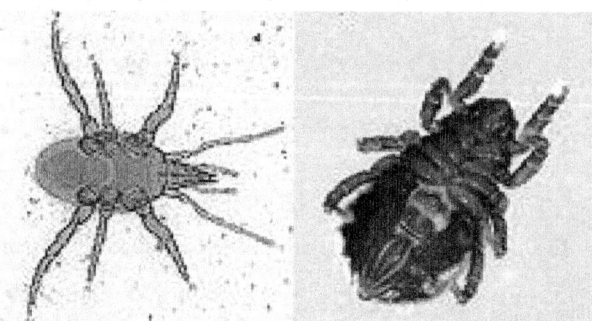

Fig. Left : A Fast Moving Predatory Mite with Long Legs. Right : a Slower Moving Decomposer Mite.

- Long, thin animals can move through the soil. Hard-bodied, rigid animals such as millipedes can push through the soil without damage while the soft-bodied earthworms can move soil particles gently aside by alternately lengthening and contracting their bodies. Earthworms lengthen and push between soil particles. They then dig tiny bristles on their bodies into the burrow walls, which helps anchor themselves before pulling the rest of their bodies forward. Earthworms and millipedes also create burrows by simply eating their way through the soil. They are called "tunnelers".

Fig. Long Slender Bodies of an Earthworm (above) and Millipede (below).

- Some insects have developed different bits of their anatomy which act as shovels. Dung beetles have flanges on their legs and heads which help move soil particles from their burrows to the soil surface — a bit like a mini-bulldozer. Mole crickets have front legs with wide flanges and spurs and the first thoracic segment behind the head is enlarged and which protects the head while digging. These animals are called "excavators" — likening their anatomy and digging activities to large earthmoving equipment.

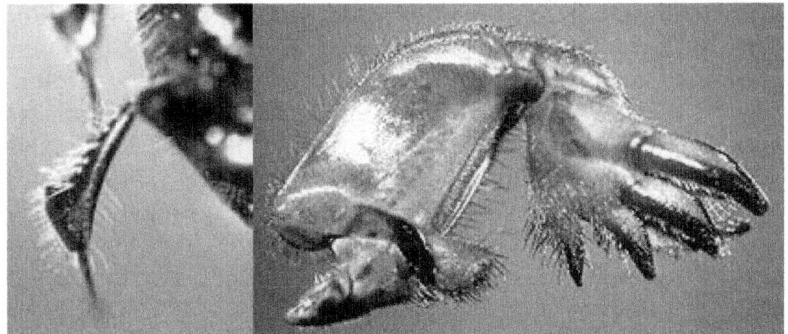

Fig. Digging Legs of Dung Beetle (left) and Mole Cricket (Right).

- Earthworms secrete a mucous layer to lubricate their bodies against abrasive surfaces in the soil.

- Many animals that live in the surface soil and in litter, dung and logs have a flattened body shape so they can slide easily under the litter layer, stones and logs for protection (*e.g.* centipedes, cockroaches).

- Many true soil animals are white as they don't need pigment to protect them against ultra-violet light (*e.g.* springtails, millipedes).

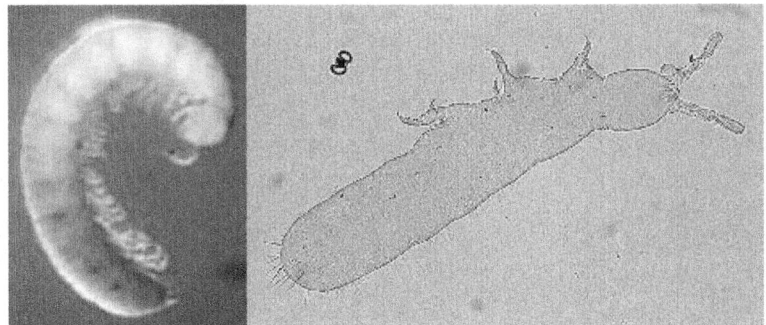

Fig. White Millipede (left) and Springtail (right) which Live Deep in the soil.

Adaptations of Behaviour

- Many soil animals, particularly those that move up into the surface layers of litter, or onto the soil surface, are active in the hours of dark (*e.g.* spiders and snails). By being nocturnal, they can avoid predators (*e.g.* birds) and they can avoid the hotter, drier hours of the day. Many soil animals will move away from a light source, preferring darker parts of a light gradient (*e.g.* woodlice).

- Some social insects can create their own habitat by constructing nests where the micro-climate is carefully controlled. Termites and ants excavate soil cavities and move soil particles about to create living spaces and foraging galleries. Termites nests can be sealed or opened to the air depending on prevailing climatic conditions.

- Predators in soil can have a lie-in-wait strategy (antlions, trapdoor spiders, scorpions) while other predators are active hunters. The "lie-in-wait" animals, use strategies that conceal them in the soil (*e.g.* antlion pits, trapdoors into underground tunnels of trapdoor spiders) and then spring out and capture their prey as it comes too near their lair. The "active hunters" leave their daytime homes in the soil for the night shift (*e.g.* wolf spiders and centipedes). Protected by the night from things that eat them, they roam about on the soil surface hunting for prey (wolf spiders, centipedes). These animals generally have long , "all the better to catch you with", legs.

- Large soil animals have various ways of moving soil about so that they can progress through it (*e.g.* shoveling it aside, eating through it). Although smaller invertebrate animals, such as springtails and mites, are not strong enough to move soil particles aside, they are tiny enough to use the existing pore spaces and channels between soil particles, to move from place to place.

Chapter 7

SOIL MICRO-ORGANISMS

INTRODUCTION

Pesticides are the chemical substances that kill pests and herbicides are the chemicals that kill weeds. In the context of soil, pests are fungi, bacteria insects, worms, and nematodes etc., that cause damage to field crops. Thus, in broad sense pesticides are insecticides, fungicides, bactericides, herbicides and nematicides that are used to control or inhibit plant diseases and insect pests. Although wide-scale application of pesticides and herbicides is an essential part of augmenting crop yields; excessive use of these chemicals leads to the microbial imbalance, environmental pollution and health hazards. An ideal pesticide should have the ability to destroy target pest quickly and should be able to degrade non-toxic substances as quickly as possible.

The ultimate "sink" of the pesticides applied in agriculture and public health care is soil. Soil being the storehouse of multitudes of microbes, in quantity and quality, receives the chemicals in various forms and acts as a scavenger of harmful substances. The efficiency and the competence to handle the chemicals vary with the soil and its physical, chemical and biological characteristics.

Effects of Pesticides

Pesticides reaching the soil in significant quantities have direct effect on soil micro-biological aspects, which in turn influence plant growth.

Some of the most important effects caused by pesticides are :

- Alterations hi ecological balance of the soil micro-flora,
- Continued application of large quantities of pesticides may cause ever lasting changes in the soil micro-flora,
- Adverse effect on soil fertility and crop productivity,

- Inhibition of N2 fixing soil micro-organisms such as *Rhizobium, Azotobacter, Azospirillum* etc., and cellulolytic and phosphate solubilising micro-organisms,

- Suppression of nitrifying bacteria, *Nitrosomonas* and *Nitrobacter* by soil fumigants ethylene bromide, Telone, and vapam have also been reported,

- Alterations in nitrogen balance of the soil,

- Interference with ammonification in soil,

- Adverse effect on mycorrhizal symbioses in plants and nodulation in legumes, and

- Alterations in the rhizosphere micro-flora, both quantitatively and qualitatively.

Persistence of Pesticides in Soil

How long an insecticide, fungicide, or herbicide persists in soil is of great importance in relation to pest management and environmental pollution. Persistence of pesticides in soil for longer period is undesirable because of the reasons : a) Accumulation of the chemicals in soil to highly toxic levels, b) May be assimilated by the plants and get accumulated in edible plant products, c) Accumulation in the edible portions of the root crops, d) To be get eroded with soil particles and may enter into the water streams, and finally leading to the soil, water and air pollutions. The effective persistence of pesticides in soil varies from a week to several years depending upon structure and properties of the constituents in the pesticide and availability of moisture in soil. For instance, the highly toxic phosphates do not persist for more than three months while chlorinated hydrocarbon insecticides (*e.g.* DOT, aldrin, chlordane etc.) are known to persist at least for 4–5 years and some times more than 15 years.

From the agricultural point of view, longer persistence of pesticides leading to accumulation of residues in soil may result into the increased absorption of such toxic chemicals by plants to the level at which the consumption of plant products may prove deleterious/hazardous to human beings as well as livestock's. There is a chronic problem of agricultural chemicals, having entered in food chain at highly inadmissible levels in India, Pakistan, Bangladesh and several other developing countries in the world. For example, intensive use of DDT to control insect pests and mercurial fungicides to control diseases in agriculture had been known to persist for longer period and thereby got accumulated in the food chain leading to food contamination and health hazards. Therefore, DDT and mercurial fungicides has been, banned to use in agriculture as well as in public health department.

Biodegradation of Pesticides in Soil

Pesticides reaching to the soil are acted upon by several physical, chemical, and biological forces. However, physical and chemical forces are acting upon/ degrading the pesticides to some extent, micro-organism's plays major role in the degradation of pesticides. Many soil micro-organisms have the ability to act upon pesticides and convert them into simpler non-toxic compounds. This process of degradation of pesticides and conversion into non-toxic compounds by micro-organisms is known as "biodegradation". Not all pesticides reaching to the soil are biodegradable and such chemicals that show complete resistance to biodegradation are called "recalcitrant".

The chemical reactions leading to biodegradation of pesticides fall into several broad categories which are discussed in brief in the following paragraphs :

- *Detoxification* : Conversion of the pesticide molecule to a non-toxic compound. Detoxification is not synonymous with degradation. Since a single chance in the side chain of a complex molecule may render the chemical non-toxic.

- *Degradation* : The breaking down/transformation of a complex substrate into simpler products leading finally to mineralisation. Degradation is often considered to be synonymous with mineralisation, *e.g.* Thirum (fungicide) is degraded by a strain of *Pseudomonas* and the degradation products are dimethlamine, proteins, sulpholipaids, etc.

- *Conjugation (complex formation or addition reaction)* : In which an organism make the substrate more complex or combines the pesticide with cell metabolites. Conjugation or the formation of addition product is accomplished by those organisms catalysing the reaction of addition of an amino acid, organic acid or methyl crown to the substrate, for *e.g.*, in the microbial metabolism of sodium dimethly dithiocar-bamate, the organism combines the fungicide with an amino acid molecule normally present in the cell and thereby inactivate the pesticides/chemical.

- *Activation* : It is the conversion of non-toxic substrate into a toxic molecule, for *e.g.* Herbicide, 4-butyric acid (2, 4-D B) and the insecticide Phorate are transformed and activated micro-biologically in soil to give metabolites that are toxic to weeds and insects.

- *Changing the spectrum of toxicity* : Some fungicides/pesticides are designed to control one particular group of organisms/pests, but they are metabolised to yield products inhibitory to entirely dissimilar groups of organisms, for *e.g.* the fungicide PCNB fungicide is converted in soil to chlorinated benzoic acids that kill plants.

Biodegradation of pesticides/herbicides is greatly influenced by the soil factors like moisture, temperature, PH and organic matter content, in addition to microbial population and pesticide solubility. Optimum temperature, moisture and organic matter in soil provide congenial environment for the break down or retention of any pesticide added in the soil. Most of the organic pesticides degrade within a short period (3-6 months) under tropical conditions. Metabolic activities of bacteria, fungi and actinomycetes have the significant role in the degradation of pesticides.

Criteria for Bioremediation/Biodegradation

For successful biodegradation of pesticide in soil, following aspects must be taken into consideration :

- Organisms must have necessary catabolic activity required for degradation of contaminant at fast rate to bring down the concentration of contaminant,
- The target contaminant must be bioavailability,
- Soil conditions must be congenial for microbial/plant growth and enzymatic activity and
- Cost of bioremediation must be less than other technologies of removal of contaminants.

According to Gales (1952) principal of microbial infallibility, for every naturally occurring organic compound there is a microbe/enzyme system capable its degradation.

Strategies for Bioremediation

For the successful biodegradation/bioremediation of a given contaminant following strategies are needed :

- *Passive/ intrinsic Bioremediation* : It is the natural bioremediation of contaminant by tile indigenous micro-organisms and the rate of degradation is very slow.
- *Biostimulation* : Practice of addition of nitrogen and phosphorus to stimulate indigenous micro-organisms in soil.
- *Bioventing* : Process/way of Biostimulation by which gases stimulants like oxygen and methane are added or forced into soil to stimulate microbial activity.
- *Bioaugmentation* : It is the inoculation/intro duction of micro-organisms in the contamin ated site/soil to facilitate biodegradation.

ROLE AND APPLICATION

The concept of effective micro-organisms (EM) was developed by Professor Teruo Higa, University of the Ryukyus, Okinawa, Japan (Higa, 1991; Higa and Wididana, 1991a). EM consists of mixed cultures of beneficial an naturally-occurring micro-organisms that can be applied as inoculants to increase the microbial diversity of soils and plant. Research has shown that the inoculation of EM cultures to the soil/plant ecosystem can improve soil quality, soil health, and the growth, yield, and quality of crops.

EM contains selected species of micro-organisms including predominant populations of lactic acid bacteria and yeasts and smaller numbers of photosynthetic bacteria, actinomycetes and other types of organisms. All of these are mutually compatible with one another and can coexist in liquid culture.

EM is not a substitute for other management practices. It is, however, an added dimension for optimising our best soil and crop management practices such as crop rotations, use of organic amendments, conservation tillage, crop residue recycling, and bio-control of pests. If used properly, EM can significantly enhance the beneficial effects of these practices.

Throughout the discussion which follows, we will use the term "beneficial micro-organisms" In a general way to designate a large group of often unknown or ill-defined micro-organisms that interact favourably in soils and with plants to render beneficial effects which are sometimes difficult to predict. We use the term "effective micro-organisms" or EM to denote specific mixed cultures of known, beneficial micro-organisms that are being used effectively as microbial inoculants.

Conceptual design is important in developing new technologies for utilising beneficial and effective micro-organisms for a more sustainable agriculture and environment. The basis of a conceptual design is imply to first conceive an ideal or model and then to devise a strategy and method for achieving the reality. However it is necessary to carefully co-ordinate the materials, the environment, and the technologies constituting the method. Moreover one should adopt a philosophical attitude in applying microbial technologies to agricultural production and conservation systems.

There are many opinions on what an ideal agricultural system is. Many would agree that such an idealised system should produce food on a long-term sustainable basis. Many would also insist that it should maintain and improve human health, be economically and spiritually beneficial to both producers and consumers, actively preserve and protect the environment, be self-contained and regenerative, and produce enough food for an increasing world population.

Utilisation and Recycling of Energy

Agricultural production begins with the process of photo-synthesis by green plants which requires solar energy, water, and carbon dioxide. It occurs through the plants ability to utilise solar energy in "fixing" atmospheric carbon into carbohydrates. The energy obtained is used for further bio-synthesis in the plant, including essential amino acids and proteins. The materials used for agricultural production are abundantly available with little initial cost. However, when it is observed as an economic activity, the fixation of carbon by photo-synthesis has an extremely low efficiency mainly because of the low utilisation rate of solar energy by green plants. Therefore, an integrated approach is needed to increase the level of solar energy utilisation by plants so that greater amounts of atmospheric carbon can be converted into useful substrates.

Although the potential utilisation rate of solar energy by plants has been estimated theoretically at between 10 and 20 per cent , the actual utilisation rate is less than 1 per cent . Even the utilisation rate of C4 plants, such as sugar cane whose photo-synthetic efficiency is very high, barely exceeds 6 or 7 per cent during the maximum growth period. The utilisation rate is normally less than 3 per cent even for optimum crop yields.

Past studies have shown that photo-synthetic efficiency of the chloroplasts of host crop plants cannot be increased much further; this means that their bio-mass production has reached a maximum level. Therefore, the best opportunity for increasing biomass production is to somehow utilise the visible light, which chloroplasts cannot presently use, and the infrared radiation; together, these comprise about 80 per cent of the total solar energy. Also, we must explore ways of recycling organic energy contained in plant and animal residues through direct utilisation of organic molecules by plants.

Thus, it is difficult to exceed the existing limits of crop production unless the efficiency of utilising solar energy is increased, and the energy contained in existing organic molecules (amino acids, peptides and carbohydrates) is utilised either directly or indirectly by the plant. This approach could help to solve the problems of environmental pollution and degradation caused by the misuse and excessive application of chemical fertilizers and pesticides to soils. Therefore, new technologies that can enhance the economic-viability of farming systems with little or no use of chemical fertilizers and pesticides are urgently needed and should be a high priority of agricultural research both now and in the immediate future.

Preservation of Natural Resources and the Environment

The excessive erosion of topsoil from farmland caused by intensive tillage and row-crop production has caused extensive soil degradation and also contributed to the pollution of both surface and groundwater. Organic wastes from animal production, agricultural and marine processing industries, and municipal wastes (*i.e.*, sewage and garbage), have become major sources of environmental pollution in both developed and developing countries. Furthermore, the production of methane from paddy fields and ruminant animals and of carbon dioxide from the burning of fossil fuels, land clearing and organic matter decomposition have been linked to global warming as "greenhouse gases".

Chemical-based, conventional systems of agricultural production have created many sources of pollution that, either directly or indirectly, can contribute to degradation of the environment and destruction of our natural resource base. This situation would change significantly if these pollutants could be utilised in agricultural production as sources of energy.

Therefore, it is necessary that future agricultural technologies be compatible with the global ecosystem and with solutions to such problems in areas different from those of conventional agricultural technologies. An area that appears to hold the greatest promise for technological advances in crop production, crop protection, and natural resource conservation is that of beneficial and effective micro-organisms applied as soil, plant and environmental inoculants.

Beneficial and Effective Micro-organisms for a Sustainable Agriculture

Agriculture in a broad sense, is not an enterprise which leaves everything to nature without intervention. Rather it is a human activity in which the farmer attempts to integrate certain agroecological factors and production inputs for optimum crop and livestock production. Thus, it is reasonable to assume that farmers should be interested in ways and means of controlling beneficial soil micro-organisms as an important component of the agricultural environment. Nevertheless, this idea has often been rejected by naturalists and proponents of nature farming and organic agriculture. They argue that beneficial soil micro-organisms will increase naturally when organic amendments are applied to soils as carbon, energy and nutrient sources. This indeed may be true where an abundance of organic materials are readily available for recycling which often occurs in small-scale farming. However, in most cases, soil micro-organisms, beneficial or harmful, have often been controlled advantageously when crops in various agroecological zones are grown and cultivated in proper sequence (*i.e.*, crop rotations) and without the use of pesticides. This would explain why scientists have long been interested in the use of beneficial micro-organisms as soil and plant inoculants to shift the micro-biological equilibrium in a way that enhances soil quality and the yield and quality of crops.

Most would agree that a basic rule of agriculture is to ensure that specific crops are grown according to their agro-climatic and agroecological requirements. However, in many cases the agricultural economy is based on market forces that demand a stable supply of food, and thus, it becomes necessary to use farmland to its full productive potential throughout the year.

The purpose of crop breeding is to improve crop production, crop protection, and crop quality. Improved crop cultivars along with improved cultural and management practices have made it possible to grow a wide variety of agricultural and horticultural crops in areas where it once would not have been culturally or economically feasible. The cultivation of these crops in such diverse environments has contributed significantly to a stable food supply in many countries. However, it is somewhat ironic that new crop cultures are almost never selected with consideration of their nutritional quality or bio-availability after ingestion.

As will be discussed later, crop growth and development are closely related to the nature of the soil micro-flora, especially those in close proximity to plant roots, *i.e.*, the rhizosphere. Thus, it will be difficult to overcome the limitations of conventional agricultural technologies without controlling soil micro-organisms. This particular tenet is further reinforced because the evolution of most forms of life on earth and their environments are sustained by micro-organisms. Most biological activities are influenced by the state of these invisible, mi nuscule units of life. Therefore, to significantly increase food production, it is essential to develop crop cultivars with improved genetic capabilities (*i.e.*, greater yield potential, disease resistance, and nutritional quality) and with a higher level of environmental competitiveness, particularly under stress conditions (*i.e.*, low rainfall, high temperatures, nutrient deficiencies, and agressive weed growth).

To enhance the concept of controlling and utilising beneficial micro-organisms for crop production and protection, one must harmoniously integrate the essential components for plant growth and yield including light (intensity, photo-periodicity and quality), carbon dioxide, water, nutrients (organic-inorganic) soil type, and the soil micro-flora. Because of these vital inter-relationships, it is possible to envision a new technology and a more energy-efficient system of biological production.

Low agricultural production efficiency is closely related to a poor co-ordination of energy conversion which, in turn, is influenced by crop physiological factors, the environment, and other biological factors including soil micro-organisms. The soil and rhizosphere micro-flora can accelerate the growth of plants and enhance their resistance to disease and harmful insects by producing bio-active substances. These micro-organisms maintain the growth environment of plants, and may have secondary effects on crop quality. A wide range of results are possible depending on their predominance and activities at any one time.

Nevertheless, there is a growing consensus that it is possible to attain maximum economic crop yields of high quality, at higher net returns, without the application of chemical fertilizers and pesticides. Until recently, this was not thought to be a very likely possibility using conventional agricultural methods. However, it is important to recognise that the best soil and crop management practices to achieve a more sustainable agriculture will also enhance the growth, numbers and activities of beneficial soil micro-organisms that, in turn, can improve the growth, yield and quality of crops.

CONTROLLING THE SOIL MICRO-FLORA

Principles and Strategies

Principles of Natural Ecosystems and the Application of Beneficial and Effective Micro-organisms. The misuse and excessive use of chemical fertilizers and pesticides have often adversely affected the environment and created many a) food safety and quality and b) human and animal health problems. Consequently, there has been a growing interest in nature farming and organic agriculture by consumers and environmentalists as possible alternatives to chemical-based, conventional agriculture. Agricultural systems which conform to the principles of natural ecosystems are now receiving a great deal of attention in both developed and developing countries. A number of books and journals have recently been published which deal with many aspects of natural farming systems. New concepts such as alternative agriculture, sustainable agriculture, soil quality, integrated pest management, integrated nutrient management and even beneficial micro-organisms are being explored by the agricultural research establishment. Although these concepts and associated methodologies hold considerable promise, they also have limitations. For example, the main limitation in using microbial inoculants is the problem of reproducibility and lack of consistent results.

Unfortunately certain microbial cultures have been promoted by their suppliers as being effective for controlling a wide range of soil-borne plant diseases when in fact they were effective only on specific pathogens under very specific conditions. Some suppliers have suggested that their particular microbial inoculant is akin to a pesticide that would suppress the general soil microbial population while increasing the population of a specific beneficial micro-organism. Nevertheless, most of the claims for these single-culture microbial inoculants are greatly exaggerated and have not proven to be effective under field conditions. One might speculate that if all of the microbial cultures and inoculants that are available as marketed products were used some degree of success might be achieved because of the increased diversity of the soil micro-flora and stability that is associated with mixed cultures. While this, of course, is a hypothetical example, the fact remains that there is a greater likelihood of controlling the soil micro-flora by introducing mixed, compatible cultures rather than single pure cultures.

Even so, the use of mixed cultures in this approach has been criticised because it is difficult to demonstrate conclusively which micro-organisms are responsible for the observed effects, how the introduced micro-organisms interact with the indigenous species, and how these new associations affect the soil/plant environment. Thus, the use of mixed cultures of beneficial micro-organisms as soil inoculants to enhance the growth, health, yield, and quality of crops has not gained widespread acceptance by the agricultural research establishment because conclusive scientific proof is often lacking.

The use of mixed cultures of beneficial microorganisms as soil inoculants is based on the principles of natural ecosystems which are sustained by their constituents; that is, by the quality and quantity of their inhabitants and specific ecological parameters, *i.e.*, the greater the diversity and number of the inhabitants, the higher the order of their interaction and the more stable the ecosystem. The mixed culture approach is simply an effort to apply these principles to natural systems such as agricultural soils, and to shift the micro-biological equilibrium in favour of increased plant growth, production and protection.

It is important to recognise that soils can vary tremendously as to their types and numbers of micro-organisms. These can be both beneficial and harmful to plants and often the predominance of either one depends on the cultural and management practices that are applied. It should also be emphasised that most fertile and productive soils have a high content of organic matter and, generally, have large, populations of highly diverse micro-organisms (*i.e.*, both species and genetic diversity). Such soils will also usually have a wide ratio of beneficial to harmful micro-organisms.

Controlling the Soil Micro-flora for Optimum Crop Production and Protection

The idea of controlling and manipulating the soil micro-flora through the use of inoculants organic amendments and cultural and management practices to create a more favourable soil micro-biological environment for optimum crop production and protection is not new. For almost a century, micro-biologists have known that organic wastes and residues, including animal manures, crop residues, green manures, municipal wastes (both raw and composted), contain their own indigenous populations of micro-organisms often with broad physiological capabilities.

It is also known that when such organic wastes and residues are applied to soils many of these introduced micro-organisms can function as bio-control agents by controlling or suppressing soil-borne plant pathogens through their competitive and antagonistic activities. While this has been the theoretical basis for controlling the soil micro-flora, in actual practice the results have been unpredictable and inconsistent, and the role of specific micro-organisms has not been well-defined.

For, many years micro-biologists have tried to culture beneficial micro-organisms for use as soil inoculants to overcome the harmful effects of phytopathogenic organisms, including bacteria, fungi and nematodes. Such attempts have usually involved single applications of pure cultures of micro-organisms which have been largely unsuccessful for several reasons. First, it is necessary to thoroughly understand the individual growth and survival characteristics of each particular beneficial micro-organism, including their nutritional and environmental requirements. Second, we must understand their ecological relationships and interactions with other micro-organisms, including their ability to coexist in mixed cultures and after application to soils.

There are other problems and constraints that have been major obstacles to controlling the micro-flora of agricultural soils. First and foremost is the large number of types of micro-organisms that are present at any one time, their wide range of physiological capabilities, and the dramatic fluctuations in their populations that can result from man's cultural and management practices applied to a particular farming system. The diversity of the total soil micro-flora depends on the nature of the soil environment and those factors which affect the growth and activity of each individual organism including temperature, light, aeration, nutrients, organic matter, pH and water. While there are many micro-organisms that respond positively to these factors, or a combination thereof, there are many that do not. Micro-biologists have actually studied relatively few of the micro-organisms that exist in most agricultural soil, mainly because we don't know how to culture them; *i.e.*, we know very little about their growth, nutritional, and ecological requirements.

The "diversity" and "population" factors associated with the soil micro-flora have discouraged scientists from conducting research to develop control strategies. Many believe that, even when beneficial micro-organisms are cultured and inoculated into soils, their number is relatively small compared with the indigenous soil inhabitants, and they would likely be rapidly overwhelmed by the established soil micro-flora. Consequently, many would argue that even if the application of beneficial micro-organisms is successful under limited conditions (*e.g.*, in the laboratory) it would be virtually impossible to achieve the same success under actual field conditions. Such thinking still exists today, and serves as a principle constraint to the concept of controlling the soil micro-flora.

It is noteworthy that most of the micro-organisms encountered in any particular soil are harmless to plants with only a relatively few that function as plant pathogens or potential pathogens. Harmful micro-organisms become dominant if conditions develop that are favourable to their growth, activity and reproduction. Under such conditions, soil-borne pathogens (*e.g.*, fungal pathogens) can rapidly increase their populations with devastating effects on the crop. If these conditions change, the pathogen population declines just as rapidly to its original state. Conventional farming systems that tend towards the consecutive planting of the same crop (*i.e.*, mono-culture) necessitate the heavy use of chemical fertilizers and pesticides. This, in turn, generally increases the probability that harmful, disease-producing, plant pathogenic micro-organisms will become more dominant in agricultural soils.

Chemical-based conventional farming methods are not unlike symptomatic therapy. Examples of this are applying fertilizers when crops show symptoms of nutrient-deficiencies, and applying pesticides whenever crops are attacked by insects and diseases. In efforts to control the soil micro-flora some scientists feel that the introduction of beneficial micro-organisms should follow a symptomatic approach. However, we do not agree. The actual soil conditions that prevail at any point in time may be most unfavourable to the growth and establishment of laboratory-cultured, beneficial micro-organisms.

To facilitate their establishment, it may require that the farmer make certain changes in his cultural and management practices to induce conditions that will :

(a) Allow the growth and survival of the inoculated micro-organisms and

(b) Suppress the growth and activity of the indigenous plant pathogenic micro-organisms.

An example of the importance of controlling the soil micro-flora and how certain cultural and management practices can facilitate such control is useful here. Vegetable cultivars are often selected on their ability to grow and produce over a wide range of temperatures. Under cool, temperate conditions there are generally few pest and disease problems. However, with the onset of hot weather, there is a concomitant increase in the incidence of diseases and insects making it rather difficult to obtain acceptable yields without applying pesticides. With higher temperatures, the total soil microbial population increases as does certain plant pathogens such as Fusarium, which is one of the main putrefactive, fungal pathogens in soil. The incidence and destructive activity of this pathogen can be greatly minimised by adopting reduced tillage methods and by shading techniques to keep the soil cool during hot weather. Another approach is to inoculate the soil with beneficial, antagonistic, antibiotic-producing micro-organisms such as actinomycetes and certain fungi.

Application of Beneficial and Effective Micro-organisms : A New Dimension

Many micro-biologists believe that the total number of soil micro-organisms can be increased by applying organic amendments to the soil. This is generally true because most soil micro-organisms are heterotrophic, *i.e.*, they require complex organic molecules of carbon and nitrogen for metabolism and bio-synthesis.

Whether the regular addition of organic wastes and residues will greatly increase the number of beneficial soil micro-organisms in a short period of time is questionable. However, we do know that heavy applications of organic materials, such as seaweed, fish meal, and chitin from crushed crab shells, not only helps to balance the micro-nutrient content of a soil but also increases the population of beneficial antibiotic-producing actinomycetes. This changes the soil to a disease-suppressive condition within a relatively short period.

The probability that a particular beneficial micro-organism will become predominant, even with organic farming or nature farming methods, will depend on the ecosystem and environmental conditions. It can take several hundred years for various species of higher and lower plants to interact and develop into a definable and stable ecosystem. Even if the population of a specific micro-organism is increased through cultural and management practices, whether it will be beneficial to plants is another question. Thus, the likelihood of a beneficial, plant-associated micro-organism becoming predominant under conservation-based farming systems is virtually impossible to predict. Moreover, it is very unlikely that the population of useful anaerobic micro-organisms, which usually comprise only a small part of the soil micro-flora, would increase significantly even under natural farming conditions.

This information then emphasises the need to develop methods for isolating and selecting different microor ganisms for their beneficial effects on soils and plants. The ultimate goal is to select micro-organisms that are physiologically and ecologically compatible with each other and that can be introduced as mixed cultures into soil where their beneficial effects can be realised.

Application of Beneficial and Effective Micro-organisms : Fundamental Considerations

Micro-organisms are utilised in agriculture for various purposes; as important components of organic amen dments and composts, as legume inoculants for biological nitrogen fixation as a means of suppressing insects and plant diseases to Improve crop quality and yields, and for reduction of labour. All of these are closely related to each other. An important consideration in the application of beneficial micro-organisms to soils is the enhancement of their synergistic effects. This is difficult to accomplish if these micro-organisms are applied to achieve symptomatic therapy, as in the case of chemical fertilizers and pesticides.

If cultures of beneficial micro-organisms are to be effective after inoculation into soil, it is important that their initial populations be at a certain critical threshold level. This helps to ensure that the amount of bioactive substances produced by them will be sufficient to achieve the desired positive effects on crop production and/or crop protection. If these conditions are not met, the introduced micro-organisms, no matter how useful they are, will have little if any effect. At present, there are no chemical tests that can predict the probability of a particular soil-inoculated micro-organism to achieve a desired result. The most reliable approach is to inoculate the beneficial micro-organism into soil as part of a mixed culture, and at a sufficiently high inoculum density to maximise the probability of its adaptation to environmental and ecological conditions.

The application of beneficial micro-organisms to soil can help to define the structure and establishment of natural ecosystems. The greater the diversity of the cultivated plants that are grown and the more chemically complex the biomass, the greater the diversity of the soil micro-flora as to their types, numbers and activities.

The application of a wide range of different organic amen dments to soils can also help to ensure a greater microbial diversity. For example, combinations of various crop residues, animal manures, green manures, and municipal wastes applied periodically to soil will provide a higher level of microbial diversity than when only one of these materials is applied. The reason for this is that each of these organic materials has its own unique indigenous micro-flora which can greatly affect the resident soil micro-flora after they are applied, at least for a limited period.

CLASSIFICATION OF SOILS BASED ON THEIR MICRO-BIOLOGICAL PROPERTIES

Most soils are classified on the basis of their chemical and physical properties; little has been done to classify soils according to their physicochemical and microb iological properties. The reason for this is that a soil's chemical and physical properties are more readily defined and measured than their micro-biological properties.

Improved soil quality is usually characterised by increased infiltration; aeration, aggregation and organic matter content and by decreased bulk density, compaction, erosion and crusting. While these are important indicators of potential soil productivity, we must give more attention to soil biological properties because of their important relationship (though poorly understood) to crop production, plant and animal health, environmental quality, and food safety and quality. Research is needed to identify and quantify reliable and predictable biological/ecological indicators of soil quality. Possible indicators might include total species diversity or genetic diversity of beneficial soil micro-organisms as well as insects and animals.

The basic concept here is not to classify soils for the study of micro-organisms but for farmers to be able to control the soil micro-flora so that biologically-mediated processes can improve the growth, yield, and quality of crops as well as the tilth, fertility, and productivity of soils. The ultimate objective is to reduce the need for chemical fertilizers and pesticides.

Functions of Micro-organisms. Putrefaction, Fermentation, and Synthesis

Soil micro-organisms can be classified into decomposer and synthetic micro-organisms. The decomposer micro-organisms are sub-divided into groups that perform oxidative and fermentative decomposition. The fermentative group is further divided into useful fermentation (simply called fermentation) and harmful fermentation (called putrefaction). The synthetic micro-organisms can be sub-divided into groups having the physiological abilities to fix atmospheric nitrogen into amino acids and/or carbon dioxide into simple organic molecules through photo-synthesis.

Fermentation is an anaerobic process by which facultative micro-organisms (*e.g.*, yeasts) transform complex organic molecules (*e.g.*, carbohydrates) into simple organic compounds that often can be absorbed directly by plants. Fermentation yields a relatively small amount of energy compared with aerobic decomposition of the same substrate by the same group of micro-organisms.

Aerobic decomposition results in complete oxidation of a substrate and the release of large amounts of energy, gas, and heat with carbon dioxide and water as the end products. Putrefaction is the process by which facultative heterotrophic micro-organisms decompose proteins anaerobically, yielding malodorous incompletely oxidised, metabolites (*e.g.*, ammonia, mercaptans and indole) that are often toxic to plants and animals.

The term "synthesis" as used here refers to the biosynthetic capacity of certain micro-organisms to derive metabolic energy by "fixing" atmospheric nitrogen and/or carbon dioxide. In this context we refer to these as "synthetic" micro-organisms, and if they should become a predominant part of the soil micro-flora, then the soil would be termed a "synthetic" soil.

Nitrogen-fixing micro-organisms are highly diverse, ranging from "free-living" autotrophic bacteria of the genus Azotobacter to symbiotic, heterotrophic bacteria of the genus Rhizobium, and blue-green algae (now mainly classified as blue-green bacteria), all of which function aerobically. Photo-synthetic micro-organisms fix atmospheric carbon dioxide in a manner similar to that of green plants. They are also highly diverse, ranging from blue-green algae to green algae that perform complete photo-synthesis aerobically to photo-synthetic bacteria which perform incomplete photo-synthesis anaerobically.

Relationships Between Putrefaction, Fermentation, and Synthesis

The processes of putrefaction, fermentation, and synthesis proceed simultaneously according to the appropriate types and numbers of micro-organisms that are present in the soil. The impact on soil quality attributes and related soil properties is determined by the dominant process. The production of organic substances by micro-organisms results from the intake of positive ions, while decomposition serves to release these positive ions. Hydrogen ions play a pivotal role in these processes.

A problem occurs when hydrogen ions do not recombine with oxygen to form water but are utilised to produce methane, hydrogen sulfide, ammonia, mercaptans and other highly reduced putrefactive substances most of which are toxic to plants and produce malodors. If a soil is able to absorb the excess hydrogen ions during periods of soil anaerobiosis and if synthetic micro-organisms such as photo-synthetic bacteria are present, they will utilise these putrefactive substances and produce useful substrates from them which helps to maintain a healthy and productive soil.

The photo-synthetic bacteria, which perform incomplete photo-synthesis anaerobically, are highly desirable, beneficial soil micro-organisms because they are able to detoxify soils by transforming reduced, putrefactive substances such as hydrogen sulfide into useful substrates. This helps to ensure efficient utilisation of organic matter and to improve soil fertility. Photo-synthesis involves the photo-catalysed splitting of water which yields molecular oxygen as a by-product. Thus, these micro-organisms help to provide a vital source of oxygen to plant roots.

Reduced compounds such as methane and hydrogen sulfide are often produced when organic materials are decomposed under anaerobic conditions. These compo unds are toxic and can greatly suppress the activities of nitrogen-fixing micro-organisms. However, if synthetic micro-organisms, such as photo-synthetic bacteria that utilise reduced substances, are present in the soil, oxygen deficiencies are not likely to occur. Thus, nitrogen-fixing micro-organisms, coexisting in the soil with photo-synthetic bacteria, can function effectively in fixing atmospheric nitrogen even under anaerobic conditions.

Photo-synthetic bacteria not only perform photo-synthesis but can also fix-nitrogen. Moreover, it has been shown that, when they coexist, in soil with species of Azotobacter, their ability to fix nitrogen is enhanced. This then is an example of a synthetic soil. It also suggests that by recognising the role, function, and mutual compatibility of these two bacteria and utilising them effectively to their full potential, soils can be induced to a greater synthetic capacity. Perhaps the most effective synthetic soil system results from the enhancement of zymogenic and synthetic micro-organisms; this allows fermentation to become dominant over putrefaction and useful synthetic processes to proceed.

Classification of Soils Based on the Functions of Micro-organisms

As discussed earlier, soils can be characterised according to their indigenous micro-flora which perform putrefactive, fermentative, synthetic and zymogenic reactions and processes. In most soils, these three functions are going on simultaneously with the rate and extent of each determined by the types and numbers of associated micro-organisms that are actively involved at any one time.

Disease-Inducing Soils. In this type of soil, plant pathogenic micro-organisms such as Fusarium fungi can comprise 5 to 20 per cent of the total micro-flora if fresh organic matter with a high nitrogen content is applied to such a soil, incompletely oxidised products can arise that are malodorous and toxic to growing plants. Such soils tend to cause frequent infestations of disease organisms, and harmful insects. Thus, the application of fresh organic matter to these soils is often harmful to crops. Probably more than 90 per cent of the agricultural land devoted to crop production worldwide can be classified as having disease-inducing soil. Such soils generally have poor physical properties, and large amounts of energy are lost as "greenhouse" gases, particularly in the case of rice fields. Plant nutrients are also subject to immobilisation into unavailable forms.

Disease-Suppressive Soils

The micro-flora of disease-suppressive soils is usually dominated by antagonistic micro-organisms that produce copious amounts of antibiotics. These include fungi of the genera Penicillium, Trichoderma, and Aspergillus, and actinomycetes of the genus Streptomyces. The antibiotics they produce can have biostatic and biocidal effects on soil-borne plant pathogens, including Fusarium which would have an incidence in these soils of less than 5 per cent. Crops planted in these soils are rarely affected by diseases or insect pests. Even if fresh organic matter with a high nitrogen content is applied, the production of putrescent substances is very low and the soil has a pleasant earthy odour after the organic matter is decomposed.

These soils generally have excellent physical properties; for example, they readily, form water-stable aggregates and they are well-aerated, and have a high permeability to both air and water. Crop yields in the disease-suppressive soils are often slightly lower than those in synthetic soils. Highly acceptable crop yields are obtained whenever a soil has a predominance of both disease-suppressive and synthetic micro-organisms.

Zymogenic Soils

These soils are dominated by a micro-flora that can perform useful kinds of fermentations, *i.e.*, the breakdown of complex organic molecules into simple organic substances and inorganic materials. The organisms can be either obligate or facultative anaerobes. Such fermentation-producing micro-organisms often comprise the micro-flora of various organic materials, *i.e.*, crop residues, animal manures, green manures and municipal wastes including composts. After these amendments are applied to the soil, their number : and fermentative activities can increase dramatically and overwhelm the indigenous soil micro-flora for an indefinite period.

While these micro-organisms remain predominant, the soil can be classified as a zymogenic soil which is generally characterised by,

- Pleasant, fermentative odours especially after tillage,
- Favourable soil physical properties (*e.g.*, Increased aggregate stability, permeability, aeration and decreased resistance to tillage
- Large amounts of inorganic nutrients, amino acids, carbohydrates, vitamins and other bioactive substances which can directly or indirectly enhance the growth, yield and quality of crops,
- Low occupancy of Fusarium fungi which is usually less than 5 per cent, and
- Low production of greenhouse gases (*e.g.*, methane, ammonia, and carbon dioxide) from croplands, even where flooded rice is grown.

Synthetic Soils

These soils contain significant populations of micro-organisms which are able to fix atmospheric nitrogen and carbon dioxide into complex molecules such as amino acids, proteins and carbohydrates. Such micro-organisms include photosynthetic bacteria which perform incomplete photo-synthesis anaerobically, certain Phycomycetes (fungi that resemble algae), and both green algae and blue — green algae which function aerobically. All of these are photo-synthetic organisms that fix atmospheric nitrogen. If the water content of these soils is stable, their fertility can be largely maintained by regular additions of only small amounts of organic materials. These soils have a low Fusarium occupancy and they are often of the disease-suppressive type. The production of gases from fields where synthetic soils are present is minimal, even for flooded rice.

This is a somewhat simplistic classification of soils based on the functions of their predominant types of micro-organisms, and whether they are potentially beneficial or harmful to the growth and yield of crops. While these different types of soils are described here in a rather idealised manner, the fact is that in nature they are not always clearly defined because they often tend to have some of the same characteristics. Nevertheless, research has shown that a disease-inducing soil can be transformed into disease-suppressing, zymogenic and synthetic soils by inoculating the problem soil with mixed cultures of effective micro-organisms. Thus it is somewhat obvious that the most desirable agricultural soil for optimum growth, production, protection, and quality of crops would be the composite soil indicated in Fig., *i.e.*, a soil that is highly zymogenic and synthetic, and has an established disease-suppressive capacity. This then is the principle reason for seeking ways and means of controlling the micro-flora of agricultural soils.

CONTROL

Controlling the soil micro-flora to enhance the predominance of beneficial and effective micro-organisms can help to improve and maintain the soil chemical and physical properties. The proper and regular addition of organic amendments are often an important part of any strategy to exercise such control.

Previous efforts to significantly change the indigenous micro-flora of a soil by introducing single cultures of extrinsic micro-organisms have largely been unsuccessful. Even when a beneficial micro-organism is isolated from a soil, cultured in the laboratory, and reinoculated into the same soil at a very high population, it is immediately subject to competitive and antagonistic effects from the indigenous soil micro-flora and its numbers soon decline. Thus, the probability of shifting the "micro-biological equilibrium" of a soil and controlling it to favour the growth, yield and health of crops is much greater if mixed cultures of beneficial and effective micro-organisms are introduced that are physiologically and ecologically compatible with one another. When these mixed cultures become established their individual beneficial effects are often magnified in a synergistic manner.

Actually, a disease-suppressive micro-flora can be developed rather easily by selecting and culturing certain types of gram-positive bacteria that produce antibiotics and have a wide range of specific functions and capabilities; these organisms include facultative anaerobes, obligate aerobes, acidophilic and alkalophilic microbes.

These micro-organisms can be grown to high populations in a medium consisting of rice bran, oil cake and fish meal and then applied to soil along with well-cured compost that also has a large stable population of beneficial micro-organisms, especially facultative anaerobic bacteria. A soil can be readily transformed into a zymogenic/synthetic soil with disease-suppressive potential if mixed cultures of effective micro-organisms with the ability to transmit these properties are applied to that soil.

The desired effects from applying cultured beneficial and effective micro-organisms to soils can be somewhat variable, at least initially. In some soils, a single application (*i.e.*, inoculation) may be enough to produce the expected results, while for other soils even repeated applications may appear to be ineffective.

The reason for this is that in some soils it takes longer for the introduced micro-organisms to adapt to a new set of ecological and environmental conditions and to become well-established as a stable, effective and predominant part of the indigenous soil micro-flora. The important consideration here is the careful selection of a mixed culture of compatible, effective micro-organisms properly cultured and provided with acceptable organic substrates. Assuming that repeated applications are made at regular intervals during the first cropping season, there is a very high probability that the desired results will be achieved.

There are no meaningful or reliable tests for monitoring the establishment of mixed cultures of beneficial and effective micro-organisms after application to a soil. The desired effects appear only after they are established and become dominant, and remain stable and active in the soil. The inoculum densities of the mixed cultures and the frequency of application serve only as guidelines to enhance the probability of early establishment. Repeated applications, especially during the first cropping season, can markedly facilitate early establishment of the introduced effective micro-organisms.

Once the "new" micro-flora is established and stabilised, the desired effects will continue indefinitely and no further applications are necessary unless organic amendments cease to be applied, or the soil is subjected to severe drought or flooding.

Finally, it is far more likely that the micro-flora of a soil can be controlled through the application of mixed cultures of selected beneficial and effective micro-organisms than by the use of single or pure cultures. If the micro-organisms comprising the mixed culture can coexist and are physiologically compatible and mutually complementary, and if the initial inoculum density is sufficiently high, there is a high probability that these micro-organisms will become established in the soil and will be effective as an associative group, whereby such positive interactions would continue. If so, then it is also highly, probable that they will exercise considerable control over the indigenous soil micro-flora which, in due course, would likely be transformed into or replaced by a "new" soil micro-flora.

AFFECTS OF SOIL MICRO-ORGANISMS ON PLANT HEALTH AND NUTRITION

Soil micro-organisms, sometimes spelled as soil micro-organisms, are a very important element of healthy soil. Knowing what microbes in soil eat, the conditions they thrive in and the temperatures that they are most active in is important in organic gardening and organic lawn care. From a practical standpoint, it boils down to organic matter, but not just any organic matter. These facts below will help you plan your activities around the time they are most beneficial. Below is a partial list of important functions they perform.

Soil micro-organisms are responsible for :

- Ansforming raw elements from one chemical form to another. Important nutrients in the soil are released by microbial activity are Nitrogen, Phosphorus, Sulfur, Iron and others.

- Breaking down soil organic matter into a form useful to plants. This increases soil fertility by making nutrients available and raising CEC levels.

- Degradation of pesticides and other chemicals found in the soil.

- Suppression of pathogenic micro-organisms that cause diseases. The pathogens themselves are part of this group, but are highly outnumbered by beneficial microbes.

Types of Micro-organisms in Soil

There are several types of micro-organisms in soil that benefit plants. Together they make up an immense population of living organisms. One teaspoon of soil may contain millions of various types. Below is a list of common soil micro-organisms found throughout the world.

- *Bacteria* : Small, single cell organisms that make up the single most abundant type of microbe. They have a very wide range of conditions that they live in from the artic wastelands to the steaming waters of volcanic hot springs. In soils, they multiply rapidly under the proper conditions. When conditions are wrong for one species, it is right for another. This is not always a good thing since a balance is what is required.

- *Fungi* : The largest microbe group in terms of mass. Some fungi are beneficial, called mycorrhiza, that form a symbiotic relationship with plant roots, either externally or internally. Within the fungi group are pathogen fungi. These are disease causing fungi, some of which can be quite devastating to plants.

- *Protozoas* : Small single cell microbes that feed on bacteria.

- *Actinomycetes* : Necessary for the breakdown of certain components in organic matter.

- *Algae :* Beneficial groups such as blue-green algae, yellow-green algae and diatoms. Some of these can produce their own energy through photo-synthesis.

Soil micro-organisms are living, breathing organisms and, therefore, need to eat. They compete with plants for nutrients including Nitrogen, Phosphorus, Potassium and micro-nutrients as well. They also consume amino acids, vitamins, and other soil compounds. Their nutrients are primarily derived from the organic matter they feed upon. The benefit is that they also give back or perform other functions that benefit higher plant life.

Organic Matter

It is a variety of natural substances including decomposed leaves, grass clip-ping, shed roots, wood chips, etc. Humus (well decomposed organic matter) is the richest source for plant growth. Organic matter comes in many different nutrient levels, especially Nitrogen. While soil microbes need carbon (C) to live, they also need the nitrogen contained in organic matter. Therefore, the Carbon to Nitrogen ratio (C : N) is very important.

Problems with Low Nitrogen Organic Matter

Organic matter low in Nitrogen will also have a slower breakdown rate than organic matter with higher nitrogen levels. The microbes will consume the Nitrogen element first and the grass will get what is left over. So organic mat-ter high in carbon and low in Nitrogen will provide little nitrogen to the grass. However, if another N source is applied over the organic matter, it will speed up the decomposition.

Therefore, the rule of thumb is to choose an organic matter with higher levels of nitrogen. Anything lower than four per cent (4 per cent) Nitrogen with high Carbon content should be considered a soil amendment and not a fertilizer.

How Temperature Affects Soil Microbe Activity

Soil Microbe activity is dependent on soil tempe ratures. For simplicity, all essential soil microbes are classified into the three different temperature ranges they are most active in.

1. *Psychrophiles* : Active in temperatures less than 68 degrees.
2. *Mesophiles* : Active in temperatures between 77 degrees and 95 degrees. This makes up the largest group of soil microbes and the range most activity charts are based on.
3. *Thermopholes* : Active in temperatures from 115 degrees to 150 degrees. From a plant and landscape view, this group will rarely apply.

Since the primary group contains Mesophiles, this has a great influence on the degree of soil microbe activity. Areas where the temperatures are warm most of the year, organic matter can be consumed very quickly. Tropical rainforests are so lush in part because of consistently warm temperatures, which promote fast breakdown of organic matter and the release of nutrients into the soil.

Cooler areas, such as Canada and parts of Europe, that have extended periods of cool weather below 77 degrees will benefit far less from the additions of organic matter. This is because far fewer microbes are active in that temperature range. It is possible to build unhealthy levels of organic matter if you follow the example of those in warmer climates.

The scientific rule is this : With every 18 degree rise in temperature, from 32 degrees to 95 degrees, there is a 1.5 to 3.0 per cent increase in microbial activity. (Carrol/Waddington/Rieke) Remember, the food availability to microbes, the quality of organic matter, soil types, pH level, per cent of Nitrogen, etc. will also have an effect on microbial activity level.

Soil pH Factor

Most soil micro-organisms can tolerate a wide range of soil levels. However, bacteria favours a neutral to slightly alkaline soil up to 8.0. When pH drops below 6, fungi begin to dominate as bacteria finds it less favourable.

Soil Moisture

Just as temperature levels stimulate different soil microbes, so does soil moisture. Some are obvious. Persistent, damp conditions with heavy shade will promote the growth of algae while hindering microbes that thrive in sunny locations. Proper lawn watering requires deep watering so that the soil is wet 4 inches deep. Shallow watering means only the surface is wet. It drys out quickly and can greatly hinder soil microbes.

Oxygen Levels Necessary for Healthy Microbes

There is a balance to everything, including oxygen in soil. Compacted soil will have less oxygen and less water holding capacity. Clay soil consists of extremely tiny particles, even smaller than silt. Clay with proper structure can have sufficient oxygen, but it can also compact easily. Since soil micro-organisms consume oxygen, but low oxygen soils will quickly deplete what oxygen it has and lower the soil micro-organism levels. In lawns, deeply water to a level of 4 inches deep and allow it to dry before watering again. The deeper soil will remain moist at sufficient levels. Shallow watering means when the surface moisture is gone there is no moisture deeper to support a healthy microbe population.

Decomposition

Well decomposed organic matter is the oldest form of bio-stimulant and very effective. Modern advancements allow us to apply specific bacteria and other ingredients necessary for healthy plants. Other than organic matter, humic acid is a well performing bio-stimulant for lawns and gardens that has shown to stimulate soil micro-organisms.

Humic acid is known to :

- Enhance soil nutrient content.
- Increase CEC.
- Stimulate and increase microbial numbers.
- Increases soil moisture holding capacity.
- Improves soil structure.

Soil micro-organisms are one of the most important elements of a healthy soil. A good lawn care programme will take advantage of the many benefits of soil microbes and the bio-stimulants that encourage them.

Chapter 8

SOIL ECOSYSTEM

INTRODUCTION

Soils are rich ecosystems, composed of both living and non-living matter with a multitude of interaction between them. Soils play an important role in all of our natural ecological cycles–carbon, nitrogen, oxygen, water and nutrient. They also provide benefits through their contribution in a number of additional processes, called ecosystem services. These services range from waste decomposition to acting as a water filtration system to degrading environmental contaminants.

The diversity and abundance of life that exists within the soil is greater than in any other ecosystem. A handful of soil can contain billions of different organisms that play a critical role in soil quality to support plant growth. Although we understand the vital services that these organisms provide by breaking down organic debris (plants, animals, and other organic materials) and recycling nutrients, scientists have only begun to study the rich and unique diversity that is a part of the soil ecosystem.

Ecological Cycling

Each ecological cycle is unique, although similar elements can appear in more than one cycle. While most move between the atmosphere (air), hydrosphere (water), lithosphere (land) and biosphere (living things), other nutrient cycles are limited to movement between rocks and soils and plants and animals. However, even the nutrients from these limited cycles, such as potassium, calcium, phosphorus and magnesium, are essential for life.

Water and nitrogen resources, both essential to all living things, stay constant within their cycles? meaning their only change is in the forms they take. The water cycle is very dynamic as water can change from vapour to liquid to snow to ice. Soils role in this process is through infiltration, storage, and transpiration.

Nitrogen, which makes up more than three-quarters of the Earth's atmosphere, must be broken down into other forms in order to be used by living organisms. It is within the nitrogen cycle that soil bacteria converts nitrogen into usable elements (called nitrogen fixing) for plants, animals and humans before it is eventually returned to the atmosphere.

Oxygen is unique in that it not only has its own cycle, it is often integrated into elements within other ecological cycles, as water (H_2O), carbon dioxide (CO_2), iron oxide (Fe_2O_3), and many others. Within the biosphere, photo-synthesis is the key driver of theoxygen cycle as plants take in carbon dioxide and expel oxygen for animal and human use. Additionally, in water, oxygen is constantly being dissolved and consumed by micro-organisms leading to balance.

The carbon cycle is by far the cycle of greatest interest due to its importance in both climate change and global warming. Soil plays a critical role in this cycle since the majority of carbon in the atmosphere comes from biological reactions within the soil. The biological/physical carbon cycle occurs over days, weeks, months, and years and involves the absorption, conversion, and release of carbon by living organisms through photo-synthesis, respiration, and decomposition.

The geological carbon cycletakes place over hundreds of millions of years and involves the cycling of carbon through the various layers of the Earth. A large amount of organic carbon sinks to the ocean floor to be buried into the Earth's crust. It is thought that more carbon dioxide is stored in the world's soils than is circulated within the atmosphere. Throughout the Earth's history, the release of CO_2 from deep below the surface occurs as a geological event, such as a volcanic eruption.

Ecosystem Services

Aside from its participation in various bio-geochemical cycles and nutrient exchange, soil provides a number of other critical ecosystem services. These services differ from other ecosystem benefits in that there is a human demand for the natural assets and/or benefits. Several important benefits are listed below. Soil is a natural protector of seeds and plants. Within a soil ecosystem seeds can disperse and germinate. The soil provides a physical support system for plants, while both retaining and delivering nutrients to them. This, in turn, provides humans and other animals with a source of food as well as resources for potential medicinal or other goods. In addition, soil can both hold and release water, thereby providing for plant growth, flood control, and water filtration and purification services.

Soils also play a central role in the management, processing and detoxification of a variety of wastes, both natural and man-made. Soil organisms decompose many organic compounds, such as manure, remains of plants, fertilizers and pesticides, preventing them from entering water and becoming pollutants. Human activity adds a wide variety of substances to the environment, some of which are hazardous or toxic. As long as the concentration is not greater than the ecosystem's ability to handle it, micro-organisms in the soil can degrade or detoxify many of these substances, rendering them harmless to humans, animals, and the environment.

ABIOTIC SOIL COMPONENTS

Abiotic soil components include mineral matter (clay, silt, sand), water, air and organic matter. Air and water percentages vary significantly with soil texture, weather and plant water uptake.

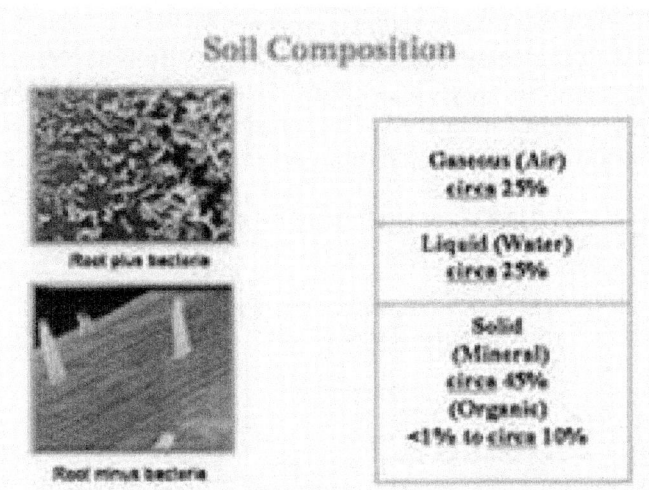

Mineral matter is composed of various proportions of sand, silt and clay particles. Sand particles are 0.05 to 2 mm in diameter, silt particles are 0.002 to 0.05 mm in diameter and clay particles are less than 0.002 mm in diameter. Because clay particles have a very large surface area to volume ratio, they can hold much more water and nutrients than larger particles.

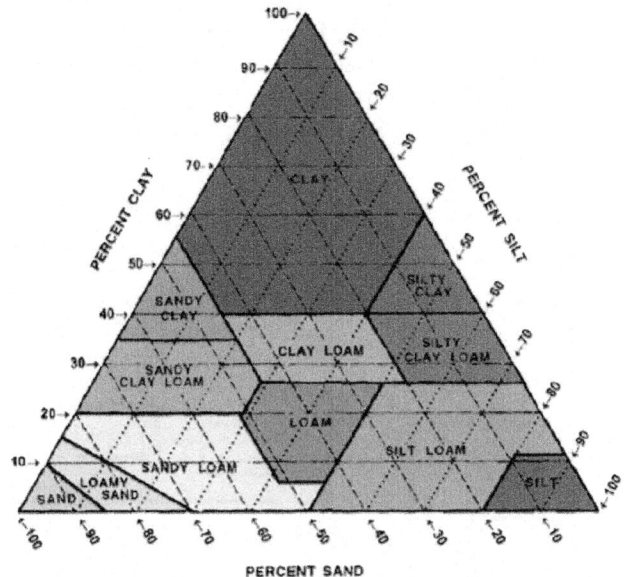

Soil texture is the proportion of sand, silt and clay in a soil. The soil texture triangle, shown here, is used to classify a soil into one of 11 different categories, each of which has different physical and chemical properties. The example shown here (10 per cent clay, 70 per cent sand and 20 per cent silt) is a sandy loam. Soil texture affects nearly every aspect of soil use and management, but is not affected by management unless significant soil erosion occurs.

Water and air: Since each size particle confers different physical and chemical properties on a soil, soil texture is an important determinant of water retention, bulk density, aeration and fertility. The aeration and water status of a soil, in turn, have important influences on soil biota activity.

SOIL PROCESSES

Mineralization-Fixation
Nitrogen Transport and Transformation

Ionic (Mineral) Forms
NH_4^+
NO_3^-
NO_2^-

Mineralization

Fixation
(assimilation)

Organic Forms
Proteins (Amino Acids)
Nucleic Acids
Microbial cell wall
Constituents (chitin and peptidoglycans

Bird, 2004

Many types of chemical and biological processes exist in soils. These include mineralisation of organic matter and fixation of atoms of mineral matter into organic compounds. The processes take place within ecosystems (groups of organism interacting with their abiotic environment).

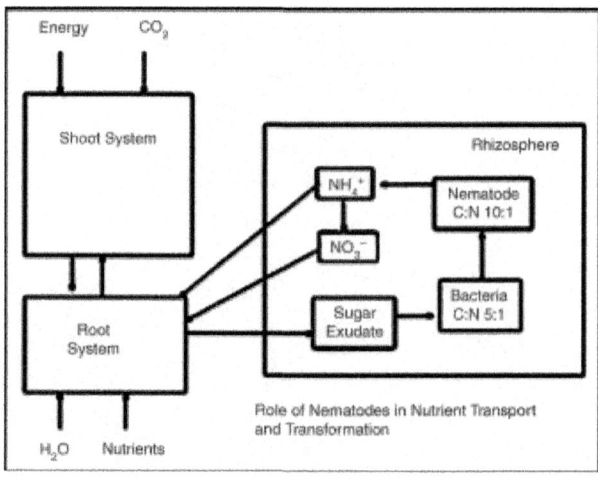

Energy CO_2

Shoot System

Rhizosphere

NH_4^+ Nematode C:N 10:1

NO_3^-

Sugar Exudate   Bacteria C:N 5:1

Root System

H_2O Nutrients

Role of Nematodes in Nutrient Transport and Transformation

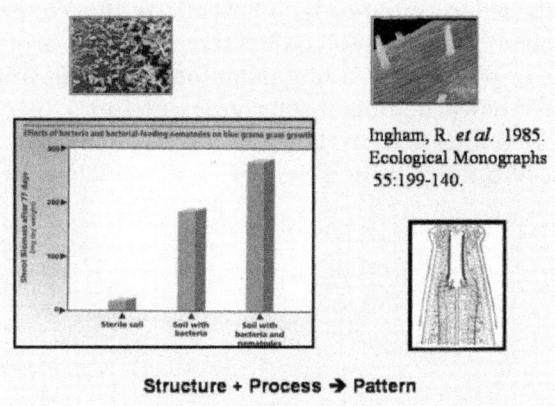

Ingham, R. *et al.* 1985.
Ecological Monographs
55:199-140.

Structure + Process → Pattern

In the presence of both bacteria and nematodes primary productivity is enhanced.

Influence of Tart Cherry Ground Cover and Nutrient Management on the Active Carbon and Nitrogen Pool Sizes.		
Management Systems	Active Carbon Pounds Peracre	Active Nitrogen
Conventional	1.054	59
Cover Crop	1.350	95
Mulch	1.409	85
Compost	1.349	108
Biosystem	1.680	110
Data are Two Year Average of Samples Taken in May 1999 and 200.For Detailed Management Information,Refer to Appendix at the end of this Chapter.Source:D.R Mulch C.E Edson TC Wilson.		

Organic matter : Soil organic matter (SOM), though usually comprising less than five person of a soil's weight, is one of the most important components of ecosystems. SOM strongly modifies soil organism habitat and provides a food source for much of the soil biota. When soil micro-organisms feed, they change the form of SOM and in the process release inorganic nutrients, especially nitrogen, phosphorus and sulfur. This process is called decomposition and is an important process in all healthy ecosystems. Because soil micro-organisms are continually consuming the SOM portion of their home, SOM must be continuously replenished to maintain soil quality.

Patterns

In addition to obtaining inorganic nutrients and water from soil, the root system serves as a host for various herbivores, including fungi, bacteria, nematodes, arthr opods and insects. Decomposers, including fungi, bacteria, actinomycetes and earthworms, mineralise labile and resistant substrates (soil organic matter).

These are referred to as first-order inter-actions. In second-order inter-actions, organisms feed on organisms involved in first order inter-actions. Numerous species of soil-borne organisms including nematodes, insects, mites, fungi, bacteria, and protozoa feed as carnivores, bacterivores or fungivores on the organisms involved in the previous activity level. Soil ecosystems seem to function very much the same as the aboveground pastures with which we are all more familiar.

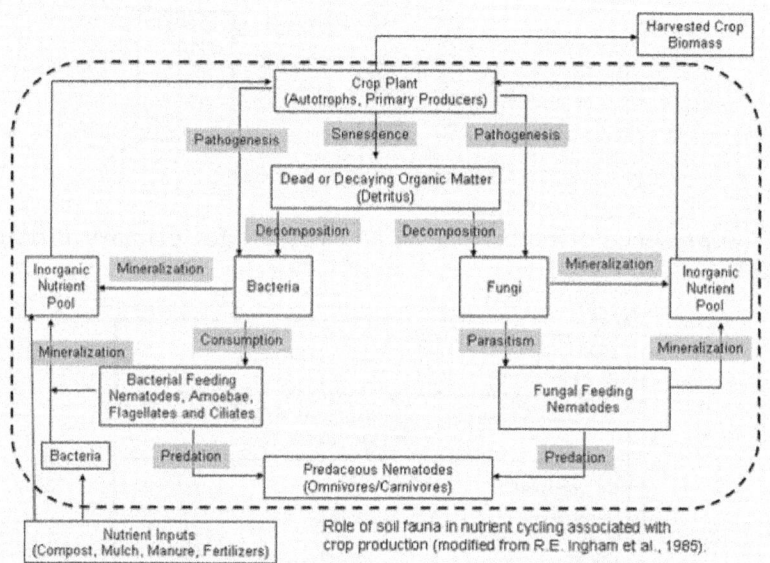

Role of soil fauna in nutrient cycling associated with crop production (modified from R.E. Ingham et al., 1985).

Soil ecosystems function in accordance with the Second Law of Thermodynamics, which states that "in any energy conversion, the final product will consist of less useable energy than the original product, because of the inevitable loss of energy in the form of heat." The amount of biomass, therefore, is less in each subsequent interaction order or trophic level.

CHEMICAL COMPOSITION OF SOIL

The industrialisation of our society has led to an increased production and emission of both xenobiotic and natural chemical substances. Many of these chemicals will end up in the soil. Various soil constituents have a great capacity to retain chemicals, especially those with apolar molecules or positively charged divalent and trivalent ions. Consequently, the soil is a net sink for all kinds of chemicals, and concentrations are often considerably higher than in any other environmental compartment. This situation may lead to smaller or larger impacts on the functioning of soil ecosystems. Important ecological functions of the soil are those associated with organic matter decomposition, mineralisation of nutrients, and synthesis of humic substances. For that reason, an increasing need exists for methods to assess the side effects of these chemicals on soil ecosystems.

An overload of chemicals will affect both abiotic soil properties and, directly and indirectly, soil biota. This paper will, therefore, start with a description of the possible sources and consequences of chemical pollution for abiotic soil properties. Subsequently, methods are described for the determination of the effects of chemicals on soil organisms.

When considering methods to assess the effects of chemicals on soil biota, two types of tests can be distinguished. The first contributes to the prediction of the potential effects of single chemicals on soil ecosystems. For that purpose, mainly single-species laboratory tests are conducted; at times, more complex micro- ecosystem, mesocosm, or field studies are carried out. This type of testing, which aims at establishing dose-response relationships and the estimation of LC_{50}, EC_{50}, or NOEC values, may be called "prognosis."

The second type of method is aimed at assessing the potential ecological risk of a certain case of soil pollution. In such a situation, several chemicals may be involved. To determine whether a specified case of soil pollution poses a real hazard for soil biota, both laboratory and field studies may be performed. This type of testing may be called "diagnosis."

Quantification of Input of Chemicals in the Soil

Before methods to assess the effects of chemicals on soil components can be presented, information must be provided about the various ways chemicals enter the soil and how these inputs can be quantified.

Major Elements

Wet Deposition

This is the major process for the input of nitrogen, sulphur, and chloride, and is significant for other elements. Excluding problems such as the definition of input by mist and fog, the measurement of wet deposition is essentially the estimation of rainfall and its elemental content.

Type and design of rain gauges can be found in *Tropical Soil Biology and Fertility : A Handbook of Methods* and their position relative to the ground surface is usually the reason for an underestimation in rainfall, depending on wind and evaporation losses in exposed areas. To avoid chemical changes in the sample, frequent sampling and rapid analysis are preferred to preservation.

Dry Deposition

This is defined as the direct transfer of gases and particles to different ecosystem surfaces (= receptors). Two methods are noted here :

* Micro-meteorological methods are indirect, and based on the assumption that transmission of chemical compounds is a process similar to transmission of heat and momentum. The method has been proven to produce useful results for SO_2 deposition to grass-covered areas, but does not fully satisfy to measure deposition to forests.

- Receptor-(or ecosystem-) oriented mass balance methods are more direct methods, but with problems and limitations. For chloride, sodium, and sulphur at high deposition, throughfall measurements have been useful to estimate dry deposition, when combined with wet deposition measurements.

Nitrogen Fixation

This method is well known as the acetylene-reduction (AR) method for measuring nitrogenase activity.

Isotope techniques, such as incubation in $^{15}N_2$-containing atmosphere are attractive; but for long-running experiments, practical problems arise. Further methods are the ^{15}N-isotope dilution technique and the classic total nitrogen difference method, based on a comparison of total N-yield in a N-fixing crop and that of a non-N fixing reference crop.

Mineral Weathering

Methods to estimate current weathering rates include mass balance of watersheds, radiometric methods, and mineral bag technique. Mass balance of watersheds or lysimeters is widely used and gives the best estimates of current weathering (*e.g.* Likens *et. al.*, 1977). The approach is indirect, leaving weathering as the residue in the mass balance equation :

$$E(rw) = [E(efflux) - E(influx)] + E(ps)$$

where $E(rw)$ is element release by weathering, E (efflux) is elemental losses mainly from leaching, E (influx) is input of elements mainly by dry and wet deposition, and $E(ps)$ is the change in elemental storage in plants and soil.

The results of mass balance studies depend on the accuracy of estimation of all possible sources and sinks for elements in the soil. Main sinks are elemental storage in biomass and humus, but also accumulation in the soil by microbial activity, redox reactions, surface exchange processes, and formation of secondary minerals. These sinks may easily turn into sources if biological or chemical conditions are altered. For instance, present-day acid deposition depletes base cations from the exchange sites in the soil profile.

Radiometric methods using $^{87}Sr/$ ^{86}Sr ratios are used to estimate calcium weathering.

For the mineral bag technique, selected soil or mineral fraction is put into a non-biodegradable mesh bag, and placed in the field for a specific period of time. The bag is then returned to the laboratory for analysis.

Organic Chemicals and Metals

Major sources for the input of organic chemicals and heavy metals in soils are agriculture, industries, and traffic. Both diffuse and point sources can be identified, and some chemicals (*e.g.*, pesticides) are applied in quite a controlled way enabling a proper prediction of the input in the soil.

To assess the deposition of pesticides that are generally applied under controlled conditions, sheets of aluminium foil or other inert substances can be placed on the soil and analysed after spraying. For other chemicals that may be released in a less controlled manner, chemical analysis of soil samples is needed to quantify the input. In all cases, rather specific analytical techniques are required to determine chemical concentrations in soils. Before analysis, complicated extraction and purification steps are often required. Generally, extraction with an organic solvent (*e.g.*, hexane, acetonitrile, toluene or acetone) is applied, followed by analysis by HPLC, GC, or GC-MS.

To determine the soil content of heavy metals, digestion of soil samples with strong acids (*e.g.*, $HNO_3/HClO_4$) is required. After that, the destruate can be analysed by atomic absorption spectrophotometry.

METHODS TO QUANTIFY EFFECTS OF CHEMICAL INPUT ON ABIOTIC SOIL CHARACTERISTICS

Nitrogen

Recent concerns over nitrogen deficiencies have led to others concerning excess nitrogen availability and the potential for forest decline and surface water pollution. High input of N can lead to N-saturation with serious environmental impacts on soil chemistry and water quality and on fluxes of radio-actively active (or "greenhouse") gases. The values for the characteristics are endpoints. To get information about the deposition level upon which these characteristics start to change, the "critical load" concept has been introduced. The definition is "the maximum deposition of elements that will not cause chemical changes leading to long-term harmful effects on ecosystem structure and function". A regional assessment of critical loads is very important to formulate optimal policies for emission reductions.

Sulphur

Sulphur is transformed in soils by processes similar to those occurring in the nitrogen cycle. Like nitrogen, sulphur can be oxidised, reduced, assimilated, or mineralised from organic matter.

The major differences between the two cycles are :

1. No process is equivalent to N-fixation and

2. Losses of gaseous sulphur from soils are not equivalent to N-losses due to denitrification.

Recent interest in soil sulphur transformations results from an increased awareness of the fertilizer value of the element and the recognition of the importance of the sulphate ion, which reaches the soil in acid rain, as the major counter-anion involved in cation leaching from soils. Sulphate adsorption by soils is an important property affecting the availability of sulphate to plants and the leaching of sulphate and associated cations. Sulphate adsorption is particularly important in soils subjected to acid precipitation, since it determines the impact of acid rain on cation mobility and leaching. Soil temperature and moisture influence sulphate adsorption, whereas desorption on waterlogging may also be an important reaction in soils exposed to atmospheric pollution.

Flowchart to Map Critical Loads and Areas where They have been Exceeded :

> Select receptor type
>
> (*e.g.*, soil ecosystems)
>
> ↓
>
> Determine critical chemical values
>
> ↓
>
> Select computation method (*i.e.*, model)
>
> ↓
>
> Quantify receptor distribution
>
> ↓
>
> Collect input data
>
> ↓
>
> Conduct critical load calculations
>
> ↓
>
> Draw maps according to procedures

Many agents have been used to extract sulphate and other sulphur ions from the soil. Important is 0.01 M $Ca(H_2PO_4)_2$ which appears to remove sulphate from the same pool of soil sulphur that is available for plants. Sulphate can be measured by methods including gravimetrically, turbidimetrically with barium chloride, spectrophotometric ally using methylene blue, by titrimetric methods, adsorption chromatography, ion exchange chromatography, ion-selective electrodes, and thin-layer or gas chromatography.

Recent studies on the damage to ecosystems caused by oxides of sulphur and acid rain involve the use of lysimeters, and collectors to measure through-fall, stem flow, and litter deposition. Unfortunately, most of these studies have omitted the microbial transformations of the element.

Recent laboratory studies have been concerned with the microbial cycling of sulphur in soils exposed to heavy atmospheric pollution from point sources. The first approach was to remove soils from the field at intervals throughout the season, and to have them analysed in the laboratory for S-ions and sulphur oxidising micro-organisms. In the latter approach, soils were sampled from sites exposed to point-source pollution and from relatively unpolluted sites that had essentially the same soil type, vegetation, and climate as the polluted sites. They were packed into plastic tubes. Several soil columns were placed in the polluted site and several in the non-polluted site.

Phosphorus

Both forests and grasslands are frequently phosphorus deficient to a variable degree, and this deficiency limits their productivity. Fertilizer application is, therefore, a principal means of increasing timber and grass production.

In forestry, tree needle analysis has been used for many years as the main guide in the assessment of phosphorus fertilizer requirements. Recent publications, however, indicate that this type of analysis is unrealistic as a predictor of fertilizer responses in commercial forest trees. Analysis of the forest or grassland soils has been proposed as an alternative.

Extraction methods for P_i rely on three different principles :

1. Anion exchange resin acts as a sink for solution P_i and thereby offsets the equilibrium between dissolved and soluble P_i. "Exchangeable" P_i as well as some of the more soluble precipitated p forms will enter the solution, bind to the resin, and can then be measured.

2. Changes in pH cause changes in the solubility of P_i. Acid will extract calcium P_i. Alkaline solutions will solubilise Al and Fe bound P_i. Different P_i compounds have different solubilities at various pH values, and this can be used to characterise soil P_i composition or to evaluate labile P_i.

3. Specific anions can bring P_i into solution by competing for adsorption sites and/or lowering the solubilities of cations that bind P_i. Fluoride has for instance been used under conditions of controlled pH, to release P from Al-bound forms, by forming insoluble aluminium fluoride. Organic anions have also been used to bind or chelate cations and release P_i into solution.

Methods to extract P_0 have employed alkaline solutions (*e.g.*, $NaHCO_3$ or NaOH) or various organic solvents (*e.g.*, acetylacetone which dissolves organic matter). Little progress has been made towards characterising P_0 extracts in terms of the mechanisms for bringing P_0 into solution, binding modes in the soil and its availability to plants.

A new, physiologically based root bioassay has been developed, that appears to be sensitive in assessing P-deficiency in plants. The bioassay relies on the negative relationship between the rate of metabolic uptake of [32]p-labelled phosphorus by roots from a standardised solution in the laboratory and the amount of phosphorus supply in the original rooting environment. This method has been successfully applied to forest stands and grasslands.

Carbon

The CO_2 concentration in the atmosphere has increased by 25 per cent over the past 100 years, and a consensus exists that a doubling of the concentration may occur by the middle of the next century. The largest terrestrial carbon sources and sinks influencing CO_2 fluxes are the forests, which account for approximately two-thirds of the photo-synthesis. The effects of a doubling of the CO_2 concentration on the growth and development of trees is known for a few species. A general finding is an increase of the tissue density of the leaves, a change in leaf structure, and an increase in the C/N ratio of the tissues.

This changed C/N ratio may reduce the decomposition rates of plant material and modify the nutrient availability (Coûteaux *et. al.*, 1991). On the other hand, increased atmospheric concentrations of CO_2, together with trace gases such as methane (CH_4), nitrous oxide (N_2O), and chlorofluorohydrocarbons (CFCs), are effecting changes in the global heat balance, resulting in significant changes in climate over the next century. Twice as much carbon is found in the top metre of soil compared to the amount in the atmosphere, and CO_2 emissions from soils will increase as organic decomposition is enhanced at higher temperatures. CO_2 emissions are particularly sensitive to temperatures between 0°C and 5°C. Based on a recent model predicting global emissions from soil organic matter, a world temperature rise of 0.3°C per decade has been estimated to result in an additional release of CO_2 from soil organic matter over the next 60 years, equivalent to about 19 per cent of that released by combustion of fossil fuels if present use of fuel were to continue unabated. These calculations suggest that increased decomposition of soil organic carbon could make an important contribution to the greenhouse effect.

Methods to measure C/N ratios have already been given. Measurements of CO_2 evolution under field conditions are described and standardised.

Organic Chemicals and Metals

Effects of organic chemicals on the abiotic soil properties are rarely mentioned in the literature. Some chemicals, such as the herbicide paraquat, are incorporated into clay particles, but the extent to which this may influence the swelling and shrinking behaviour of the clay is unknown. Other chemicals or their degradation products are incorporated into the soil organic matter. This is a physical or a biological process, which may result, in case of chlorinated organics, in chlorination of the soil organic matter. The extent to which this process occurs and the consequences for the soil characteristics are unknown. Presently, no methods exist to determine the potential impact of organic chemicals on soil abiotic properties.

Metals generally occur in the soil solution as positively charged cations, competing for negatively charged adsorption places on the soil particles. An overload of metals will affect the ionic balance of the soil; it may also lead to a release of other, less strongly bound, metals or cations from the soil. Often this process is slow, not affecting soil abiotic properties to a great extent. Only in the case of flooding a soil with salt water, containing an excess of cations, was the swelling and shrinking properties of clays shown to be severely affected. No methods can be given to measure the impact of metals on soil abiotic properties.

METHODS TO ASSESS THE POTENTIAL RISK OF CHEMICALS FOR SOIL ORGANISMS (PROGNOSIS)

A brief description is presented of single-species laboratory tests, microcosm tests, and field tests. For an extended overview of these tests, the reader is referred to Van Straalen and Van Gestel.

Single-Species Laboratory Toxicity Tests

Among soil invertebrates, only earthworms have seriously been considered as test organisms during the past decade, and some standardised test methods are available. For microfauna and mesofauna, only few tests are available, although these animals are among the most numerous and species-rich groups of soil animals. Many species, however, are promising test animals, because they are easy to culture and their size allows for small-scale experimental set-ups with many replications. Besides these soil animals, higher plants have also been considered for testing, and standardised tests with some plant species are available.

In several tests, artificial substrates (nutrient solution, agar, silica gel, filter paper) are used, the composition of which greatly affects toxicity. However, extrapolation of these test results to the field remains problematic. If concentrations in test solutions can be equated with pore water concentrations, sorption data may be used to express the toxicity per unit of soil. The validity of this extrapolation, however, still has to be investigated. The same extrapolation problems may arise for tests in which the main route of exposure is via the food. For such tests, a conversion of food concentrations to soil concentrations may be needed.

Higher Plants

A test with higher plants has been described in an international test guideline (OECD, 1984b), while some others are under discussion. The OECD guideline 208 on higher plant toxicity testing uses several plant species, representing different agricultural crops and both monocotylodoneous and dycotylodoneous species.

Protozoans and Nematodes

Protozoans and nematodes live in the soil pore water, and the best way to test them is to use methods similar to those used in aquatic toxicology. Among the protozoans the ciliates *Tetrahymena pyriformis, Colpoda cucullus,* and *Paramecium aurelia* have been considered for test animals, as have the nematode species *Caenorhabditis elegans, Panagrellus silusiae,* and *Plectus parietinus* however, an accepted test procedure is unavailable.

Lsopods and Millipedes

Isopods are an interesting group of animals in heavy metal research, because of their unique ability to concentrate extreme amounts of metals in their bodies. Their use as a test animal for soil toxicity studies, however, is restricted to a few cases, and no attempts have yet been made to arrive at standardisation. *Porcellio scaber, Oniscus asellus,* and *Trichoniscus pusillus* are three species frequently investigated. Among these, *T. pusillus* seems to be the most suitable as a test species, as it has a somewhat shorter life-cycle compared to *P. scaber* and *O. asellus.* All three species are very easy to culture, and do not require special conditions.

Usually isopods are kept on a plaster substrate, and are fed with partly decomposed leaves, either intact or ground, to which chemicals can be added. Increase in growth over several weeks is observed, but is rather variable, even for one individual. Reproduction is difficult to assess, because, after mating, females may retain the sperm for a long period before producing eggs, which are carried in a brood pouch. Tests require a minimum period of four weeks.

Millipedes (Diplopoda) are another important group of saprotrophic soil invertebrates, but they have never been considered seriously as test animals. The most widely investigated species is *Glomeris marginata.* The species *Cylindroiulus britannica* is also well suited as a test animal. Test conditions for millipedes are similar to those for isopods.

Oribatid Mites

A reproduction toxicity test using the partheno genetic oribatid mite *Platynothrus peltifer* has been described by Denneman and Van Straalen. This seems to be the only oribatid used so far in soil toxicity experiments, although oribatids comprise hundreds of species, and are usually the most numerous group of arthropods in forest soils.

In the test with *P. peltifer,* the animals are exposed to contaminated algae, and the number of eggs are counted. The test is very laborious, as the animals hide their eggs in small crevices; it is also a rather lengthy test (9 to 12 weeks) because of the low rate of egg production in this species and its long life-cycle (1 year), which are remarkable features for such a small animal (± 1 mm). *P. peltifer* appeared to be rather resistant to cadmium, copper, and lead in terms of lethality, but very susceptible in terms of egg production. Due to their peculiar habits, species such as *P. peltifer* tend to be forgotten in the development of toxicity tests. It is, however, the most sensitive soil invertebrate tested so far for cadmium, while it is more sensitive than springtails for copper and lead.

Collembola

Collembola are a relatively well investigated group of soil animals. Several species have been used frequently in toxicity experiments : *Onychiurus* spp. The first three species are parthenogenetic (thelytokous); *O. cincta* is sexual, sperm being transferred indirectly through spermatophores deposited on the substrate by the male.

Three different exposure systems have been described :

1. Through feeding on fungi grown on contaminated agar,

2. Through feeding on directly contaminated food, and,

3. Residual exposure (treated substrate, *e.g.* sand, leaves, soil).

When testing Collembola with contaminated fungi, the animals are kept on a plaster of Paris substrate in a Petri dish, and fed on a piece of agar, overgrown with hyphae (*e.g., Verticillium bulbillosum*). Egg production, growth and survival are recorded regularly throughout a period of several weeks. The advantage of this system is that substances are offered in a natural way, *i.e.*, after being taken up and possibly transformed to naturally occurring complexes. Concentration levels in the fungus, however, are difficult to maintain or to set to specific values.

When testing Collembola with directly contaminated food, chemicals are added in water or acetone solution to the food (algae, yeast, ground leaf material). Food can be offered as droplets on filter paper discs, while the animals are kept on a plaster or sand substrate. In this manner, concentrations can be manipulated easily, while growth, egg production, and survival are monitored over a period of several weeks.

The third system of testing Collembola is to use the artificial soil medium developed for earthworm toxicity tests. Juvenile Collembola *(Folsomia candida)* are placed in artificial soil with dry yeast provided for food. After 28 days, the number of remaining animals and their offspring are counted after flotation extraction. The *Folsomia* test is very easy to carry out; it requires little attention during the test; and it gives reproducible results. Another advantage is the use of artificial soil similar to the earthworm test; thus, experimental results can be compared between earthworms and springtails. The only disadvantage of the test is that reproduction cannot be observed directly, and cannot be separated from juvenile mortality and hatching success. The*Folsomia* test is now undergoing the process of international standardisation.

Enchytraeids

Enchytraeids can be cultured easily on agar and on (artificial) soil substrates, when fed rolled oats. The toxicity tests described in the literature all use species of the genus *Enchytraeus*. The well-known species *Cognettia sphagnet orum* can also be bred easily in the laboratory, but its tendency to fragmentate upon handling makes it less suitable for toxicity tests. Westheide *et. al.* described a test in which the test chemical is incorporated in 1.5 per cent nutrient agar. Two species are used, *Enchytraeus* cf. *globuliferus*and *E. minutus*. Reproduction, measured as the number of cocoons and juveniles produced, is the endpoint studied in this test. This method seems to provide an easy and reproducible test, but, because of the use of an agar substrate, results cannot be translated to real soil.

Römbke (1989) described a test with *Enchytraeus albidus*, using the OECD artificial soil prescribed for earthworm toxicity tests. Adult *E. albidus* are exposed for 28 days to different concentrations of the test chemical, mixed homogeneously through the artificial soil. Survival of adult worms and the number of juveniles produced are the endpoints studied. In this way, both acute and sublethal effects are combined in one test. The reproducibility of the method cannot be judged and, because only one chemical was tested, no conclusions can be drawn with respect to the sensitivity of the sublethal endpoint. An important positive aspect of this test is that the substrate used is the same artificial soil used in the internationally accepted earthworm toxicity tests.

Lumbricids

In the guidelines of OECD (1984a) and EEC (1985), *Eisenia fetida* and its sibling species *E. andrei* are recommended. Both species are commonly found in compost and dung heaps, and can be cultured easily in the laboratory on a substrate of horse dung or cow dung. According to the existing guidelines on acute toxicity testing with earthworms, other real soil dwelling species may also be used. Such species are, however, hard to culture in the laboratory, because they have long generation times and need large volumes of soil. So, for practical reasons, the use of the two *Eisenia* species is recommended.

Three acute toxicity tests exist. In the filter paper contact test (OECD, 1984a), adult earthworms of the species *Eisenia* spp. are exposed to filter paper wetted with a solution of the test substance. Mortality is assessed, and the 48-hour LC_{50} value is expressed as µg per cm². The method has been shown to be easy, fast, and highly reproducible. Several authors including Heimbach have demonstrated, however, that this test has no predictive value for the effect of chemicals on earthworms in the soil; it can only be used to rank chemicals.

In the artificial soil test, adult earthworms of the species *Eisenia* spp. are exposed to the test chemical, which is mixed through an artificial soil substrate for 14 days. This artificial soil is made up by mixing (dry weight) 10 per cent sphagnum peat, 20 per cent kaolin clay, 70 per cent quartz sand, while some $CaCO_3$ is added to adjust the pH to 6.0±0.5. The moisture content of the substrate is adjusted to about 55 per cent (w/w) or to 40-60 per cent of the water holding capacity. Mortality is the only test parameter, and LC_{50} values are expressed as mg per kg. dry soil. Van Gestel and Ma (1990) have demonstrated that results obtained in this artificial soil can easily be translated to natural soils by using sorption data. For this reason, the use of the artificial soil is acceptable, and the test can be concluded to have enough predictive value with respect to effects that occur in the field. The test has been shown to be reproducible.

In the Artisol test adult earthworms of the species *Eisenia* spp. are exposed to chemicals mixed through a substrate of amorphous silica gel (Artisol) for fourteen days. Survival is the only test parameter, and results are expressed in terms of LC_{50} values. The silica gel substrate does not bear any resemblance to natural soil; thus for reasons of ecological realism and extrapolation towards natural soil, this test cannot be recommended.

Recently, two sublethal toxicity tests have been described. In both tests, the OECD artificial soil and the earthworm species *Eisenia* spp. are used. In the first test (Van Gestel *et. al.,* 1989), chemicals are mixed homogeneously through the artificial soil, and after three weeks of exposure effects on the growth and cocoon production by adult earthworms are determined. The worms are fed by supplying a small amount of (untreated) cow dung in a small hole in the middle of the soil. By incubating cocoons produced for five weeks in untreated artificial soil, effects on hatchability (per cent fertile cocoons, number of juveniles per cocoon) and the total number of offspring per adult worm can be determined.

The second method was developed to determine the sublethal effects of pesticides on earthworms. The pesticide is sprayed onto the soil surface, and earthworms are fed by applying about 0.5 g cow dung per animal to the soil surface once a week. Pesticides are applied in two treatment levels, corresponding with the recommended dose and a five fold dose. After six weeks incubation, adult worms are removed from the substrate and weighed. The test substrate containing cocoons and juveniles is incubated for another four weeks. Food is given when required. After ten weeks, the juveniles are extracted from the substrate by hand-sorting or heat extraction and counted. Effects on earthworm growth and on the total number of offspring produced per tray are determined. The method has been subjected to a (German) ring test, and will be revised on the basis of the results of it. The method is only applicable for pesticides, and the recovery of all juveniles from the artificial soil by hand-sorting is difficult. This hinders comparison of this method with that of Van Gestel *et. al.*. Furthermore, the method of pesticide application is not standardised, which may be the reason for the variability observed in the first ring test.

Molluscs

In the limited number of toxicity tests using terrestrial molluscs, exposure was via the food. Russel *et. al.* (1981) described a method for toxicity experiments with the garden snail *Helix aspersa*. The snails were kept in polyethylene boxes, filled with a substrate of moist quartz sand covered by a piece of woven glass towel. The snails are fed a diet of ground Purina Lab-Chow (formulation for rats, mice, and hamsters) supplemented with $CaCO_3$. Parameters affected include survival, reproductive behaviour, dormant state, new shell growth, and food consumption. Similar test methods using the snail *Helix pomatia* or the slug *Arion ater* have been described by other authors.

Beneficial Arthropods

Arthropods that may improve the production of agricultural products are designated as "beneficials," and commercial interest exists in designing and applying pesticides in such a way that beneficials are least affected. The working group on "Pesticides and Beneficial Orga nisms" of the International Organisation for Biological and Integrated Control of Noxious Animals and Plants (IOBC) has contributed significantly to designing ecotoxicological test methods and decision schemes to evaluate the hazard of pesticides.

The hymenopteran groups Ichneumonidae, Braconidae, and Chalcidoidea contain a large number of parasitoid species. The female insect deposits an egg in or on a host (usually an insect egg or larva), which is then gradually eaten as the offspring develop. The host selection process and the life-cycle of the parasitoid are finely tuned to the host, and many species will attack only a single or a few host species. Furthermore, other hymenopteran species used in toxicity tests include-*Diaeretiella rapae*, an internal parasite of aphids such as *Myzus persicae, Phygadeuon trichops,* a parasite of *Delia* species (bulb flies), *Coccygomimus (=Pimpla) turionellae,* a polyphagous parasite of Lepidoptera (Tortricidae, Geometridae, Noctuidae), and *Opius* sp., a parasite of leaf mining insects. The methods used for these species are similar to those described for *Trichogramma* and *Encarsia.*

Within the order of the Coleoptera, the families Carabidae (ground beetles), Staphylinidae (rove beetles), and Coccinellidae (lady birds) contain representatives that are common in agricultural fields and are recognised for their predation of pests.

Among the various arthropod groups, spiders seem to be particularly sensitive. This phenomenon often appears in field tests with pesticides, where catches of surface active spiders are reduced in a manner similar to that of predatory mites following pesticide application. The families Erigonidae and Linyphiidae (money spiders) are important groups with a great species richness. The recommendations made by the International Commission for Plant Bee Relations (ICPBR) have been included in a guideline of the European and Mediter ranean Plant Protection Organisation (EPPO) to evaluate the hazards of pesticides to the honey bee *Apis mellifera.* Several countries have slightly different national guidelines to test pesticides on honey bees.

Microcosm Tests Including Those on Soil Micro-flora

Microcosm Tests

Single-species tests are carried out under rather artificial conditions, and disregard ecological inter-actions between different species. To evaluate effects of chemicals under more natural conditions, model ecosystems, microcosms or micro-ecosystems have been designed that simulate certain aspects of real ecosystems, and are yet simple enough for experimental use. Decomposing invertebrates have been considered for such systems, because their activities can be assessed conveniently in terms of system functions such as leaf litter fragmentation and nutrient conversions.

Several terrestrial model ecosystems have been described, without attempt to arrive at standardisation. The system may either be closed or open to the ambient air, and contains intact core samples from a natural habitat or a more or less standardised soil (*e.g.,* Bond *et. al.,* 1976). For ecotoxicological tests, the use of standardised soils seems to be most appropriate, since it allows the chemical to be mixed homogeneously through the soil, and it minimises experimental variation between replicate units. The effects of various pretreatments, such as drying, sterilising, inoculation, litter type, age of the litter, however, have a significant impact on the behaviour of the system and need to be investigated thoroughly.

Natural rainfall may be simulated, and leachate can be collected. Various chemical analyses of the leachate solution may indicate aspects of decomposer activity : dissolved organic carbon, NH_4, NO_3, pH, and Ca. The advantage of this procedure is that repeated sampling in time from the same soil column is possible.

A disadvantage of the leaching procedure is that the humidity of soil and litter is unstable, and is difficult to standardise. Moreover, toxicants added to the system may be displaced through the column or leached out. Microbial respiration can be estimated by measuring CO_2 production in the microcosms. For that purpose, CO_2-free air (20.8 per cent O_2; 79.2 per cent N_2) is guided through the soil column, and the resulting CO_2 is subsequently measured by infrared gas analysis.

Verhoef and Dorel, Verhoef *et. al.* and Verhoef and Meintser have used these types of microcosms, filled with pine litter, to study the effects of gaseous (NH_3) and wet ($(NH_4)_2SO_4$) atmospheric deposition. N-deposition eliminates the stimulation of mineral leaching by the collembolan *Tomocerus minor*. Neither survival nor growth of the animals are affected. Reproduction, however, is negatively influenced. In pine litter, which has been confronted with high N input for several decades, *T. minor* slows down mineral leaching by stimulating microbial growth.

Van Wensem and Van Wensem *et. al.* added chemicals to poplar leaf litter, which is incubated for 4 weeks, after which some replicates are terminated to determine DOC, NH_4, NO_3, and pH. The remaining replicates are incubated for another four weeks, with eight isopods *(Porcellio scaber)* added to each system. Survival and growth of the isopods can be assessed, as well as particle size distribution and concentrations of minerals of the remaining litter. The organotin fungicide triphenyltin hydroxide increased the concentration of soluble ammonium in the litter, due partly to excretion by isopods and partly to stimulating effects on the micro-flora.

In systems with isopods, the organotin decreased ammoniation in treatment levels higher than 10 µg per g, but in systems without isopods the organotin had no significant effect. The addition of isopods in this case, therefore, made the system quite sensitive, which was unexpected (triphenyltin is a fungicide), and would not have been noticed in a single-species test using isopods.

Mothes-Wagner *et. al.*described a more complex microcosm system, consisting of 25 litres of natural or standardised soil that are inoculated with nematodes *(Pelodera strongyloides)* and enchytraeides *(Enchytraeus coronatus)* and sown with bush beans *(Phaseolus vulgaris)*. After emergence of the beans, spider mites *(Tetranychus urticae)* are introduced. After a preincubation period of about six months in the laboratory, in a greenhouse, or in the field, the systems can be treated with the test chemical. Test parameters are survival, reproduction, and population growth of the introduced organisms. Further more, measurement of several histological and enzymatic parameters in these organisms is recommended. As no substantial test results are available, the predictive value and sensitivity of this system cannot be evaluated.

Tests on Soil Microbial Processes

Tests on single species of isolated micro-organisms in artificial substrates are not regarded as representative for the soil ecosystem, and will, therefore, not be considered here.

Based on a series of workshops held during the 1970s and 1980s, Somerville and Greaves formulated several recommended tests to assess the side effects of pesticides on the soil micro-flora. For all microbial tests in soil, the use of freshly sampled soil containing an active micro-flora was considered essential. Prolonged storage and drying of the soil should be avoided.

For a proper assessment of the effect of chemicals, at least two different soil types should be used. A short description is given here of several tests on microbial processes related to the conversion of nutrients in soil. Unless stated otherwise, all tests are carried out in the dark at a temperature of 20±2°C, and the test chemicals are mixed homogeneously through the soil. Generally, soils are tested at a moisture content corresponding to field capacity or to 40-60 per cent of the water holding capacity.

Test for Soil Respiration and Mineralisation of Substrates

In these tests the production of CO_2 from small soil samples (100 g) treated with the test chemical is measured continuously or semi-continuously. The tests should run for a minimum of 30 days. Tests may be performed in either unamended soil or in soil amended with a substrate. For this purpose mostly 0.5 per cent (w/w) lucerne or horn meal is used. The disadvantage of this soil respiration test is that the activity of the total soil micro-flora is determined. When certain species are affected by the test chemical, this will often not be noticed, as other (less sensitive) species may take over the activity of the sensitive ones.

During the past decade some new test methods have been developed which aim to determine chemical effects on more specific groups of soil micro-organisms. One is the addition of a readily degradable substrate and the determination of the short-term respiration rate. Such a test was described by Haanstra and Doelman using glutamic acid as a substrate. Soils are amended with glutamic acid and the CO_2-production is measured.

Glucose may be used as a substrate. The duration of the test is no longer than 100-120 hours. The test appeared to be quite sensitive to heavy metals. These short-term respiration tests may be combined with a biomass determination, and seem to be more sensitive than the traditional respiration tests. Another alternative may be found in the addition of more persistent substrates such as lignin or cellulose. Only a few soil micro-organisms are capable of degrading these substrates, and, when they are affected by the test chemical, no others can take over their activity.

The disadvantage of the previously described soil respiration or substrate degradation methods is that less sensitive species of micro-organisms may grow on the substrate during the test. This results in a shift among the micro-flora towards more resistant species, masking the possible elimination of sensitive species. For this reason, Van Beelen *et. al.* developed test methods using the mineralisation of low concentrations of ^{14}C–acetate, ^{14}C–chloroform or other labelled substrates. The amount of substrate applied is very low (1 μg per L) to ensure that no growth of the micro-flora will occur.

This amount of substrate is added to a slurry of the test soil, prepared by mixing the homogenised soil with an equal weight of ground water. The test chemicals are added to the slurry in the desired concentration levels. Samples are incubated at 10°C. The test duration depends on the capacity of the micro-flora in the soil sample to mineralise the test substrate, and is chosen depending on the half-life of the acetate mineralisation. Acetate mineralisation is measured by determining the amount of $^{14}CO_2$ released from the sample and by determining the amount of ^{14}C remaining in the suspension at the end of the test.

Test for Ammoniation and Nitrification

In ammoniation tests, the release of inorganic nitrogen from soil organic matter or a substrate (*e.g.*, plant material or horn meal) is studied in a way comparable to soil respiration tests. The influence of nitrification, *i.e.*, the conversion of ammonia into nitrate, may also be studied in these tests. Ammoniation is performed by a wide variety of soil micro-organisms, and is, therefore, relatively insensitive to perturbation.

The advantages of nitrification are :

- That fewer species of micro-organisms are involved in this process and,

- That the process is considered to be of ecological and agricultural importance. Therefore, either combining these parameters in one test or running a separate test on nitrification is recommended. Nitrification tests can be performed in soil amended with either $(NH_4)_2SO_4$ or with organic substrates such as lucerne or horn meal. For this purpose, substrate equivalent to approximately 100 mg N per kg. soil is added, and the disappearance of NH_4^+ and the appearance of NO_3^- is monitored. In case the rate of NO_3^- formation does not follow the disappearance rate of the NH_4^+, the soil should also be checked for the formation of NO_2. To check whether the test soil is capable of nitrification and whether the organic matter amendment is suitable for ammoniation and nitrification, studies are also recommended.

Test for Nitrogen Fixation

Tests on both symbiotic and asymbiotic nitrogen fixation can be identified. Tests on symbiotic nitrogen fixation in fact consider the unique relationship between the host plant and *Rhizobium,* and, therefore, include in one test effects on both the plant and the bacteria. These experiments are conducted in a soil suitable for growth of the plant. The plant, seeds, or soil can be inoculated with *Rhizobium* if no suitable bacteria are present in the soil. Effects on plant growth and the degree of nodulation should be included. In tests on asymbiotic nitrogen fixation, the degree of acetylene reduction (or formation of ethylene from acetylene) by soil samples is determined in relation to the addition of the test chemical.

Test for Denitrification

Denitrification is the conversion of nitrate to atmospheric nitrogen, and will especially take place under anaerobic conditions. This process may be relevant in soil, as microsites may become anaerobic. In this test, soils are generally flooded with a layer of water, and nitrate is supplied as a substrate. Additionally, an organic substrate such as glucose is added to the soil. Since the process cannot be quantified, the formation of nitrogen gas, the disappearance of nitrate, and the formation of nitrite are measured as test parameters.

Tests on Enzyme Activity in Soil

As in the tests on microbial processes, chemicals are also mixed homogeneously through the soil in tests on enzyme activity. Moisture content is adjusted to field capacity or to 40-60 per cent of water holding capacity, and all incubations are done in the dark at 20°C. Several soil enzymes, relevant to microbial processes in soil, can be used as test parameters, as noted below.

Soil Enzyme : Urease

At several intervals, small soil samples (6-7g) are taken, and incubated with 5 ml of demineralised water and 1.0 ml of a solution containing 60 mM urea. Incubation is at 35°C for 5 hours on a shaking water bath. A phenylm ercury acetate solution in 2 M KCl is added to the soil samples to stop the urease reaction. After 10 minutes of shaking, the soil suspensions are filtered. The filtrates are analysed photometrically at 525 nm for urea concentrations.

Soil Enzyme : Dehydrogenase

Soil samples (5–10g) are incubated with a solution of TTC (2,3,5-triphenyl tetrazolium chloride) in 0.1 M tris buffer solution (pH 7.6), and incubated for 24 hours at 30 or 37°C. The reduced triphenyl formazan formed is extracted with methanol and quantified by measuring the absorbance at 485 nm. Dehydrogenase reflects a broad range of microbial oxidative activities, and does not consistently correlate to microbial numbers, CO_2 evolution or O_2-consumption. Additionally, dehydrogenase activity may depend upon the nature and concentration of amended C-substrates and alternative electron acceptors. Rossell and Tarradellas concluded that short-term (substrate-induced) dehydrogenase activity may reflect the impact of chemicals on the physiologically active biomass of the soil micro-flora.

Soil Enzyme : Phosphatase

At several intervals, 0.5 g soil samples are taken, and incubated with 5 mM *p*-nitrophenylphosphate (*p*–NPP) for 1 hour in a shaker at 20°C. Phosphatase activity is measured as the amount of *p*-nitrophenol formed using a spectrophotometer. Phosphatase is said to bear little relation to total phosphate availability in soils. Its relevance for microbial activity in soil may, therefore, be questionable.

Somerville and Greaves stated that soil enzyme activities would be of little value to monitor side effects of pesticides on micro-flora.

The main reasons for this were :

- The total enzymatic activity of the soil is made up of various fractions, and quantifying the contribution of each to the catalysis of a particular substrate is extremely difficult; furthermore, many enzymes are formed extracellularly, and will still be active when the micro-organisms responsible for their production have been eliminated.

- There is no universally agreed methodology, and almost any result can be achieved by varying assay conditions (temperature, pH, substrate). Although tests on enzyme activity have been described by many authors, few data are available to judge the reproducibility of these methods. Also soil animals, such as collembola and isopods, significantly influence the activity of several enzymes, such as urease, dehyd rogenase, and cellulase. Therefore, discriminating between direct and indirect effects of the tested chemicals on micro-organisms is difficult. Other enzymes that are more or less frequently used as test parameters for microbial activity in soil are: arylsulphatase, b-glucosidase, b-acetylgluco-saminidase, saccharase, galactosidase, protease, and phosphodiesterase.

Field Tests

The reliability of microcosm studies in the laboratory to interpret field conditions is much debated. Microcosms differ from the field situation concerning the influence of temperature and moisture dynamics, the influence of root presence, and the composition of the soil biota community. A recent study compared microcosm studies in the laboratory with mesocosm studies and direct field measurements concerning microbial respiration, enzyme activities, and availability of macro-nutrients in interaction with soil animals; these soil process variables appeared to be of the same order of magnitude.

The tests described in the preceding sections provide only a rough estimate of the possible hazard imposed by a chemical in the environment. In many cases, this degree of precision is sufficient. Usually the laboratory test is considered to be a "worst case" situation, since test animals are exposed to a constant concentration that is relatively available, because the test substrate is prepared freshly. Under field conditions, exposure may be lower since the chemical is not distributed unifonnly over the habitat, and bio-availability will often be lower due to various sorption processes. By contrast, the laboratory test considers the test organism under optimal conditions, without secondary stresses, such as those of food shortage, drought, and cold. The uncertainties attached to the laboratory-to-field extrapolation can be avoided by conducting experiments under semi-field or field conditions.

Various organisations have recommended test protocols for field investigations. In several cases, guidelines for field tests are part of the national registration procedures for pesticides. Furthermore, considerable scientific research has been done in which side-effects of pesticides have been published. Some attempts have also been made to develop standardised procedures for field tests to assess the effects of pesticides on earthworms.

Cage Tests Using Selected Arthropod Species

Some of the arthropods used as laboratory test species can also be exposed to chemicals under semi-field conditions, while exposed in cages. Hassan *et. al.* lists protocols developed for *Trichogramma cacoeciae, Phygadeuon trichops, Coccygomimus turionellae, Phytoseiulus persimilis, Aleochara bilineata, Chrysoperla camea,* and *Drino inconspicua.* The usual procedure is to treat a group of plants with a spray of the chemical. A cage is then put over the plants after which test animals, with hosts or food, are introduced. The cage is installed either in a greenhouse or outdoors under a cover to provide shelter from rain and excessive sunshine. After an adequate duration of exposure, the performance of beneficials is compared with water-treated controls.

Cage tests have been described in detail for testing with pollinators. Bees *(Apis mellifera)* from small colonies are made to forage on a flowering crop in cages measuring minimally $2 \times 2 \times 3$ m, with a 3 mm mesh netting. The product is applied to the plants, and not to the cage walls, by spraying. The EPPO-guideline does not require replication of the treatment. Effects are recorded at several intervals, preferably 0, 1, 2, 4, and 7 days after treatment. Observations are made on the number of dead bees, on foraging activity, and on behaviour. The results are compared with a blank control (usually water-sprayed) and a positive control (a reference product known to be hazardous to bees, *e.g.,* parathion).

Honey Bee Field Test

OEPP/EPPO also provides a guideline for field tests on honey bees. A chemical to be tested is applied to a plot of at least 1500 m^2, with the crop for which the chemical is intended, or another crop attractive to bees (rape, *Phacelia*), in full flower. Per treatment, three colonies of honey bees are placed in or on the edge of the plot. Test plots should be separated by at least 500 to 1000 m^2 to avoid bees foraging on the wrong plot. Replication of the treatment is considered desirable, but is not required in the EPPO guideline. A blank control (untreated, or treated with a reference product known to present a low hazard to bees), as well as a positive control (*e.g.,* parathion, dimethoate) are applied to separate test plots.

After treatment, observations are made at several intervals, preferably after 0, 1, 2,4,7, and 14 days. Meteorological data are recorded during the entire period of the trial. Several parameters are estimated such as the number of foraging bees in the crop, behaviour of bees on the crop and around hives, mortality of bees (using dead bee traps), pollen collection (using pollen traps), pollen in collected honey, number of bees on frames, brood status in frames, and residues in dead bees, pollen wax, and honey. For the test to be valid, mortality in the negative control should not exceed 15 per cent, while mortality in the positive control should be statistically significant.

Arthropod Fauna in Arable Crops

Hassan *et. al.* summarise the recommendations made by the IOBC Working Group Pesticides and Beneficial Organisms for full-scale field tests. The methods are suitable for a variety of crops, but have been applied mostly to winter wheat.

The trial is laid out in a replicated block design : three large fields with similar agronomic history are each divided into three treatment areas, where each treatment area covers at least 3 ha. The treatments are a spray with the product to be tested, a blank treatment (water spray), and a positive control (*e.g.*, dimethoate).

Sampling is planned on seven occasions : 10 and 5 days before treatment, 2, 5, 10, and 20 days after treatment, and just before harvest. Sampling activities are concentrated in the central parts of each treatment area. Crop foliage fauna is collected with a suction net sampler (*e.g.*, the Dietrich vacuum sampler). Soil surface fauna are sampled using pitfall traps, left in the field for 5 days. Visual inspection, water traps, and sticky traps can provide additional information on those arthropods sampled inefficiently by pitfalls or suction samplers.

The fauna collected are identified to at least the family level. Some groups where species can be recognised easily can be further subdivided (*e.g.*, carabid beetles).

In addition to the large-scale experiments suggested for arthropods in winter wheat, smaller set-ups have been suggested by Edwards and Thompson and Eijsackers and Van de Bund. Treatment plots of 3 × 3 m are recommended for microarthropods (Collembola, mites), while 10 × 10 m plots are suitable for studies on beetles and spiders. In all cases, however, the plots should be fenced, preferably using a polythene sheet, protruding 15 cm below ground and 40 cm above. The barrier should limit the immigration from neighbouring plots by surface-active arthropods such as beetles.

The statistical treatment of data from field experiments is not harmonised. Yet, such harmonisation may be important since the probability of finding effects of the treatment will depend on the power of the statistical analysis. Stewart-Oaten *et al.* (1986) suggested the use of "pseudoreplication in time," to allow for a detailed evaluation of the effects of a treatment in relation to a control. In this design, also called BACI (Before After Control Impact comparison), the correlation between the observations from control plots and treatment plots before treatment is used to assess the effects in the treatment plots as deviations from the expectations made on the basis of the control plots.

Arthropod Fauna in Orchards

Hassan *et. al.* summarise the standardised methods developed by the IOBC Working Group "Integrated Protection in Orchards." The methods consist of catching the fauna in collectors placed under the trees to which a chemical has just been applied. The collectors can be trays, canvas sheets, or funnels ("Steiner funnels"), with at least 0.5 m² of collecting area. In each trial, both the control (water spray) and the test treatment are followed by a "cleaning" treatment, 48 hours after the initial treatment. The "cleaning" treatment consists of dichlorvos at double the recommended dose, which will remove all beneficials present. The effectiveness of the treatment can thus be expressed in relation to the total population of beneficials present in the treated tree. The use of a reference chemical with each treatment, *e.g.*, phosalone, is also recommended.

The design of the trial is a complete randomised block design, or a balanced incomplete block design. Each replicate is represented by one tree with one or more fauna collectors; six to eight replicates per treatment are recommended. The trees should be separated by at least one untreated tree. The fauna are gathered from the collectors 24 hours and 48 hours after the treatment, as well as 24 hours after the "cleaning" treatment. Only the arthropods, of which there is at least an average of ten individuals per collector, are considered. The fauna are identified to the species or the family level.

Earthworm Field Tests

Although no standardised guidelines exist yet for the study of pesticide side-effects on earthworms, recomm endations for the performance of such a test have been formulated by Kula.

Field studies with earthworms can be performed on arable land or on permanent grass; in both cases, a minimum number (100 individuals per m² of earthworms is required, and the relevant species *(Aporrectodea caliginosa* and *Lumbricus terrestris)* must be present. Minimum plot size should be 100 m², and at least four replicate plots should be used per treatment. A study should include a control, the highest recommended dose, and a toxic standard (benomyl). In many cases also a manyfold (*e.g.*, fivefold) of the recommended dose should be studied. At each sampling time, at least two samples (sampling area 0.25 m² should be taken from each replicate plot. Preferably treatment should take place in the spring, and samples should be taken 1, 4 to 6, and 12 months after application. For the sampling of earthworms, the formaldehyde method and electrical sampling methods seem to be the most useful.

METHODS TO ASSESS THE IMPACT OF SOIL CONTAMINATION ON SOIL ORGANISMS (DIAGNOSIS)

Laboratory and Field Bioassays

The main characteristic of laboratory bioassays is that the potential toxicity of samples of field soil is studied in the laboratory using laboratory-bred test organisms. So, in such a test, well known organisms having similar characteristics are used, and the test can be performed under controlled conditions ruling out other possible disturbing influences. The advantage of bioassays is that they provide a direct indication of the toxicity of a specific soil, and integrate the effect of all substances present. The disadvantage is that the specific chemicals causing the observed effects cannot be identified. Bioassays can also be applied to determine the bioavailability of pollutants in soils as an indication of the potential risk for higher trophic levels.

Laboratory bioassays with earthworms *(Eisenia fetida)* and higher plants *(Cyperus esculentus)* have been described by Marquenie and Simmers and Van Gestel *et. al.* the latter authors used these bioassays to study the influence of soil clean-up on the bio-availability of metals. In these methods, cylinders with a diameter of about 18 cm are filled with a 20 cm layer of soil. The cylinders have a perforated bottom and are placed in a dish filled with water. After one week of preincubation, test organisms are introduced. After 4 weeks, earthworms are sorted out of the soil; after incubation on wet filter paper for 24-48 hours to void the gut, they are analysed for metal content. Plants are harvested, and shoots are analysed. Van Gestel *et. al.*described similar bioassays, using lettuce *(Lactuca sativa)* and radish *(Raphanus sativus)*, to determine metal bio-availability in soils. For these organisms, smaller amounts of soil are needed (about 400 mg), and as in the bioassays using *C. esculentus*, some nutrient solution was added to the soils to stimulate plant growth.

Bioassays using the earthworm species Lumbricus terrestris and E. fetida have been described by Menzie *et. al.* They studied survival of the earthworms after exposure to contaminated soil for 28 and 14 days, respectively, and also determined the uptake of some selected chemicals in the earthworms.

Menzie *et. al.* and Callahan *et. al.* studied the potential risk of contaminated soils using *in situ* bioassays with earthworms. Contaminated soil was placed in plastic buckets placed in the ground from which the soil was taken. The buckets were constructed to allow for free exchange of air and water. Adult earthworms of the species *Lumbricus terrestris* were placed in the buckets, and observations were made after 1 and 7 days. Survival and morbidity (burrowing, coiling, shortening, swelling, lesions) were the test parameters, and earthworms (including gut content) were analysed to determine bioaccumulation of the main pollutants. This bioassay proved to give a good indication of the possible risk of polluted soil.

Two observations merit comment. First, the bioassays last only 7 days, which might be too short to allow for an equilibrium in the uptake of highly lipophilic chemicals. Thus, uptake of these chemicals by the earthworms may be under-estimated. Second, uptake might be misjudged, because of the presence of contaminated soil in the earthworm gut. From the authors' experience, the gut content of an earthworm may account for about 50 per cent of its dry weight.

Kopezski used small enclosures (3 cm long; 4.8 cm in diameter) to study the impact of acidification on population growth and decomposition activity of the collembola species *Folsomia candida* and *Heteromurus nitidus* in a forest soil. The enclosures were filled with 1 g of hazel leaf litter and 1g of wafers, and buried into the soil. By using cheese cloth for the bottom and top ends of the enclosures, free contact with the surrounding soil was ensured. After 6 months, samples were analysed. A significant correlation appeared between animal numbers and soil pH, with *H. nitidus* being most sensitive. Decomposition showed a somewhat weaker correlation with soil pH.

To study the effects of N-deposition on the inter-actions between soil fauna and micro-flora and the effects on mineralisation in coniferous forest soils, lysimeters have been used by Berg and Verhoef. Intact soil cores with total soil fauna or absence of mesofauna are treated with three $(NH_4)_2SO_4$ concentrations (10, 50, and 200 kg N per ha per g). The lysimeters are defaunated by means of microwave treatment. Before installation, they are incubated with a soil-spore suspension. Faunal groups extracted from the soil are added to the different treatments.

Migration of fauna between the lysimeter and the surrounding soil can occur through holes in the lysimeter. The lysimeters are covered by a gauze lid and covered just above the top by a plexiglass roof preventing rain input. The lysimeters are watered every two weeks by hand.

Mutagenicity Tests

Besides the bioassays using invertebrates or higher plants, special methods may be necessary to assess the potential risk of contaminated soils. Some compounds present in soil as contaminants (*e.g.*, PAHs) are known mutagens, and their effects can be assessed by genotoxicity tests developed for drinking water, surface water, and sediments. These tests may be applicable for industrially contaminated sites, waste disposals, or sewage sludge-amended soils.

The usual approach to assess mutagenicity is to record the number of re-vertants in a Salmonella thyphimurium strain, plated with the test substance on a histidine deficient medium. To include those mutagens that require metabolic activation, a rat liver microsome suspension is added. In addition to this test, several other procedures have been proposed, which are often more sensitive than this test.

To apply mutagenicity tests to contaminated soils, an extract must be obtained. The extraction may be crucial to the validity of the results : extractions with solvents such as methanol or dimethylsulphoxide often induce a stronger mutagenic response in the Salmonella test, compared to water leachates; this difference in potency is especially true for superficial soil horizons.

The ecological relevance of mutagenicity test results for soils is difficult to evaluate, as this field of research is still underdeveloped. Mutagenicity is detected not only at contaminated sites but also in uncontaminated soil and often bears no clear relationship with the levels of known chemical mutagens in soil.

Field Studies (Biomonitoring)

Biomonitoring studies in soil often deal with the study of the effect of chemical stress on the structure and function of entire ecosystems or at least at the community level. For that purpose, studies are performed to determine the impact on litter decomposition in forest ecosystems or on communities of soil meso-fauna. Foodweb models for soil organisms have been constructed for agro-ecosystems. and are developed for a coniferous forest soil. based on stratified litterbag experiments. With these models effects of excessive N-input on the soil ecosystem can be estimated. No standardised guidelines exist for such studies.

Also certain species of organisms can be selected and followed in time to detect certain deviations that may be due to the impact of chemical contamination, a process called biomonitoring. Isopods are for instance, excellent bio-indicators for metal contamination because of their ability to concentrate metals in their body tissues.

Tolsma *et. al.* recommended inclusion of primary producers (the plant species Urtica dioica and Holcus lanatus), detritivores (earthworms and isopods), and carnivores (mice or moles) in a biomonitoring system for the terrestrial environment. These organisms were selected because of their capability to accumulate metals. PAH. and organochlorine pesticides.

CONCLUSIONS

The array of methods used in testing terrestrial invertebrates is diverse mainly because different tests have been developed with different aims. Many methods are still poorly described, especially in relation to the medium to which the chemical is applied, and the consequences for bio-availability.

The potential for standardisation of a test system is important when one strives for international use of test methods. For some species, a standardised method may not be possible; this holds for the oribatid mite Platynothrus peltifer that has a very long life cycle and a low reproduction rate not allowing for a proper determination of effects on reproduction within a reasonable test duration. For others, such as the earthworms and the collembola Folsomia candida, tests have already been standardised at an international level, as is the case for many tests on beneficial arthropods. The usefulness of a test system to derive environmental quality criteria also comprises the substrate used. When real or artificial soils are used as a test substrate, test results may be applied directly to derive soil quality criteria; however, this is not the case for tests on soil organisms using other substrates or exposure routes, such as water, agar, or nutrient solutions used in tests with protozoans and nematodes. Results of such tests cannot be translated directly to soil quality criteria. This conclusion holds also for exposure routes used in tests on beneficial arthropods; such tests can only be useful for the risk assessment of pesticides when test results can be related to natural exposure routes.

By one approach, each species should be tested under its optimal conditions, and inter-species harmonisation of conditions (*e.g.*, by using the same test substrate) is not to be recommended. The usefulness of more than one test on the same chemical is, however, very limited when the results for two species cannot be compared to each other. Standardisation of the test substrate should, therefore, be considered in the further development of test methods.

Only a few methods are available to determine the potential risk of contaminated soils. Because of the complex nature of contaminated soil, in which often a mixture of chemicals is involved, such methods are urgently needed. Also more knowledge is needed on the toxicity of combinations of toxicants.

For a suitable risk assessment, a battery of tests should be available. Such a test battery should contain organisms representing different taxonomic groups as well as the soil community. For that purpose, the Health Council of the Netherlands (1991) selected 24 parameters to be applied for both diagnosis and prognosis of chemical effects in terrestrial ecosystems and sediments. A conclusion was that, especially for higher levels of organisation, tests are lacking. For an initial screening of possible effects, a test system should consider higher plants, decomposition capacity of the soil, invertebrates, and vertebrates (the latter being outside the scope of this paper).

Chapter 9

SOIL MICROBIAL ACTIVITY

SOIL FERTILITY AND NUTRITION

Profitable canola production relies heavily on adequate plant nutrition, which in turn is affected by management of soil fertility. In addition, the nutritional level of the plant will affect the crop response to stress factors such as disease and adverse weather. Balanced, effective fertilizer management not only contributes to profitable canola yield but also helps to maintain the productivity of the soil resource.

Basic Plant Nutrition

The living plant depends on a number of basic factors for normal growth :

- Light
- Air
- Water
- Nutrients
- Physical support.

Soil plays an important role in all these factors except for light. If any of these basic factors are limiting, plant growth will be reduced or the life cycle may not be completed this is called the principle of limiting factors. In other words, plant growth potential is limited by the factor in shortest supply.

Yield may be reduced when one nutrient reaches excessive levels that cause toxicity, so the proper balance of nutrients is important. Also, other factors such as improper management or pests can lower yield. Therefore, a systems approach is necessary to integrate all the factors in the best combination to achieve the most economic yield.

Essential Plant Nutrients

The plant mineral composition does not simply reflect the elements needed for growth. Plants can selectively absorb required elements for their growth. But they also can take up elements not needed for growth.

The terms essential plant nutrient or essential mineral element were formed to describe the minerals needed by plants to grow and complete life cycles. Essential plant nutrients must be directly involved in some aspect of the plant metabolism such as structural material, enzymes or hormones, and they must not be totally replaceable by another mineral element. For higher plants such as canola, there are 14 essential nutrients (besides CO_2, oxygen and water).

Table. Essential Plant Nutrients

Macro-nutrients	N,	P,	K,	S,	Mg,	Ca
Micro-nutrients	Fe,	Mn,	Zn,	Cu,	B, Mo, Cl and Ni	

Table also indicates that plant nutrients are often classified by the relative amounts needed. Macro-nutrients are needed in large amounts relative to micro-nutrients. Table shows the relative amounts of nutrients contained in a typical canola crop. Some nutrients can be accumulated in plants much higher than necessary for growth.

Table. Approximate Amounts of Nutrients in the Above- Ground Portion of a 1,960 kg/ha (35 bu/ac) Canola Crop

Element	kg/ha	lb/ac
Nitrogen (N)	112–134	100–120
Phosphorus (P)	1–28	15–25*
Potassium (K)	67–134	60–120*
Sulphur (S)	22–28	20–25
Calcium (Ca)	45–67	40–60
Magnesium (Mg)	13–20	12–18
Iron (Fe)	~1	~1
Chlorine (Cl)	~0.8	~0.7
Manganese (Mn)	~0.2	~0.2
Zinc (Zn)	~0.2	~0.2
Boron (B)	~0.2	~0.2
Copper (Cu)	~0.7	~0.06
Nickel (Ni)	~0.004	~0.004
Molybdenum (Mo)	~0.004	~0.004

Note :

* P X 2.3=P_2O_5; K X 1.2=K_2O

**Crop uptake of nutrients is greatly affected by conditions in the soil or weather (dry, wet, cold, compaction, nutrient imbalances, salinity, etc.).

General Nutrient Uptake

The following discussion outlines nutrient movement into and through the canola plant. The level of most nutrients in the plant sap is much higher than in the water surrounding the roots. For example, a typical N content in a canola plant at the rosette stage would be 5 to 6 per cent N, whereas a fertile soil in the spring would contain about 0.0002 per cent N in a plant available form on a dry weight basis. The N level in the soil solution would be in the range of 0.00002 per cent . Therefore, plant nutrient uptake must be highly selective.

Nutrient uptake begins when plant available forms move from the soil water through pores in the root skin (exodermis) into the free space of the roots. This free space comprises about 5 to 10 per cent of the root internal volume. This movement is a passive process (doesn't require energy from the plant) driven either by diffusion (movement due to differences in concentration) or mass flow (simply carried by water flowing into the roots). The movement is selective since pores into the free space act as a size filter. Many nutrient ion diameters are much smaller than the pores. For example, potassium and calcium are only 10 to 20 per cent of the pore size, and have easy access to the free space. Large diameter substances such as metal chelates, viruses and fungi are restricted from entry by the small pore size.

As plant roots grow, the soil volume and surface area explored increases, which increases the capacity for nutrient absorption. In addition, roots possess a cation exchange capacity (CEC) due to negative charges in cell walls. This root CEC attracts positive ions (cations) like ammonium (NH_4+) but repels negative ions (anions) such as nitrate (NO_3-). (For details on CEC, see section âœSoil Properties that Affect Plant Nutrition.)

After entry into the free space, nutrients move into the cell interior by crossing a plasma membrane found on the inside of cell walls. Another similar membrane is found surrounding a large central storage compartment (vacuole) that usually fills more than 80 per cent of the total cell volume. The plasma and vacuole membranes are effective barriers and are the main sites for nutrient uptake selectivity. These membranes contain carrier systems or ion pumps that transport certain nutrients. Such systems are called active since they require energy from the plant to work. This energy demand for ion uptake by roots is considerable, taking up to 1/3 of the energy during rapid growth. The energy for root activity arises from respiration, which requires carbohydrates and oxygen. This explains why nutrient uptake often stops in flooded soils there is a lack of oxygen.

Some active uptake systems are constant while others have a rate that can be regulated. As the plant level of nutrients and related compounds increases, the root uptake rate can decrease (negative feedback). In contrast, as plants build tissue, the level of nutrient âœ building blocks decreases, and the roots are signalled to increase the uptake rate (positive feedback) for nutrients. Passive ion channels through the membranes allow for selective nutrient movement.

The selectivity of the various transport systems across the membranes is not absolute. There is often competition between ions of similar size and charge. For example, chloride (Cl-) competes with nitrate (NO_3-). This competition between

Cl– and NO_3 - is important in certain saline soils with Cl- as a major component of the salt. Most prairie soils contain salt with sulphate as the main anion.

Since cation and anion uptake are regulated differently, plants must be able to compensate for differences in electrical charges that arise from disproportionate uptake of cations and anions. Plant cells maintain a pH in the range 7.3 to 7.6 by either releasing or consuming hydrogen cations (H+), which is achieved by formation or removal of organic acids.

The nutrient journey continues in a path from cell to cell through tiny connecting tubes (plasmodesmata), although some nutrients can continue to move between cells through the free space. The next barrier occurs at the waxy layer (Casparian band) that surrounds the central vascular tissue (phloem and xylem). The phloem and xylem are special tissues that act like highways for nutrient transport from roots to leaves (xylem) and from leaves to growing points and roots (phloem). In young root tips, the Casparian strip is not well formed and thus is an incomplete barrier.

The mechanism how ions pass through the Casparian strip and into the xylem (âœxylem loading) is not well understood. There is probably a combination of active (ion pumps) and passive channels for ion movement into the xylem. Xylem loading is regulated separately from root uptake, thus creating a control system for nutrient movement. Nutrients are carried by water up through the xylem. Water flows up through xylem tissues due to a suction-like force created when water evaporates from the leaves, and from slight pressure produced by roots. Once inside the xylem sap, nutrients can be unloaded and reloaded before reaching the end growing points.

Once nutrients reach their targets, there often is considerable recycling, especially for the mobile nutrients such as N. For example, a normal feature of plants appears to be simultaneous import and export of nutrients from leaves. This dynamic nutrient cycling is termed remobilisation or retranslocation. In young vegetative plants, nutrient recycling occurs from the mature leaves to roots and young leaves through the phloem. The remobilisation ability of different nutrients affects where deficiency symptoms occur. Deficiency symptoms of mobile nutrients such as N will first appear in old tissues. In contrast, deficiency symptoms of nutrients with limited mobility such as sulphur and copper will occur in young tissue, and can hinder flower/seed development.

Nutrient remobilisation is particularly important when seeds are forming. At this stage, mobile nutrients are being exported from ageing leaves while nutrient imports are decreasing. Also, root activity and nutrient uptake generally decrease by this stage due to drying soils, nutrient depletion in the soil and a relative shift in the energy supply from roots to developing pods and seeds. Plant parts with a strong energy or nutrient demand are called sinks. As a result, old leaves are sacrificed to supply pod and seed growth. Mobile nutrients in the seed have mostly been transferred from other plant tissue" in canola the sources are pods, stems and leaves.

Roots are not the only sites where nutrient uptake can occur. Some nutrients can be absorbed by leaves and other above ground plant parts. Nutrients in the gas form NH_3 (ammonia), NO_2 (nitrogen dioxide) and SO_2 (sulphur dioxide) can enter leaves through leaf pores (stomata) and then be changed into organic forms. These gases are major air pollution components and in some areas contribute considerably to plant nutrition. In areas with intensive livestock operations, NH_3 uptake can contribute 10 to 20 per cent of the nitrogen for adjacent crops. SO_2 is readily absorbed by leaves. In a European field experiment, almost half of the total sulphur (S) taken up by vegetative rapeseed came from atmospheric S compounds, probably SO_2. This may partly explain why S deficiencies have increased in western Canada after environmental regulations enforced clean-up of S emissions from gas plants.

Soil Properties That Affect Plant Nutrition

Soil is a complex mixture of non-living substances (minerals, organic matter, gases and liquids) and living organisms (bacteria, fungi, insects, worms, etc.). These factors influence soil fertility either directly or indirectly.

Soil solids consist of mineral particles, organic matter in varying stages of decomposition and living organisms. Solids make up about half the soil volume, while water and gases make up the other half in the pore space.

Soil mineral particles vary widely in size and are classified by size :

- Rocks are larger than 2 mm in diameter (0.08″)
- Sand particles range from 0.05 to 2 mm (0.002 to 0.08″) in diameter
- Silt particles range from 0.002 to 0.05 mm (0.00008 to 0.002″) in diameter
- Clay particles are smaller than 0.002 mm in diameter″(0.00008″) in diameter.

These particles are made from various mineral types with different elemental composition, which affects weathering processes and thus the release of certain nutrients. Two soils with identical texture could be drastically different in fertility due to differences in mineral composition. Potassium is an example of a plant nutrient whose supply arises from mineral weathering in soil.

The soil colloidal fraction refers to microscopic particles of clay and organic matter. The surface of the colloidal fraction is where most soil chemical reactions occur and it is very important in nutrient supply.

The proportion of sand, silt and clay determines the texture of a soil. Soil texture is grouped into five or more classes. The texture influences fertility by affecting moisture holding capacity, air exchange and the CEC. Adequate moisture is key to fertilizer response and potential yield for canola in western Canada.

The CEC is an important property that influences the soil storage of many plant nutrients. Most nutrients are present in the soil water as positively charged cations. A few are negatively charged anions. The CEC indicates a soils ability to hold or store cations. Prairie soil particles typically have a negative charge.

The process of electrical attraction that holds cations to negative surfaces of soil colloids is called adsorption (not absorption). The cations are not permanently stuck to the colloidal surface and can be exchanged with other cations. With time certain cations may become âœ fixed into forms that are not easily removed from the exchange complexes. Adsorbed cations are not removed by water moving through the soil and can be accessed by plant roots. Cations with a higher positive charge (for example Ca^{+2}) are held more tightly than those with a lower charge (for example K+).

Soil negative charges arise due to substitutions in the mineral crystals by elements with smaller positive charge, and due to reactions at the edges. Organic particles also contain a significant number of negative charges. The total particle surface area in a soil increases as particle sizes get smaller. Therefore, a soil high in clay has a much greater surface area than a sandy soil.

A high clay soil also has a bearing on the surface area and negative charge. Clay minerals are microscopic layers of aluminium and silicon crystals formed by weathering of other minerals. Thus clays are called secondary minerals. The type of clay depends on the original minerals and the weathering extent. Fairly âœ young clays common in western Canada (such as montmorillonite) have a 2 : 1 arrangement of silica : alumina crystal sheets, while older clays have a 1 : 1 arrangement. Generally, 2 : 1 clays have 10 to 100 times more surface area, negative charges and, consequently, a higher CEC than 1 : 1 clays. Organic colloids have 10 to 100 times more negative charges and higher CECs than the 1 : 1 clays. Therefore, soil organic matter levels greatly influence the CEC.

The CEC strongly influences soil fertility. A higher CEC means that more cations, including plant nutrients, can be loosely stored in a plant available form, giving the plant a greater pool of nutrients to draw from. Since most cations are not highly soluble, only small quantities can be dissolved in the soil solution at one time.

The CEC soil property allows a reservoir of nutrients to be stored then released to plant roots. This continuous replenish ment of nutrients in soil water is very important for several nutrients, including potassium. A high CEC also means that fewer cations will be lost through leaching out of the root zone.

Since soils are predominantly negatively charged, anions [such as nitrate (NO$_3^-$) and sulphate (SO$_4^{-2}$)] are repelled by soil colloids and tend to stay in the soil water. They will flow with water and are potentially subject to leaching loss.

Soil organic matter (OM) plays an important role in soil fertility as a plant nutrient storehouse. Not only does OM adsorb many cations due to a high CEC, it also stores nutrients as part of its structure. As the OM is decomposed by soil microbes, nutrients are released from the organic structure into plant available forms this process is called mineralisation.

Mineralisation from OM is the primary natural source of plant available N and S in prairie soils, and also influences P availability. Minera lisation of individual nutrients will be described in later sections. Soil OM also plays a secondary role in soil fertility by improving physical properties such as water holding capacity, infiltration, aggregation (tilth) and buffering pH.

Soil Testing for Nutrient Content

Plant nutrient content in soil varies over years, between fields and even within fields that appear uniform. Soil sampling and analysis methods (soil testing) were developed to assess the fertility level and to predict crop response to applied fertilizer or manure. Soil testing is not an exact science due to nutrient variability inherent in most fields and the inability to predict growing season weather. Although soil testing is not exact, it can help estimate soil fertility and give reasonable guides for profitable fertilizer application.

Spatial nutrient variability in fields creates problems for soil testing and fertilizer application. The variability makes it difficult to obtain representative soil samples. Using single fertilizer rates across variable fields results in over-fertilized and under-fertilized areas within the field. Although variable rate fertilization is being researched and developed, most fields still are fertilized with a single rate. In addition, fertilizer response calibrations developed from research sites with low variability will under-predict the optimum fertilizer rate for larger farm fields with more variability. For meaningful soil test results proper soil sampling is necessary. A sampling error in the field is usually much greater than the analytical error in the lab. Ensure soil samples accurately reflect the overall field. However, intensive soil sampling is not convenient, cost effective or practical.

Research shows that the accuracy of a composite soil sample increases with the number of sub-samples taken (see Figure). Accuracy refers to how similar the soil sample value is to the true field average (is the sample representative). Precision describes how often the same value can be obtained when repeating the procedure (reproducibility). The common recommendation to sample a field 20 times and mix all the samples into one composite, produces an accuracy of about Â ±17 per cent for NO_3^-, assuming 80 per cent precision. This is an acceptable level of accuracy and precision for fertilizer recommendation purposes in most cases.

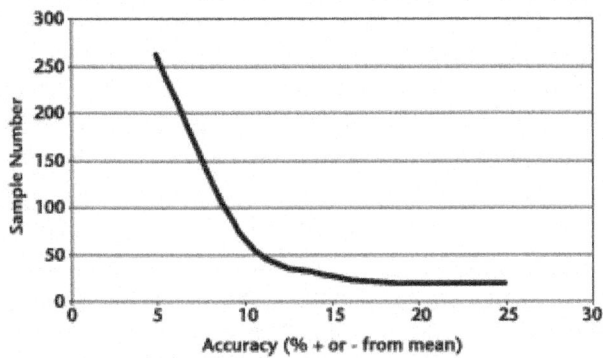

Fig. Effect of Sub-Sample Number on per cent Accuracy of Composite Sample for Nitrate-N with 80 per cent Precision.

In addition to adequate sample numbers, proper soil sampling techniques include :

- Where and how to sample each field (sampling plan)
- Proper equipment
- Proper sampling time
- Correct depth
- Proper sample handling

Soil Sampling Plan

Sample individual fields separately. The first step is to assess the field variability and identify representative areas. The type and level of variability can influence the choice of sampling plan.

Consider the four basic sampling plans :

1. Random soil sampling uses a random pattern across a field, generally avoiding unusual or problem areas such as hilltops and potholes. Bulk together 20 soil cores into one composite sample, air-dry then send to a soil test lab. This common method is adequate for smaller, relatively uniform fields.

2. Topographic sampling involves separating sets of samples based on topography. Identify the dominant topographic features such as hilltop, midslope and bottom slope, and take 20 core samples for each type. Bulk sub-samples from each type into one composite sample for each landscape type. Send several samples to the testing lab. This method can provide more meaningful results for variable fields, but at additional expense and labour. You must be willing and able to apply different fertilizer rates in the separate landscape areas.

3. Benchmark sampling expands on the topographic sampling concept by considering unique areas based on topography, soil texture and type and typical crop growth. Once these unique areas are identified, each year the samples are taken from the same spot within each benchmark area. There may be one or more benchmark sites within a field, depending on the variability. Each benchmark area then becomes a reference area on which fertilizer recommendations are based. The benchmark sampling area is much smaller than the whole field, and together with sampling in the same spot from year to year, this is assumed to reduce variability of the test results. If the benchmarks are carefully selected to represent the majority of the field, then good soil test results can be obtained.

4. Grid or systematic sampling follows an organised grid pattern, perhaps every 0.2 to 2 ha (0.5 to 5 acres). This method can reveal field nutrient variability and allow for variable rate fertilization and precision farming techniques. However, the sampling and analysis costs are not economic with current field crop prices on the prairies.

Proper Soil Sampling Equipment

Soil sampling to a 60 cm (2′) depth can be done with a probe or auger. Do not use flight or screw sampling augers if samples need to be separated by depth since mixing will occur. If testing for micro-nutrients, ensure the sampling tool is chrome plated or stainless steel and rust-free. Do not bulk samples in metal pails. Clean plastic pails labelled for location and depth work well.

Proper Sampling Date

The most accurate sampling time is just prior to seeding. However, this is not practical because time is needed to purchase and perhaps place the fertilizer before seeding. Sampling is commonly done in the spring or fall. Spring sampling is done after the soil has thawed and is no longer saturated from snow melt. Fall sampling can begin once the soil has cooled to 5 to 7Â°C. This helps reduce nutrient content changes due to microbial activity.

The old assumption that N availability does not change over the winter has been proven wrong. Research in Alberta found that available N increased by 56 kg/ha (50 lb/ac) in stubble fields from fall to late winter while the soil was frozen. Summerfallow fields increased by 73 kg/ha (65 lb/ac). This overwinter gain in available N is apparently due to death of soil microbes and subsequent release of available N forms from their ruptured cells. However, the available N gained over the winter was temporary as it was mostly lost in the spring, probably through denitrification. The stubble fields lost 45 kg/ha (40 lb N/ac) in the spring while summer fallow fields lost 73 kg/ha (65 lb/ac). Therefore, the net change from late fall to late spring was minimal. Late fall sampling tends to more accurately reflect spring NO_3- (nitrate) contents than early fall sampling, especially for Black soils. Alberta research in the 1980s compared soil samples taken in the fall (early October and early November) 'to spring samples for 26 stubble fields. Early fall samples averaged 34 kg/ha (30 lb/ac) less nitrate N than spring samples, while late fall samples averaged only about17 kg/ha (15 lb/ac) less. The early fall samples were also more variable in relation to spring samples. Overall, late fall samples more accurately predicted spring nitrate contents and grain yields than early fall. Spring samples were slightly better than late fall samples for predicting grain yield and N uptake.

In contrast, recent research in Manitoba measured very little change in soil nitrate levels in cereal stubble from early September to freeze-up. North and South Dakota extension soil scientists recommend early fall sampling in view of time constraints, but reduce the N recommendation by 0.2 kg (0.5 lb) for each sampling day prior to September 15.

Another experiment in central Alberta compared the effects of sample timing on phosphorus (P) soil test values. On average over 27 sites, the extractable P increased from 28 kg/ha (25 lb/ac) in early October to 49 kg/ha (44 lb/ac) by early November, and to 50 kg (45 lb) by spring (late April to early May).

The relationship between early fall and spring P values was not close and, therefore, it was not possible to simply correct the early fall values. In contrast, research on a Brown soil in Saskatchewan over 24 years found both overwinter increases and decreases in soil P tests, but relatively few were significant. An experiment on irrigated alfalfa on a Dark Brown soil near Let bridge, AB found significant overwinter increases in organic P. The conflicting results may be related to the differences in soil organic matter content and biological activity, and, therefore, potential for microbial changes to the plant available P pool.

In conclusion, early fall sampling can create higher than necessary fertilizer N and P recommendations due to an underestimation of spring nitrate N and available P, especially in the Black and Gray soil zones. Fertilizer response curves have been calibrated only against spring nutrient contents.

One disadvantage to late fall sampling after soil has cooled to 5 to 7Â°C is that fall fertilizer banding opportunities become more limited. By the time the samples are taken, dried, sent to the lab, analysed and results returned, the soil may have become frozen or covered with snow. On average over the prairies, soil cools by 1Â°C every five days in the fall.

Correct Sampling Depth

The appropriate sampling depth depends on the nutrients to be tested. For mobile nutrients such as N and sulphur (S), sampling to the 60 cm (2′) depth is usually the most accurate according to research conducted in the 1960s. The fertilizer response database for N was developed with 0 to 60 cm (0 to 24″) samples.

However, recent research in Saskatchewan and North Dakota indicates that the 0 to 30 cm (0 to 12″) depth may be more accurate for N than either 0 to 15 cm (0 to 6″) or 60 cm samples. This recent research and the fact that sampling 60 cm is considerably more difficult, supports the 0 to 30 cm depth as a reasonable recommendation for these areas. In contrast, research in Manitoba has documented that the 0 to 15 cm depth is inferior. Separate samples for 0 to 15 cm, 15 to 30 cm and 30 to 60 cm was often recommended in the past, but was rarely done due to additional expense and time.

For immobile nutrients such as P and potassium (K) and most micro-nutrients, the ideal depth is 0 to 15 cm because the fertilizer response calibrations are based on that depth. If only the 0 to 30 cm depth is sampled, the soil testing labs must use a correction factor to estimate the value for the 0 to 15 cm depth. Since these nutrients are relatively immobile, they tend to remain at the fertilizer application depth. Therefore, the 0 to 30 cm depth may under-estimate these nutrients and lead to high fertilizer recommendations.

Proper Sample Handling

Handle samples carefully to prevent accidental mixing and contamination. Mix each sample and spread on separate pieces of clean paper to dry at room temperature. Use a fan if required and avoid additional heat sources. Once dry, fill soil sample cartons or bags with about 0.5 kg (1 lb) of soil, and label with field number and depth.

Consistency of Soil Test Lab Recommendations

Soil sampling, analysis and interpretation is not an exact science. However, reasonable precision and accuracy is needed in order to make costly fertilizer decisions. Over the years, various growers, agencies or companies have sent duplicate samples to different labs to compare their analysis and recommendations. Unfortunately, widely differing results and recommendations have occurred, causing growers to question the credibility of the labs or the soil testing practice.

Two fundamental reasons contribute to differences in results. First, labs may be using different analytical procedures to measure soil nutrient content. For example, there are several methods to extract soil phosphorus. Or the technique may differ slightly when using a particular method. Over time, labs are harmonising their test methods and in time this problem may disappear.

Variations in soil test recommendations arise mainly due to the differences in each labels interpretation. The recommended fertilizer amounts at the same soil test level can vary significantly from lab to lab.

This may be due to using :

- Different critical (deficient) soil levels being used
- Regional fertilizer response (calibration) data but modifying recommendations to fit a particular philosophy of fertilizer use or economic payback
- Recommendations from other regions or countries
- A unique system of fertilizer recommendation not based on regional calibration data or economics.

Although consistency among labs has been improving since the first comparison studies in the 1980s, further improvements are needed. Calibration data must continue to be collected to account for changes in fertilizer application techniques and changes in other agronomic practices.

Overall, soil testing is a useful agronomic practice. Use labs that base fertilizer recommendations on economics using regional calibration data. Be prepared to question unusual recommendations based on experience and the local knowledge of qualified agronomists. Keep in mind that the accuracy of fertilizer recommendations will always be limited by sampling challenges and the inability to predict the weather of the upcoming growing season.

Plant and Tissue Testing

Crop nutritional status also can be assessed by plant and tissue analyses. These methods can supplement, but not replace, soil testing. Plant and tissue analyses measure the nutrient content of above ground plant parts during growth. The values are compared to established ranges for inadequate, adequate and excess levels.

Plant tissue testing is suitable for diagnosing crop problems that may be nutritionally related and to identify any nutrients that may be limiting yields. Plant analysis can determine if the fertilizer rate and method of application were adequate to meet crop needs.

The disadvantage of tissue sampling is timing"after tissue samples are taken and analysed, it may be too late to correct deficiencies in the current crop. No reliable interpretative criteria exist for nutrient ranges in seedling canola. Also, nutrient contents usually differ greatly between different plant parts and ages. Therefore, the proper part must be sampled at the proper growth stage.

An adequate sample will contain 50 to 80 plants, depending on the nutrients to be tested and plant part/age. Avoid unusual, dead or stressed plants, as well as those covered with soil or recent sprays. Cut samples with a clean, rust-free knife or scissors. Dry the plant samples on clean paper or plastic at room temperature (do not oven dry). After drying, keep the samples in a paper bag. The following table shows sufficiency levels for most plant nutrients in flowering canola.

Table. Plant Tissue Analysis Interpretative Criteria for Canola (whole Above Ground Plant at Flowering).

Nitrogen (N) per cent	> 2.4
Phosphorus (P) per cent	> 0.24
Potassium (K) per cent	> 1.4
Sulphur (S) per cent	> 0.24
Calcium (Ca) per cent	> 0.49
Magnesium (Mg) per cent	> 0.19
Zinc (Zn) ppm	> 14
Copper (Cu) ppm	> 2.6
Iron (Fe) ppm	> 19
Manganese (Mn) ppm	> 14
Boron (B) ppm	> 29
Molybdenum (Mo) ppm	> 0.02

Comparing the plant analysis results from two areas of a field that differ visibly in growth can be difficult to interpret because nutrient content differences can be confounded by growth differences. If the two areas differ mainly in deficiency symptoms, then comparative sampling can be useful. In this case, collect the samples soon after the symptoms appear and before major differences in growth and maturity occur. Plant and tissue analyses need to be interpreted by experienced indi viduals.

Diagnose Nutrient Deficiency Symptoms

In moderate to severe nutrient deficiencies, visible symptoms can indicate the specific nutrient that is lacking. Nutrients that are only slightly limiting often do not show visible symptoms, a situation that has been termed hidden hunger. A systematic diagnosis of visible symptoms is needed to correctly identify limiting nutrients. Symptoms usually appear on either old or young leaves depending on the mobility of the nutrient in question. Chlorosis (loss of green colour, yellowing) and necrosis (death of plant tissue, often leading to white or brown colour) are important visible symptoms. Diagnosis under field situations can be complicated by high field variability, multiple deficiencies, and other causes such as weather, pests and herbicide injury. For example, a sulphur deficiency can easily be confused with Group 2 herbicide injury due to similar symptoms.

Nitrogen (N)

Nitrogen is the most common limiting nutrient (other than water) for canola production. Therefore, a good understanding of this nutrient is needed to efficiently manage fertilizer N and maximise economic returns.

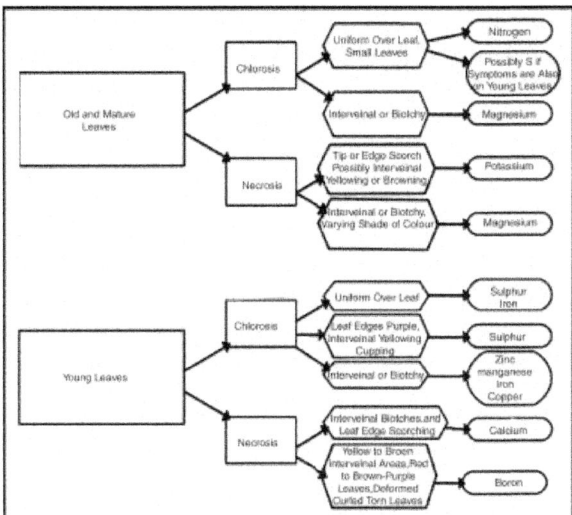

Fig. Diagnosing Nutrient Deficiency Symptoms.

Role of Nitrogen in the Canola Plant

As shown in the previous section (Table), canola, like most crops, contains large amounts of N. Nitrogen is a part of many critical plant components : amino acids and proteins (which form enzymes); genetic material (nucleotides and nucleic acids); and other components found in membranes (such as amines), co-enzymes and others. The majority of the N in green plant tissue is present as enzyme protein in chloroplasts where chlorophyll is located. By harvest, the majority of the N in a canola plant is found as seed protein. The relative N proportions in the plant changes over time and growth stage. The N proportioning closely resembles the dry matter partitioning as shown in Figure.

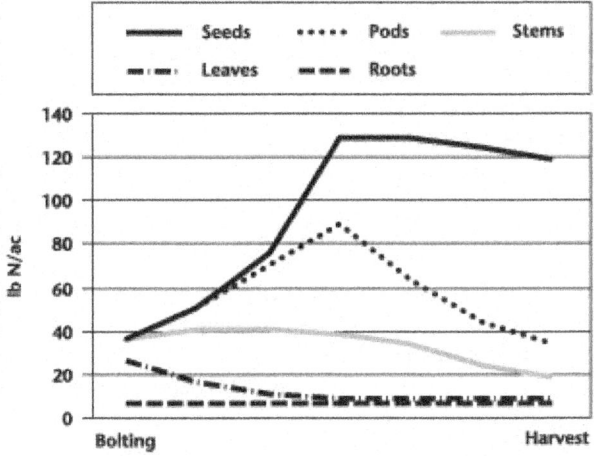

Fig. Nitrogen Partitioning in Canola.

The N level in canola plants is highest in the early seedling stage when young leaves are the majority of the plant dry matter. As the plant grows into flowering stages, the overall N level declines due to stem material and leaf loss. By maturity, canola straw contains just 0.5 to 1.5 per cent N while the seed contains 3.4 to over 4 per cent N.

Nitrogen Effects on Canola Growth

The most obvious N effect is an overall increase in plant growth (height and dry matter). This stimulation occurs early in the vegetative stage and continues into the reproductive stage. Research in England illustrates the canola leaf stimulation, leaf area and pods/seed production by fertilizer N (Figures).

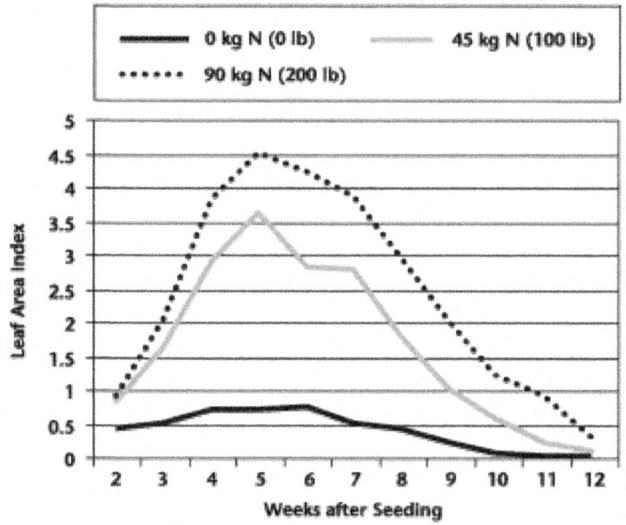

Fig. Canola Leaf Area Response.

Other research generally confirms that N fertilizer mainly increases canola leaf area index, leaf duration, plant weight, growth rates, number of flowering branches, plant height, number of flowers, number and weight of pods and seed yield. Therefore, good N fertility is necessary to produce a large, photo-synthetically efficient leaf area that will support high numbers of flowers, pods and seed yield.

Canola Nitrogen Deficiency Symptoms

Healthy canola plants with adequate N have dark green leaves. Nitrogen is mobile within the plant and can be moved from older to younger leaves and pods. Therefore, N deficiency symptoms first show up in older leaves as pale green to yellow colouring, and sometimes purpling. These older leaves tend to die early, turn brown and drop off prematurely.

Overall plant growth is slow, with short thin stems, small leaves, and few branches. The amount and time of flowering is restricted, and pod numbers are low. Nitrogen-deficient canola pictures are shown in Figure. In healthy canola, plant tissue tests of above ground material at flowering will show more than 2.5 per cent N.

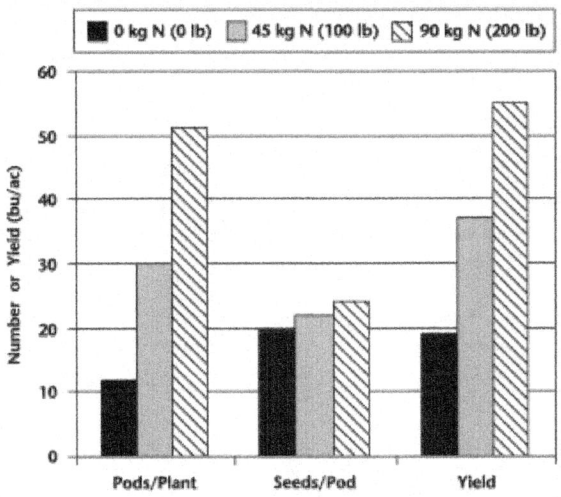

Fig. N Fertilizer Effect on Pod, Seed Number and Yield.

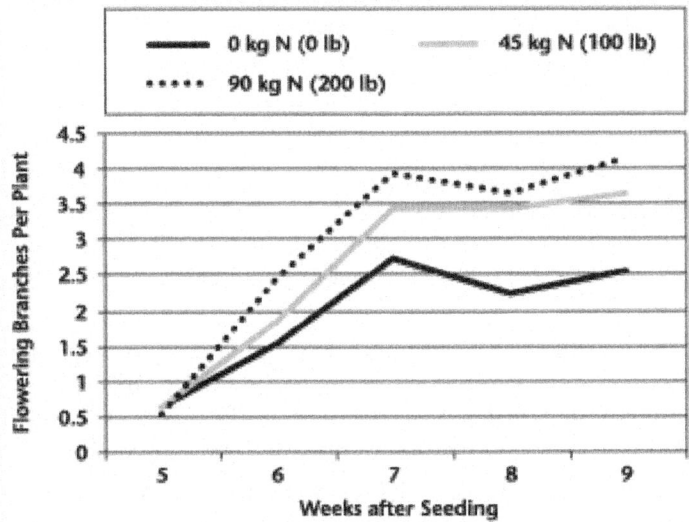

Fig. Effect of N on Flowering Branches.

Canola Response to Fertilizer Nitrogen

Canola responds well to applied fertilizer N on deficient soils. Most stubble fields have insufficient N for high canola production and thus require fertilizer or manure application. Research data has been collected that describes crop response to fertilizer in relation to initial soil reserves in western Canada and around the world. This data is used by soil testing labs to predict fertilizer requirements. Research on the prairies has found that profitable dryland canola yield response to fertilizer N is unlikely when the soil contains more than 34 to 45 kg nitrate N/ ha (75 to 100 lb nitrate N/ac) in the top 60 cm (2′).

Higher yielding winter canola types typically respond to more N fertilizer than winter wheat. In contrast, spring canola types have a similar N fertilizer response to high yielding CPS wheat in western Canada.

An example of typical canola response to fertilizer N on a deficient stubble black soil in Alberta is shown in Figure. The vertical line represents the economical level of N fertilizer under medium moisture with $265/tonne ($6/bu) canola, $0.88 kg ($0.40 lb) N and a 2 :1 risk ratio.

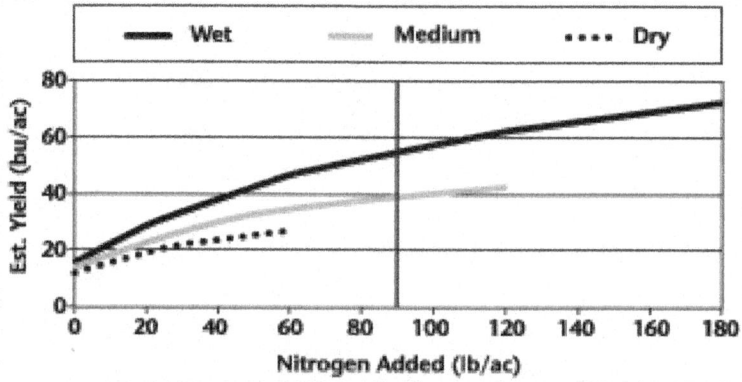

Fig. Canola Response to N Fertilizer.

The previous crop influences the crop response to N fertilizer. Different preceding crops vary in the amount of available N removed from the soil and the amount released from crop residue. In addition, the disease break with different rotational crops can affect the yield potential and, therefore, the economic response to fertilizer.

Moisture availability greatly affects the yield response to N fertilizer. Conversely, adequate N is needed for crops to respond to moisture. Canola yield is reduced under extremely dry or wet conditions. Research at Outlook, SK illustrates the synergistic response of canola to moisture and N over eight years (Figure).

Under dry soil conditions, root growth and activity are reduced, resulting in less N uptake. In addition, soil microbial growth is slower, which reduces N release from soil organic matter, but also reduces temporary tie-up by soil microbes. Normally, more plant available N is left in the soil after a dry growing season than after wet seasons.

Drought and high temperature stress near flowering and podding dramatically reduce both water and N fertilizer efficiency. This is true for both stubble

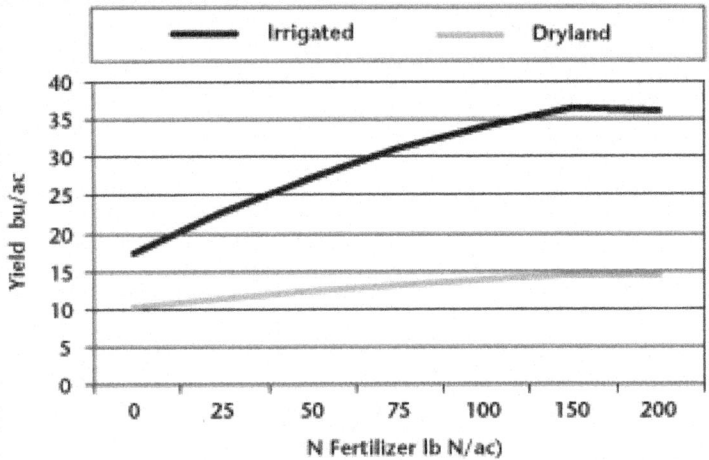

Fig. Canola Response to N and Moisture at Outlook, SK.

and fallow crops, although moisture is more often limiting for stubble crops. Unfortunately, crops use the most water during the flowering and podding period when soil moisture reserves and rainfall are usually low. Therefore, the amount and timing of rainfall are important. Studies in western Canada generally find that growing season precipitation increases grain yield two to three times more than equivalent amounts of stored soil moisture.

Heavy N fertilization can reduce canola yields when excellent spring moisture conditions are followed by drought. Under this condition, the N stimulates larger leaves, increases transpiration and moisture use. As a result, soil moisture can be depleted, leaving little for flowering, podding and seed fill. Excessive N fertilization can also reduce yields by promoting lodging, delaying maturity that may increase fall frost damage, and increase foliar disease due to the dense canopy and lodging.

Effect of Nitrogen Fertilizer on Canola Quantity

N fertilization generally increases the protein content of canola seed and meal. However, when N fertilizer is added under conditions of S deficiency, there may not be a protein increase but a rise in free amino acids due to hampered protein synthesis.

In contrast, N fertilization may slightly decrease the canola seed oil content, especially at higher rates. Plant breeders report that oil and protein contents are often inversely related any attempt to change one causes an opposite change in the other component. Although the seed oil content may decrease at high N rates, the total oil yield per acre still increases because yield increases more than oil level decreases. The overall effect of N fertilization on yield and canola quality based on research in Manitoba is illustrated in Figure.

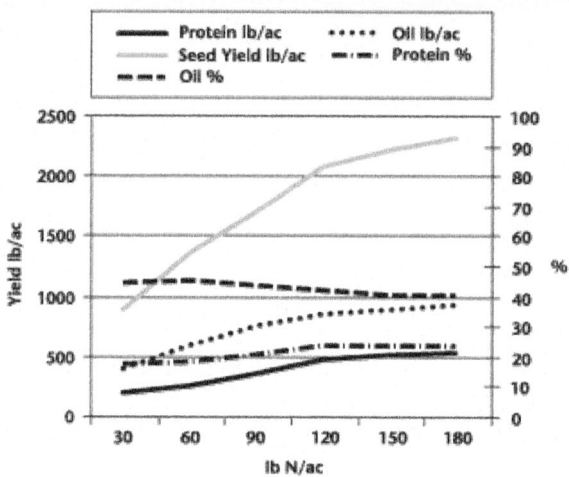

Fig. Canola Response to N and Moisture at Outlook, SK.

Nitrogen Cycle and Transformations

Nitrogen is subject to many different processes in soil due to microbial activity and physical forces (Figure).

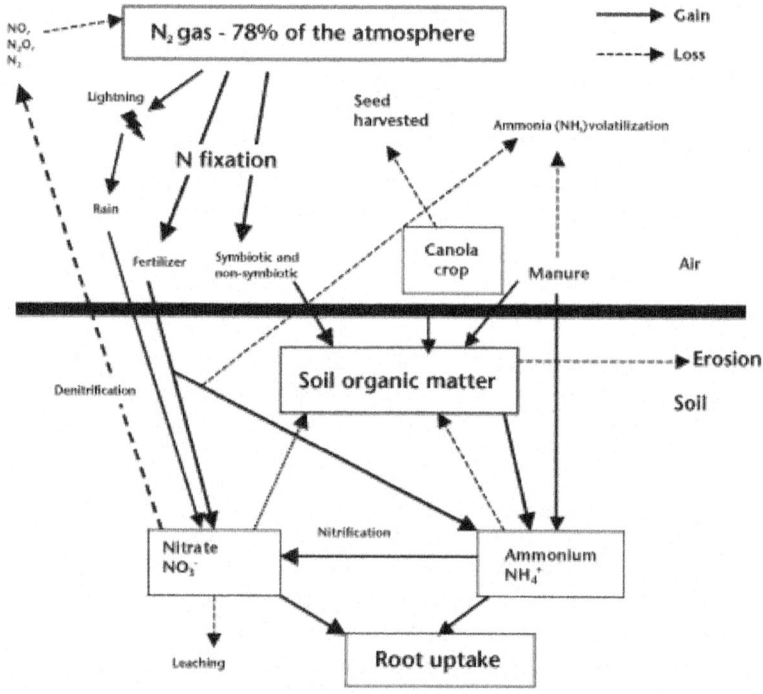

Fig. Nitrogen Cycle and Transformations.

Supply of Nitrogen by Soil and Fertilizer to Canola

Canola plants obtain N from several sources :

- Soil supply of nitrate and ammonium arising from soil organic matter decomposition and fertilizer residues of previous years.

- Nitrate and ammonium released through mineralisation during the current growing season.

- Fertilizer or manure N additions to the current crop.

- Other minor sources such as available forms deposited during rainfall and lightning, non-symbiotic N fixation in the root zone, and ammonia uptake by foliage.

Since canola plants obtain most of their N from the soil environment, understanding the N cycle is very important to managing N fertilizer effectively.

Transformations that Supply Plant Available Nitrogen

Soil organic matter is a major N reservoir containing several thousand pounds of organic N per acre. This large organic N storehouse needs to be decomposed by soil microbes before becoming available for root uptake. The decomposition process is called mineralisation.

The decomposition rate is fairly slow and variable, ranging from about 0.4 to 5 per cent per year. As shown in Figure 10, organic N is first slowly changed to ammonium by bacteria. Then different bacteria rapidly change the ammonium to nitrate, in a two-step process called nitrification.

The ammonium is first changed to nitrite (NO_2-), then to nitrate. Under normal soil conditions, the bacterial oxidation of ammonium to nitrite is much slower than nitrite to nitrate. Therefore, very little nitrite is normally found in soil. This is fortunate since nitrite is toxic to plants. The result of these rate differences is that most of the plant available N in the soil tends to be nitrate. This is also why most soil testing labs analyse soil for nitrate and dont include ammonium.

Since N transformations result from soil microbial activity, soil conditions such as temperature, moisture and acidity will strongly affect the rates at which these processes occur. If soil is cold, saturated or very acidic, then mineralisation and other microbial activities will proceed slowly. Cultivation stimulates organic matter decomposition and mineralisation of N by improving aeration and physical mixing that gives soil microbes access to new organic matter supplies.

There are several sources of plant available N to the soil system, in addition to soil organic matter decomposition and fertilizer additions (Figure). The atmosphere contains 78 per cent N_2 and thus is a huge source of N. But it first must be changed to plant available forms through biological or industrial processes called N_2 fixation. Biological fixation of atmospheric N_2 is performed by certain bacteria or blue-green algae species.

Biological N fixation in soil falls into three general types :

• Symbiosis with legumes

• Associative

• Free-living N fixing bacteria

Symbiotic N fixation is much larger relative to the other types. Canola is not a legume and cannot form the symbiosis with rhizobia to fix atmospheric N. However, canola can benefit from residual N fixed by previous legume crops.

Associative N fixation occurs when bacteria just inside the root or on the root surface use root exudates for energy to fix atmospheric N. The plant benefits indirectly when the bacteria dies and its N is released through mineralisation. The results of many experiments in Russia and throughout the world on cereal inoculation with associative N fixing bacteria have shown varying and unpredictable responses. This suggests that the inter-actions between plants and the bacteria are complex, unstable and can vary greatly depending on genotypes.

Much less research has been done on canola inoculation with associative N fixing bacteria. The limited research does show that some strains will success-fully colonise the roots of Brassica species but often do not significantly influence harvest dry weight or N accumulation.

Transformations that Reduce Plant Available Nitrogen

Several different mechanisms contribute to N loss from soil, and, therefore, lost opportunity for increasing yield. Understanding the conditions that promote such losses can be valuable in avoiding such conditions in the field and improving the N fertilizer efficiency.

Denitrification

One major N loss from soil occurs through a microbial process called denitrifi-cation. As shown in Figure, nitrate can be changed by certain bacteria to gases such as N_2O and N_2, which escape back to the atmosphere. These soil bacteria have the ability to switch their respiration from using oxygen to nitrate. Since respiration is more efficient using oxygen, these denitrifying bacteria will only switch to nitrate if oxygen is absent, such as in waterlogged soil. Therefore, denitrification becomes significant when soils become saturated. Other secondary factors that encourage denitrification include carbon availability (crop residues), warm soil, and neutral to alkaline pH.

Denitrification accounts for 10 to 50 per cent of the available N losses in prairie soil. Research on the prairies has shown that considerable N denitrification losses occur during spring thaw. For example, research near Edmonton, AB found that 16 to 60 per cent of annual denitrification loss occurred immediately following snowmelt. During spring thaw on the prairies, frozen sub-soil is often a barrier for water drainage and the overlying thawed soil becomes saturated.

The second major period of denitrification loss occurs in late spring and early summer during rainfall events that cause soil saturation. Remember that denitrification occurs regardless of the nitrate source from fertilizer, manure or from decomposition. Effective N fertilizer management strives to avoid having large amounts of nitrate present during spring melts. Summerfallow is especially prone to large denitrification losses since large amounts of nitrate and moisture are stored during the fallow year, which increases the denitrification potential in the next spring thaw. Summerfallow also is a major contributor to leaching losses of nitrate.

Immobilisation

Immobilisation is the second major N transformation that reduces the plant available N supply. Figure indicates that soil bacteria may use either nitrate or ammonium for their own growth, temporarily tying up the N in the soil organic N storehouse.

Immobilisation essentially is the reverse of mineralisation, and occurs when residues with low N content (like cereal straw) are being decomposed. Since these residues don't contain enough N for the microbes to make their own protein, they need to use the nitrate and ammonium. Soil microbes thus compete with plant roots for the available N, and plant growth suffers when N supplies are inadequate for both microbial and plant growth needs. The poor crop growth in heavy chaff rows is due in part to immobilisation of N and other nutrients by the decomposing microbes. One effective fertilization strategy is to place the fertilizer away from residues and thus avoid immobilisation losses. The remaining transformations that reduce plant available N are usually relatively minor.

Table. Seedbed Utilization (SBU) of Various Openers[a]

	Spread Width of Fertilizer in Seed Row											
	2.5 cm (1") Spoon or Sweep			5 cm (2")			7.6 cm (3")			10 cm (4") Sweep Hoe		Disc or knife
Row spacing cm	15	23	30	15	23	30	15	23	30	15	23	30
Row spacing "	6	9	12	6	9	12	6	9	12	6	9	12
SBU %	17	11	8	33	22	17	50	33	25	67	44	33

Note : *Although some openers also vertically spread seed and fertilizer, this is not considered in the table since seed should be placed at a consistent depth for uniform germination and emergence. The actual spread width varies with air flow, soil type, moisture, residue and speed, and should be checked under prevailing field conditions.

Leaching

Leaching of nitrate can occur since this form is not adsorbed to the soil and moves readily with soil water. Leaching losses can be significant in sandy soils in high rainfall areas or under summerfallow, but overall leaching probably contributes to less than 10 per cent of the available N losses on the prairies. To reduce leaching losses time fertiliser applications to avoid prolonged exposure to wet conditions, and consider band placement to delay the conversion to the vulnerable nitrate form.

Volatilisation

Volatilisation occurs when ammonia escapes from the soil to the atmosphere. Such losses happen in a variety of ways. One obvious loss occurs when anhydrous ammonia fertilizer is improperly applied (too shallow or into a too dry or wet soil). Broadcasting urea fertilizer on the surface without incorporation can also lead to significant volatilisation losses if significant rainfall (more than 6 mm (1/4″) does not occur soon after application. All ammonium based fertilizers are subject to volatilisation if broadcast on the surface of soils with high pH, surface lime salts, low soil organic matter, warm temperatures and dry conditions. To reduce volatilisation loss, ensure proper fertilizer placement into the soil.

Weeds

Weeds can contribute to poor N fertilizer efficiency by competing with crops for uptake. The competitive ability of the crop for fertilizer uptake can be improved by placing the fertilizer near crop roots rather than broadcasting or random banding.

Erosion

Erosion of topsoil carries significant N and other nutrients away from the field. Use soil conservation techniques to minimise such losses.

A final minor loss mechanism occurs when ammonium is fixed into the crystal structure of certain clays. Some soils contain expanding type clays that allow ammonium to enter within the plates of the crystal structure and become fixed. Such ammonium trapped within the crystal lattice is held tightly and unavailable for root uptake.

NITROGEN FERTILIZER

Due to the various N losses described in the previous section, N fertilizer use efficiency cannot approach 100 per cent . Generally, research in western Canada has found that N fertilizer use efficiency (fertilizer N recovered in seed) rarely exceeds 50 per cent and often is less than 20 per cent . In the latter case, this means that only 20 per cent of the fertilizer N made it into the seed. Although a small amount of fertilizer N remains in plant parts other than seed, a significant portion is lost. Fertilizer management strives to increase efficiency by increasing crop uptake and decreasing the losses. Fertilizer N management uses two main tools : placement and timing.

Seed Row Placement

Although seed row placement of N fertilizer is an efficient method for uptake, canola is sensitive to seed row N and this limits application rates. Canola seedlings are injured by excessive seed row N by the salt effect that reduces water uptake by the seed, and by ammonia toxicity. Greenhouse research at the Agriculture and Agri-Food Canada (AAFC) Beaverlodge, AB Research Centre in 1960 showed that rapeseed was sensitive to seed-placed N (see Figure). Subsequent field research confirmed that rapeseed was more sensitive to seed-placed N than cereals. During this period, seed row placement was limited to P and very low rates of N. The common drills during this time were double disc and hoe press drills, which give a very narrow seed spread.

The adoption of conservation tillage seeding systems and air-seeders has greatly influenced fertilizer placement. Development of pneumatic delivery implements has facilitated both dry fertilizer banding and direct seeding. Conservation seeding systems limit tillage passes in order to retain surface residues and this reduces the options for fertilizer application. However, newer machines designed for direct seeding have resolved the fertilizer placement issue by either placing the fertilizer away from the seed or increasing the seed-bed utilisation. Seed-bed utilisation is the spread width of fertilizer and seed relative to the row spacing. For example, a 7.6 cm (3") spread with 15 cm (6") row spacing creates 50 per cent seed-bed utilisation. Table outlines the seed-bed utilisation (SBU) obtained with different openers and row spacing.

Numerous trials have examined the safe seed row N amounts with various openers and configurations. Research conducted in Alberta from 1992-96 illustrates the effect of seed row N on canola emergence and yield. The research involved 32 site years with canola at various Alberta locations. The highest N rate and least SBU reduced canola emergence and yield 90 and 45 per cent of the time, respectively. Sites with limited moisture (due to sandy texture, low seed-bed moisture or dry conditions) two weeks after seeding experienced the greatest reduction in emergence and yield with 101 kg N/ha (90 lb N/ac) as urea and low SBU.

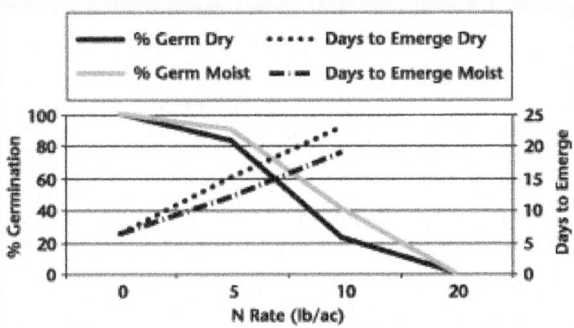

Fig. Seed Row N Fertilizer Effect on Canola Emergence.

Figure shows the approximate safe rates of seed row granular N fertilizer based on prairie research to date. In canola, there is no significant difference in seed row safety between urea (46–0–0), ammonium sulphate (21–0–0–24) or ammonium nitrate (34–0–0). Anhydrous ammonia (82–0–0) must be placed separately from the seed. If moisture conditions are dry, reduce the safe amount of seed row N by half. The N rates are in addition to N contained in seed row phosphate fertilizer.

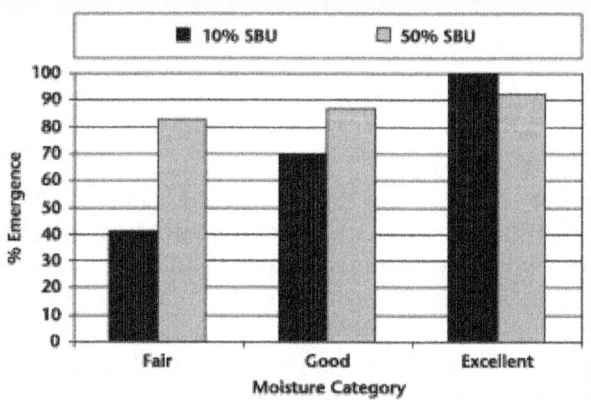

Fig. Effect of Moisture and SBU on Canola Emergence
[101 kg N/ha (90 lb N/ac)] in Alberta.

Band Placement

Band placement away from the seed row can be used to avoid toxicity and to improve fertilizer use efficiency. Banded N fertilizer is usually more efficient than broadcast incorporation because concentrating fertilizer in the band reduces the contact with soil and microbes, and reduces losses due to denitrification and immobilisation.

The banding benefit varies between soils and years mainly due to differences in moisture and susceptibility to loss. Also, if the band is located near the seed row rather than random, fertilizer loss due to weed uptake is reduced. Pre-plant banding involves placing granular, liquid or gaseous fertilizer N in a ribbon several inches below the soil surface before seeding. Banding is often done in the fall on fields that tend to be wet in the spring. Fall banding can spread the workload without significantly lowering N efficiency, and often allows growers to buy fertilizer at lower cost than in the spring. The banding depth often is 8 to 10 cm (3 to 4").

Under dry or sandy soil conditions, ensure anhydrous ammonia is placed deep enough to prevent visible gaseous loss, and ensure the soil flows well around the openers to permit a good seal behind the shanks. Spring banding can be shallower due to better moisture and tilth. However, in dry springs the banding operation can reduce seed-bed moisture and quality. Make spring band spacings narrower than in the fall.

Spring banded anhydrous ammonia can be immediately followed by seeding, providing there are several inches of vertical separation between the injection point and seed depth. Canola emergence directly over the bands may be slightly reduced, but yields are generally not affected.

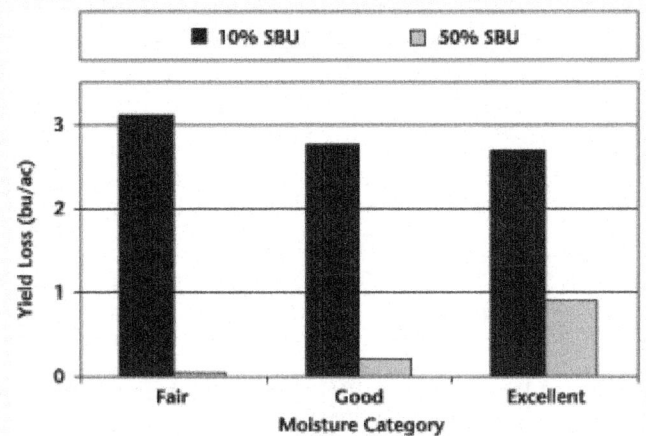

Fig. Effect of Moisture and SBU on Canola Yield [101 kg N/ha (90 lb N/ac)] in Alberta.

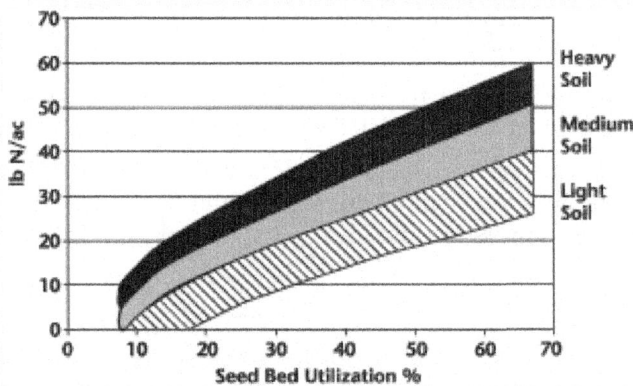

Fig. Approximate Maximum Rates for Seed Row Placement in Canola with
Good to Excellent Seed-bed Moisture.

Figure illustrates research conducted at the Agriculture and Agri-Food Canada (AAFC) Scott, SK Research Centre on the safety of seeding directly after banding anhydrous ammonia 10 to 15 cm (4 to 6") deep. Yields varied between seeding dates following anhydrous banding over the years with no relationship between yields and plant stands. Proper soil packing over the seed row to firm the soil disrupted by the banding operation was deemed more important to avoid stand reduction than was potential injury from the banded ammonia. No difference was found between banding ammonia parallel to and perpendicular to the seed row.

Side banding involves placing the fertilizer band to the side and often below the seed during the seeding operation. This method has good N use efficiency and avoids seed row toxicity if the separation is maintained under field conditions. However, side banding at seeding does involve more complicated, costly openers, increased draught and wear.

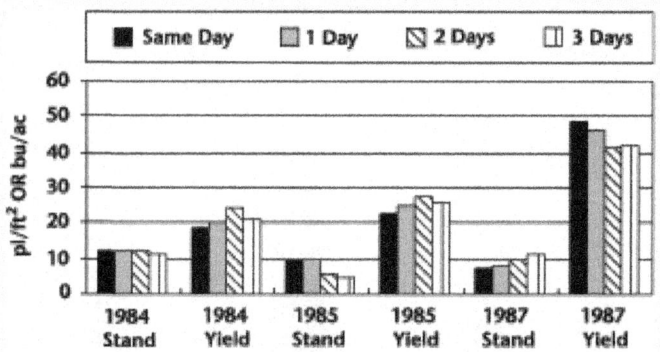

Fig. Effect of Waiting Period after Anhydrous Ammonia Banding on
Canola Stand and Yield.

Mid-row banding involves placing fertilizer between every second seed row or between a paired row during the seeding operation. This method also has good N use efficiency and avoids seed row toxicity if the separation can be maintained under field conditions.

Both side and midrow banding can improve N efficiency and yield response by favouring crop versus weed access to the fertilizer.

In recent years, openers have been designed that allow anhydrous ammonia to be side-row or mid-row banded during seeding. Ensure the anhydrous ammonia is horizontally separated from the seed by at least 5 cm (2″).

Broadcast Incorporation

N fertilizer can be spread onto the soil surface then incorporated into the soil with a tillage implement. Although this method can be time and labour saving, fertilizer efficiency is usually sacrificed. Under dry conditions, broadcast-incorporated fertilizer can be stranded in dry surface soil, and not accessible by plant roots growing down into moisture. In wet conditions, broadcast-incorporated fertilizer is more vulnerable to losses due to denitrification, immobilisation and leaching than banded fertilizer. Broadcasting without incorporation is the least efficient method of applying N fertilizer due to increased loss due to run-off, erosion and volatilisation. Broadcasting after crop emergence or topdressing generally has low efficiency, but it can serve as a rescue treatment when poor fertility was not corrected prior to seeding.

Foliar Application

Foliar application is possible, but only a limited amount of fertilizer N can enter through leaves without significant leaf burn. The only practical instance of foliar fertilization for canola on the prairies occurs under irrigation where up to 20 per cent of applied N can be supplied in irrigation water early in the growing season.

Time of Nitrogen Fertilizer Application

Nitrogen fertilizer efficiency is greatly affected by the placement method and the application date. However, differences between placement methods or timing varies widely between years or fields due to variability in weather, soil type and drainage. Generally, N fertilizer applied at or near seeding is the most efficient.

The major disadvantages of applying all the fertilizer N at seeding are :

- Isncreased time required to seed and fertilize during the short seeding season
- Higher fertilizer prices during the spring season compared to fall
- Increased risk of reduced seed-bed quality due to extra tillage/soil disturbance at or near seeding
- Seed row N limits
- Fertilizer applicators being unavailable during the busy spring season

Considerable fertilizer placement/timing research has been conducted on the prairies. Table gives approximate relative efficiencies of the various placement and timing methods.

Table. N Fertilizer Effect on Pod, Seed Number and Yield

Time and Placement	Relative Efficiency (per cent)		
	Dry*	Medium	Wet
Spring broadcast-incorporated	100	100	100
Spring branded	120**	110	105
Fall broadcast-incorporated	80	75	65
Fall banded	120	110	85

Note : *These are soil-climate categories based on typical conditions expected in the spring. **Extra tillage associated with spring banding can dry out the seed-bed, reduce emergence and yield in some cases.

An extensive survey conducted by Alberta Agriculture Food and Rural Development in 1982 on the practices of above average growers confirmed the banding benefit (top yields were associated with farmers who fall-banded fertilizer N).

Date of Fall Banding

Nitrogen fertilizer banded too early in the fall increases potential for losses on wet, poorly drained soils. Delay banding until soil temperatures have cooled to less than 10Â°C on well drained sites, and less than 5Â°C on poorly drained sites.

This will significantly reduce losses by reducing the rate of change from ammonium to the more vulnerable nitrate form. Unfortunately, the opportunity to fall band fertilizer into cold soil is quite short since snow or frozen soil can occur without much warning. On average, fall soil temperatures on the prairies drop 1°C every 5 days. In the Black and Gray soil zones, fall soil temperatures usually decline to 10°C by the last week of September. In these zones, since the banding benefit is significant due to typically wet spring conditions and moderately high N rates, begin fall banding in late September. This is earlier than the normal practice.

Nitrification Inhibitors

Since only the nitrate form is susceptible to losses through denitrification and leaching, methods to delay the conversion from ammonium to nitrate (nitrification) would be beneficial. Banding achieves this delay to some extent. However, interest in developing chemical inhibitors of nitrification has stimulated research programmes for many years. A variety of chemicals have been tested under prairie conditions, but none has achieved commercial success. For example, nitrapyrin (N-Serve) was recognised as a nitrification inhibitor in the early 1960s. Nitrification inhibitors have not been successful to date for various reasons, including cost, potential toxic effects on the soil environment and inconsistency.

Urease Inhibitors

The widespread adoption of direct seeding has resulted in interest in seed-placed fertilizer. While increasing seed-bed utilisation increases the safety of seed-placed fertilizer, the higher disturbance and draft is not always desirable. Openers with banding capability add cost and draft. Even with sidebanding, seedling damage can sometimes occur, particularly with wide row spacing, high rates of N application or insufficient separation between seed and fertilizer. Therefore, there is interest in œsafening urea fertilizer so that seedling injury is reduced, allowing more freedom for seedplaced N.

Agrotain (N-n-butyl-thiophosphoric triamide) urea by inactivating urease enzymes in the soil adjacent to the granule. This slows the breakdown of urea to ammonia, reducing the potential for seedling damage from seed-placed or side-banded applications of urea or urea ammonium nitrate. As long as N remains in the urea form, the risk of damage is minimal. In addition, since Agrotain delays the release of ammonia, there is more time for the uncharged urea to move away from the seed-row in the soil water or with rainfall. Movement of the urea away from the seed, combined with a slower release of ammonia from the urea, will decrease the concentration of ammonia in contact with the germinating seedling, thereby reducing seedling damage.

In field studies, Agrotain was effective in reducing seedling damage from side-banded urea and urea ammonium nitrate, where soil and environmental conditions led to seedling damage from the untreated fertilizer. The improved stand did not always lead to a higher crop yield because canola has the ability to compensate for reduced stands. The studies also showed that canola oil and chlorophyll content were often improved by using Agrotain.

Fig. Canola Seedlings with Sufficient N (left) Versus those with an N Deficiency.

While Agrotain appears effective for use with side-banded N applications, more information is needed to determine if the safening effect is great enough to allow for seed row placement of the full rates of urea or urea ammonium nitrate needed for a high-yielding canola crop.

Phosphorus (P)

Phosphorus is an important plant macro-nutrient, but it is required in smaller amounts than nitrogen. Western Canadian soils are commonly P deficient and fertilization usually increases yield and economic returns. Good P fertilizer management is important to optimising canola production.

Role of Phosphorus in the Canola Plant

Phosphorus functions in the plant as a structural element and also in energy transfer. The structural components that rely on P include nucleic acids (the building blocks of DNA) and phospholipids (fats and oils), which are important membrane constituents.

Phosphorus plays a significant role in energy transfer in all living organisms. The P energy transfer compounds are phosphate esters about 50 different esters have been identified. ATP (adenosine triphosphate) is the principal phosphate energy compound used for starch synthesis and nutrient uptake.

Fig. Prsogressively Less N Deficient Leaves (Left to Right).

Fig. N-Sufficient (Left) Flowering Plants and Deficient Plants.

Fig. N-sufficient Raceme (Left) and Deficient Raceme.

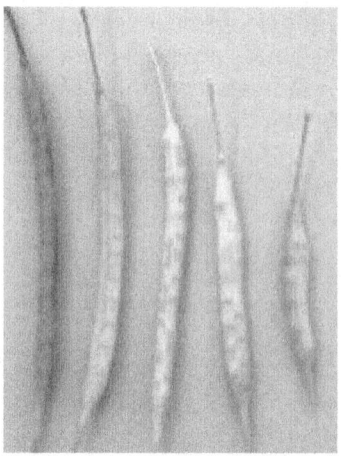

Fig. N-sufficient Pods (Left) Progressing to Deficient Pods.

Energy produced during respiration and photo-synthesis is captured by these phosphate compo unds, which then are transported to areas that are building plant tissue. The energy stored in the phosphate compound is released, and the molecule is recycled back to be âœre charged. This recycling of phosphate energy compounds is accomplished at extremely fast rates, and a small amount can satisfy the plant energy needs.

CHARACTERISTICS OF PHOSPHORUS

The main P forms taken up by roots from the soil solution are the primary and secondary phosphate ions (H_2PO_4–and HPO_4^{-2}). These phosphate anions exist transiently in the soil solution due to rapid removal by roots and microbes, or reaction with other soil minerals. The P level is highest in young vegetative material and in the canola seed. Figure illustrates the P uptake and level over the 1998 season at the AAFC Melfort, SK Research Centre. Canola seedlings take up P rapidly during early growth, but not as rapidly as N. Studies conducted in Manitoba in the 1960s showed that canola P uptake in early growth stages was more rapid than oats, flax and soybeans.

The P level remains fairly high in the leaves (0.3 to 0.4 per cent) until late flowering when significant translocation occurs into developing pods and seeds. By maturity, 75 to 80 per cent of the P in above ground dry matter is in the seed. Canola seed contains 0.7 to 0.8 per cent P, about double that of cereal grains. Canola stems and pods at harvest contain only 0.1 to 0.2 per cent P.

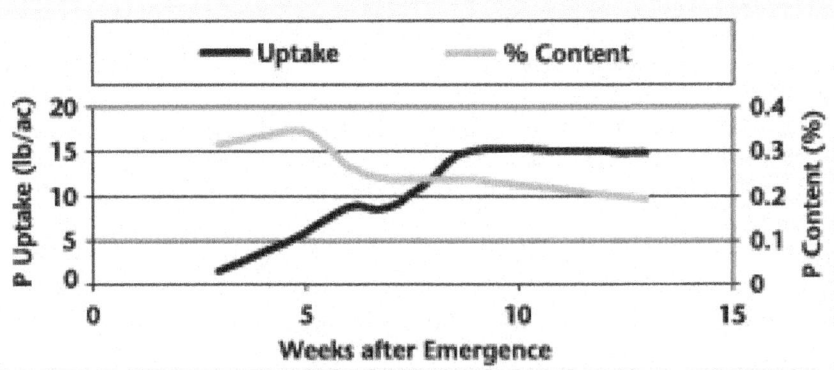

Fig. Phosphorus Content (per cent) and Uptake by Canola Over the 1998
Growing Season, Melfort, SK.

Canola is an efficient scavenger of soil P even though Brassica species are non-mycorrhizal (mycorrhizae are symbiotic associations between certain soil fungi and plant roots where the fungi contribute to the P nutrition of the plant). Many cereal crops can form these beneficial relationships. In spite of canola being non-mycorrhizal, research has shown that canola takes up more P than cereals. Canola has several mechanisms to achieve this efficient P uptake.

Canola has abundant fine roots with the ability to branch and proliferate in zones of higher nutrient content such as around fertilizer bands or granules. In addition to root proliferation in fertilizer zones due to branching, canola roots can increase the root hair number and length in response to low P conditions.

The second mechanism in canola roots that enhances P uptake is solubilisation of relatively insoluble mineral P forms. Canola has the ability to acidify the rhizosphere just behind the root tip near the zone of root hair formation. In a recent western Canada growth chamber experiment, the pH of the canola rhizosphere fell up to 0.8 units over five weeks compared to a drop of less than 0.4 units for wheat rhizosphere. Canola absorbed more of the relatively insoluble P forms than wheat. The acid generated by canola roots is predominantly caused by exudation of organic acids such as malic and citric acid. Canola roots also release enzymes (phosphatases) that mineralise phosphate from organic P pools. Cation-anion uptake imbalance may also contribute to rhizosphere acidification when the main form of N uptake is ammonium. However, under western Canadian field conditions, canola takes up the majority of N in the nitrate form.

After phosphate enters the root, there are three barriers to cross before reaching the xylem system that feeds aboveground growth. The rate of phosphate transport across these membranes is affected by the plants P status. As the plant P content increases, the P transport rate decreases (feed back regulation). The P uptake rate is often more related to shoot than to root P level. This regulated transport system requires energy. Factors that influence root respiration will affect root P uptake. For example, cold soil or low oxygen content in a saturated soil reduces root respiration and consequently P uptake. There is competition for the phosphate transport system by arsenate. This can impact P nutrition in soils high in arsenate.

The xylem loading system is usually regulated separately from the systems at the plasma and vacuole membranes. Phosphate ions typically are rapidly transported from the roots to the shoots. Unlike N, P is absorbed and transported throughout the plant in the inorganic form (mainly H_2PO_4). Similar to N, phosphate is readily remobilised from aging tissue such as leaves to more active growing points. Phosphate stored in cell vacuoles can also be readily mobilised. Immature plants adequately supplied with P have 85 to 95 per cent of the total inorganic phosphate stored in the vacuoles. In contrast, in P deficient plants, almost all the phosphate in leaves is found in active pools (cytoplasm and chloroplasts). By maturity, most of the plant P is stored in organic form as phytate in the grain.

Phytate serves as a readily accessible P source for the germinating seedling. Animal nutritionists are interested in seed phytate (including canola meal) since these compounds interfere with absorption of minerals such as zinc, iron and calcium. Considerable attention has been given to reducing phytate levels in grains, including canola, and some success is being reported.

Phosphorus Effects on Canola Growth and Deficiency Symptoms

Canola plants suffering from strong P deficiency can experience slow leaf expansion, smaller and fewer leaves. Deficiency symptoms appear by the second week of growth since canola seedlings are able to obtain sufficient P from seed reserves for the first week of growth. Figure (of field research results from five sites in western Canada in 1991) illustrates the significant increase in early season growth with P fertilization.

Phosphorus deficient leaves may have a dark green, bluish green to purplish colour since chlorophyll and protein formation are less affected than cell and leaf expansion. Under severe P deficiency, purple colouration arises from accumulation of anthocyanin pigments. Mildly deficient plants may look normal but are small. Above-ground plant P content at flowering should be above 0.24 per cent .

Root growth is less affected by P deficiency than shoot growth, leading to a typical decrease in the shoot-root ratio. With a more severe deficiency, root development is restricted, but not as dramatically as stem and leaf growth. Although overall root branching is restricted in P deficient soils, root hair length and density usually increase.

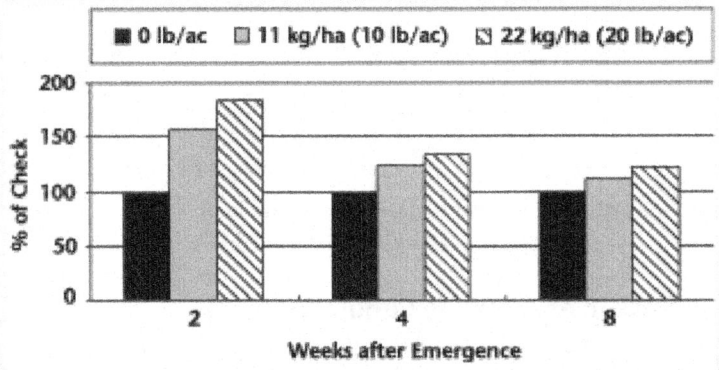

Fig. Effect of P Fertilizer (P2O5) on Canola Dry Matter after Emergence.

P deficiency affects the maturity and development of reproductive tissue. Even a mild P deficiency can result in maturity delays of several days compared to plants with adequate P. In addition to a flowering delay, a P deficiency can reduce the number of flowers and seeds per pod. Also, a P deficiency can cause leaves to die and drop early, which contributes to the overall yield loss.

Canola Response to Phosphorous Fertilizer

Most agricultural soils in Canada have inadequate P for producing canola crops. However, the canola yield response to P fertilizer on deficient soils usually is much less than the average response to N fertilizer on N deficient soils. Research in the 1960s showed that rapeseed often responded more to P fertilizer than wheat or flax. Subsequent research established that yield response could be predicted from soil test values. Central Alberta research in the 1980s found that 23 of 48 sites responded to P fertilizer.

A recent Alberta P study from 1991-1993 found a statistically significant response to P fertilizer at 42 site-years while 81 site-years had no response. Economic analysis of the results suggested that 70 per cent of the canola sites responded to 7 kg (15 lb) P_2O_5/ac and 53 per cent responded economically to 14 kg (30 lb) given canola at \$352/tonne (\$8/bu) and \$0.75/kg (\$0.34/lb) P_2O_5. Data from such fertilizer experiments are compiled into databases to predict fertilizer response. Figure 23 is an example of canola response to P fertilizer based on soil test P in Alberta.

Canola response to P fertilizer depends mainly on the amount of plant available P in the soil but is also influ enced by moisture and temperature. In cold soil, P availability and movement is reduced. Canola response to P fertilizer is greater under these conditions. Phosphorus fertilization often slightly advances maturity of canola crops by one or two days. This slight difference may be important in short growing seasons.

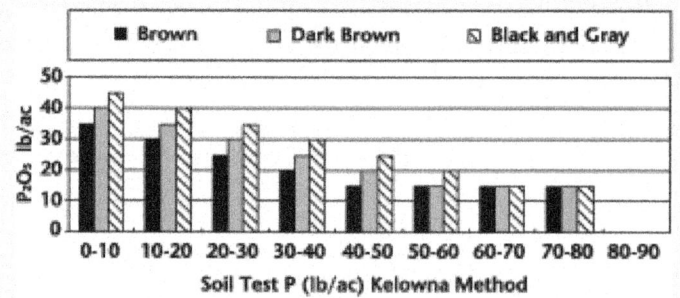

Fig. Phosphate Fertilizer Recommendations for Canola on Medium to Fine Textured, Neutral Soil under Medium Moisture.

Phosphorus Fertilizer Effect on Canola Quantity

P fertilization generally has negligible effects on canola quality. Experiments in western Canada have found that P fertilizer increased, decreased or did not affect oil content. Canola protein content has occasionally been slightly raised by P fertilization. A recent field experiment in Manitoba on two very deficient sites found that P fertilizer significantly increased both protein and oil content.

Phosphorus Cycle

Prairie soils contain significant amounts of total P"450 to 907 kg P/ac (1,000 to 2,000 lb P/ac). However, most of this soil P is relatively insoluble with limited availability to plants. Canola roots obtain P by absorbing phosphate dissolved in the soil water. Since the amount of phosphate dissolved in the soil water is very small at any given moment, there must be constant replenishment into the soil water from the insoluble forms. This replenishment of soil solution P around roots arises from slightly soluble minerals, P desorption from surfaces, organic P minerali sation and fertilizer.

Soil phosphate supply is usually highest in the pH range of 6.5 to 7.0. At high pH levels (>7.5), calcium and magnesium cations can precipitate with phosphate to form salts with low solubility. In contrast, in acidic soils (pH<6), iron and aluminium cations react with the phosphate to form insoluble compounds. Phosphate is not a mobile nutrient in soil due to these soil constituent reactions.

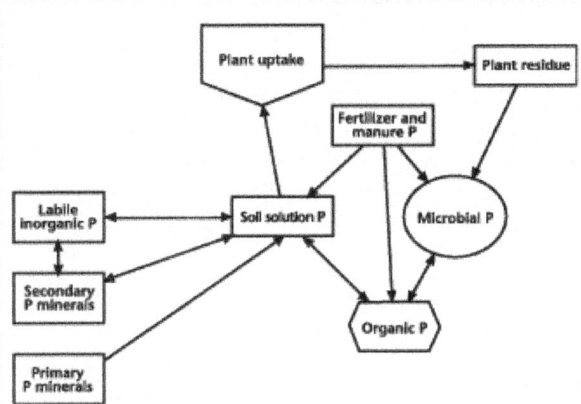

Fig. The Soil Phosphorous Cycle.

The natural soil weathering process causes acidification and this encourages the eventual conversion of primary P to secondary minerals and unavailable forms (occluded P). This transformation to unavailable forms takes centuries.

As phosphate is removed from the soil solution, the lower level stimulates phosphate release from exchangeable and labile inorganic pools. As labile pools are depleted, non-labile secondary P minerals slowly dissolve to maintain the labile and solution pools.

The organic P pool also contributes to the maintenance of phosphate in the soil solution. Organic P in prairie surface soil constitutes about 25 to 55 per cent of total P and is a large pool of potential plant available P. Microbial processes drive the organic section of the P cycle. Phosphate from organic matter can be released through decomposition and then incorporated into new microbial biomass or enter into the soil solution. Most organic P compounds released during decomposition are quickly degraded and exist briefly in soil. Some organic P compounds can be stabilised in soil through adsorption to soil constituents or by physical isolation within aggregates. Tillage decreases the soil organic P content by exposing stabilised forms to new or more vigourous microbial attack. Organic P can be degraded to phosphate by enzymes (phosphatases) released by soil microbes and by canola roots.

Significant seasonal fluctuations occur in both organic and inorganic P pools. However, reports conflict on the direction and magnitude of the fluctuations. For example, several experiments reported decreases in organic P during the summer growing season and gains over the winter, while another experiment measured major declines over the winter. Inorganic P (soil test extractable P) can also vary widely and inconsistently between fall and spring.

Phosphorus Fertilizer Management for Canola

The majority of P fertilizer is not absorbed by canola in the application year. Instead, most of the fertilizer P reacts with soil constituents to form relatively insoluble salts or stabilised P compounds. Timing and placement are two management strategies used to maximise P fertilizer uptake and yield response.

Timing of Phosphorus Fertilization

Since phosphate reacts with soil constituents to form insoluble compounds over time, P fertilizer efficiency can be increased by limiting the time from application to crop uptake. However, the P fertilizer application date affects canola yield much less than placement method. Most growers apply P fertilizer at seeding, minimising P availa bility losses by reducing reaction time. Research in western Canada has shown the effectiveness of fall and spring banding of P fertilizer is similar.

Phosphorus Fertilizer Placement

Phosphorus supply during the first two to six weeks of canola growth is critical to achieve optimal yield. Therefore, place P fertilizer to maximise early season access.

Seed row placement is an effective placement method when soil P levels are low to moderate and spring soil conditions are cold. Cold soil decreases phosphate solubility and diffusion in the soil solution, slowing P movement to roots and root uptake rates. This condition increases the likelihood of response to readily accessible seed row P. Unfortunately, canola seedlings are sensitive to seed row fertilizer and this limits seed row P rates.

The maximum safe rate of seed row P fertilizer for canola depends on seed-bed utilisation and soil moisture conditions. Pot experiment results (Figure) conducted at AAFC Beaverlodge, AB Research Centre in the 1960s illustrate canola sensitivity to seed-placed P fertilizer. The seed-bed utilisation in this experiment was very restricted (3 per cent), even less than a double-disc press drill. The dry and moist soil corresponded to 30 and 50 per cent of available water in a sandy loam.

Subsequent field research has confirmed that excessive P fertilizer placed in the seed row can reduce plant populations and yield. High seed-placed P fertilizer rates have lowered plant populations in some cases but did not affect yield. However, lower plant populations due to excessive seed row fertilizer will increase yield variability and usually lowers high yield potential under optimal growing conditions.

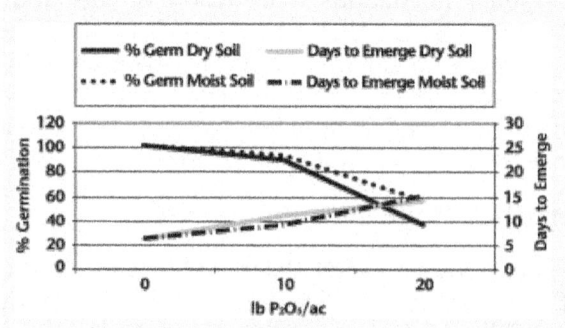

Fig. Effect of Seed-Placed P Fertilizer on Canola Emergence.

Under dry soil moisture conditions with low seed-bed utilisation (such as disc opener), the maximum safe P2O5 seed-placed rate is approximately 22 kg/ha (20 lb/ac). The rate can be safely increased to 28 kg/ha (25 lb/ac) under good moisture conditions with low seed-bed utilisation. As seed-bed utilisation increases, proportionally increase seedp laced P fertilizer rates. Some research suggests that the larger seed of B. napus will tolerate slightly more seed-placed P than B. rapa. Due to the significant emergence and yield reductions caused by moderate to high rates of seedplaced P fertilizer, place these rates separately from the seed. This fertilizer/seed separation can be achieved by increasing the spread width in the seed row or by placing the fertilizer in bands away from the seed row.

Pre-plant band placement is an effective method since it reduces fertilizer contact with sensitive canola seed and with soil constituents that will fix P over time. Also, the deeper fertilizer placement tends to be more accessible to roots as they normally grow down to moist soil. Banding fertilizer prior to seeding reduces the fertilizer handled during seeding and can provide time and labour benefits. Pre-plant band placement is currently a common method for placing all fertilizer. Phosphorus fertilizer can be banded in late fall or spring prior to seeding.

Side-banding places fertilizer near the seed row during seeding. The fertilizer normally is banded 2.5 to 5 cm (1 to 2″) below and beside the seed row. Several direct seeding machines use a mid-row or paired-row method of banding fertilizer. Air seeders with shovels or knives are used with shank spacings ranging from 20 to 36 cm (8 to 14″). The fertilizer is usually banded to a depth of 5 to 13 cm (2 to 5″). No consistent agronomic benefits accrue to banding deeper than 8 cm (3″) and fuel costs increase significantly with deeper depths. In high P-fixing soils, place fall P bands deeper than subsequent tillage depths to avoid mixing the band with soil.

Split application methods refer to combinations of band and seed row placement. Split application takes advantage of the consistent benefit of seed-placed P fertilizer up to 22 kg/ha (20 lb P_2O_5/ac), and avoids seedling injury by placing the remainder of the P fertilizer in a band (usually with N and S).

Broadcast-incorporated placement involves spreading P fertilizer on the surface followed by cultivation to work it into the soil. This method is significantly less effective than seed-placed or banded P fertilizer due to increased contact between the P and reactive soil constituents. Application rates with broadcast-incorporated P fertilizer usually have to be two to four times seed-placed or banded rates to get an equal response. Therefore, broadcast-incorporated methods are less economical.

Research comparisons of P fertilizer placement methods at typical rates show that highest yields are frequently obtained with seed-placed and split applications, followed closely by pre-plant band methods. Broadcast-incorporated methods produce significantly lower yield responses.

Research by Westco Ag research illustrates the relative usefulness of various P fertilizer placements. Differences between placement methods are largest under conditions of low soil test P (such as after forage breaking), as well as cold spring soil. On typical prairie farms that have received P fertilizer for many years, canola yield response differences between seed-placed and banded methods tend to be minimal. Phosphorus fertilizer plac ement issues have been largely resolved by ground opener development with increased seed row spread or side-band capability. Split P application between a band and seed row appears to be the most consistent method due to reduced seed row toxicity, less P uptake interference from high N bands, and a dual location that hedges against poor access in either cold or dry surface soil conditions.

The application of both N and P fertilizer in a single band is called dual banding. At low to moderate rates of N, the uptake efficiency of P is sometimes increased from a dual band. At higher N rates"above 90 kg/ha (80 lb/ac)"the concentrated N in the band can reduce early season P uptake due to ammonia and nitrite toxicity that hinders root entry into the band. This P uptake interference appears to be strongest in recent band applications, and could be a problem with dual spring banded N+P fertilizer immediately before or during seeding.

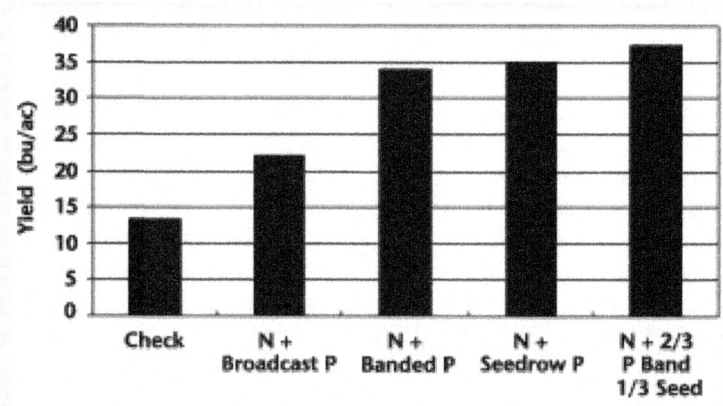

Fig. Canola Yield Response to Different P Fertilizer Placements.

Non-Traditional Sources of Phosphorus Nutrients

Phosphorus deficiencies on the prairies are normally corrected with annual applications of commercially refined P fertilizer"either dry blends of mono-ammonium phosphate (12–51–0) or liquid blends of ammonium polyphosphate (10–34–0). Manure also serves as a traditional source of P and other nutrients.

Rock Phosphate

Canola has an ability to absorb native soil P through acidification of the rhizosphere. Pot experiments have demonstrated that canola can utilise more rock phosphate than other crops, apparently due to the rhizosphere acidification. This has prompted promotion of rock phosp hate as a viable alternative P fertilizer for canola. Rock phosphate is the relatively insoluble, gray-black powdery material that is refined by fertilizer manufacturing plants into soluble phosphate fertilizer.

Idaho is a common source of rock phosphate marketed in western Canada. Research on the prairies indicates that rock phosphates do not perform satisfactorily compared to fertilizer phosphate. The poor performance is due to poor solubility, lower P_2O_5 content, and the predominance of neutral, calcareous soils on the prairies. While high rates of rock phosphate do slightly improve canola yields on some soils, this is not cost effective compared to fertilizer phosphate. Typically, rock phosphate application rates need to be six to eight times that of fertilizer phosphate for equivalent yield response.

Research by Alberta Agriculture Food and Rural Development at Ellerslie, AB illustrates the poor performance of rock phosphate compared to fertilizer P. All the P sources were seed placed and 112 kg N/ha (100 lb N/ac) was pre-plant banded.

Another non-traditional means of P nutrition recently developed for canola is biologically based. Although canola does not form symbiotic associations with mycorrhizae that improve P nutrition in other crops, other rhizosphere microbes exist that increase P solubilisation and subsequent plant P uptake.

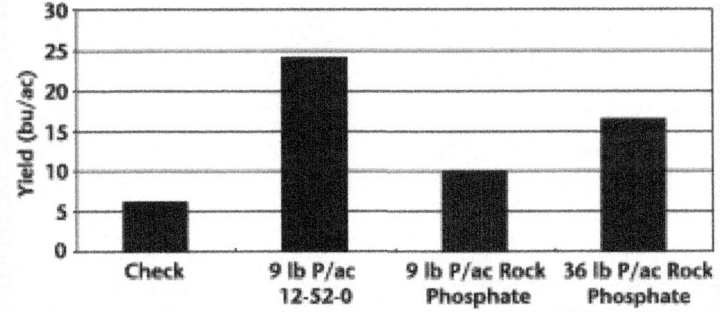

Fig. B. Rapa Yield Response to Rock Phosphate and P Fertilizer.

AAFC, Lethbridge, AB researchers identified an organism (Penicillium bilaii) that solubilised P minerals and improved the P uptake of cereals and canola. This organism was then commercialised as a seed inoculant. Field experiments conducted on the prairies have shown that inoculating canola with Provide increases early season P uptake and vegetative growth, and results in higher yield with and without P fertilizer. The canola yield response to inoculation with Provide at 15 P-responsive sites in western Canada is summarised in Figure.

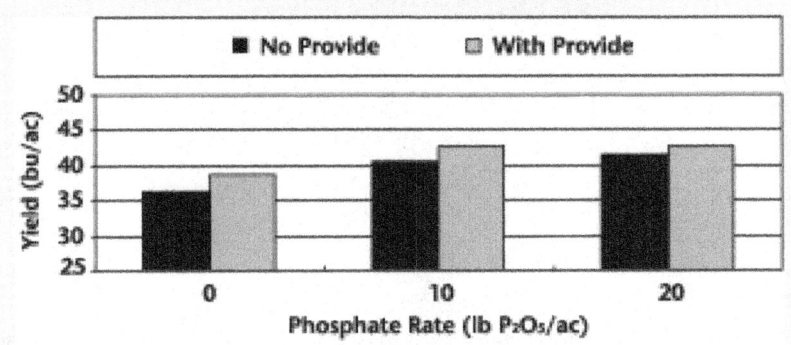

Fig. Canola Yield Response to Inoculation with Provide at P-Responsive Sites.

On average, growers can expect to apply 11 kg less P_2O_5/ha (10 lb less P_2O_5/ ac) when canola is inoculated with Provide. The adoption of Provide inoculation by canola growers has been limited, perhaps due to the inconvenience of inoculation, the short viable period after inoculation, inconsistency and cost relative to simply using more P fertilizer. In the future, biological fertility enhancing microbes will likely become more common.

Fig. P-sufficient Canola (Left) Compared to P-deficient Canola.

Fig. P-sufficient Canola at Flowering (Left) Compared to P-deficient Canola.

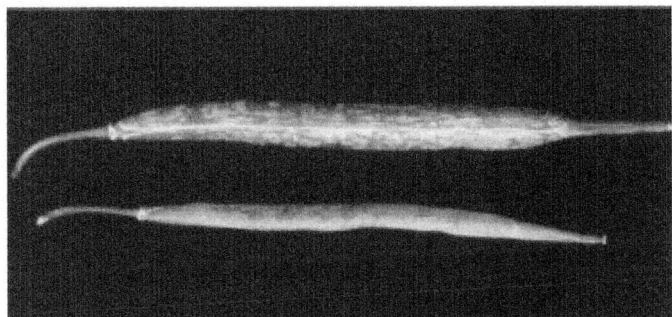

Fig. A P-sufficient Canola Pod (Top) Compared to a P-deficient Canola Pod.

Potassium (K)

The macro-nutrient potassium (K) is required in large amounts by canola similar to nitrogen (see Table). In spite of the large requirement, canola yield responses to K fertilizer (potash) are infrequent, due to ample soil K reserves on the prairies, and canola strong ability to absorb K.

Role of Potassium

Potassium is different from most other essential nutrients since it does not become part of structural components in the plant. Instead, most of the K in plants remains dissolved in the cell sap and performs several major functions.

One major function for K is that of enzyme activation. Enzymes are protein complexes that catalyse chemical reactions. More than 60 enzymes need to be activated by K. This activation occurs when potassium cations (K+) bind to the enzyme surface, changing the enzyme shape, and allowing the enzymes active site to attach to its substrate more rapidly or accurately. For example, K stimulates the activity of an enzyme (starch synthase) that catalyses starch formation from glucose. While other cations can also stimulate this enzyme, K+ is the most effective. In K deficient plants, the lack of stimulation of the starch synthase results in an accumulation of soluble sugars and N compounds, and a decrease in starch.

Another major function of K is in water relations. Potassium helps to maintain a favourable water status in plants in several different ways. Potassium cations dissolved in cell sap perform major osmotic functions. Osmosis is the tendency for water levels to equalise between different areas separated by a porous membrane. Dissolved ions such as K+ attract water and thus are osmotically active substances. Potassium is the major dissolved ion in cell sap and provides most of the osmotic œpull that draws water into roots.

Potassium cations also maintain the water relations in plants through their crucial role in regulating water loss (called transpiration) from pores (stomata) in the leaves. Although the stomata must open to allow movement of carbon dioxide and oxygen in and out of the leaves, water loss also occurs. This transpiration creates a gradient that pulls water and nutrients up through the xylem to the leaves.

However, plants cannot afford excessive water loss and need to regulate the stomata opening. For example, photo-synthesis stops during darkness, and the need for nutrients and water decreases greatly during night. Plants have developed a system that closes stomata during the dark or during drought. Potassium cations, in combination with chlorine, calcium and certain hormones, are responsible for governing the opening and closing of the stomata. Upon receiving a â€œsignal induced by darkness, K+ and Cl- are pumped from the two guard cells surroun ding the stomata, which causes a loss of turgidity of the guard cells and thus allows the pore to close. Potassium deficient plants often have higher transpiration rates and display wilting.

Potassiums osmotic activity also provides the physical force that expands cells during growth. New cells accumulate K+ and associated anions like Cl- in the large central vacuole that occupy 80 to 90 per cent of the cell volume. The K+ ions attract water and inflate the cell, stretching it to a new larger size. Potassium-deficient plants can exhibit low growth rates and small cells.

Energy relations in the plant are influenced by K. Potassium affects photo-synthesis at several levels. K+ is the main ion that counterbalances the H+ flux during photo-synthesis in the chloroplasts. Potassium also maintains a favourable pH gradient in the chloroplasts for making phosphate energy compounds. Potassium helps the translocation of photo-synthate sugars by maintaining a high pH in phloem tubes needed for loading, and by maintaining osmotic gradients needed for sap flow.

Potassium is needed for N uptake and protein synthesis. K+ cations are the major counter ions that balance nitrate during transport and storage in vacuoles. Many steps of protein synthesis require high K+ levels.

The K level is highest in seedling canola, then declines steadily up to maturity as shown in Figure 32,B. napus canola grown at the AAFC Research Centre in Melfort, SK in 1998. Canola K uptake is rapid during the early growth stages and tapers off by the end of flowering. Under high K fertility and good growth, canola can absorb more K than apparently needed, a situation termed luxury consumption.â€ As canola matures, the K level in leaves declines while the stem level increases. By harvest, the stem and straw material contain about 1 to 2 per cent K. In contrast to N and P, the K content of the seed (0.8 to 1 per cent K) is low relative to the stem. Unlike K+ in the vegetative parts, seed K is probably complexed with phytate as a salt.

Potassium Effects on Canola Growth and Deficiency Symptoms

Potassium deficiency reduces overall canola growth but to a lesser degree than N or P deficiency. Since K is mobile within the plant, deficiencies are first visible in older leaves. The edges and areas between veins of older leaves tend to turn pale green or yellow, followed by withering. The yellowing can occur first in middle leaves before older ones if observed at bolting to flowering stages. In severe cases, leaves die but remain attached to the stem. Small white spots can develop on leaves. Plants are prone to wilting during midday. Potassium deficiency symptoms in canola are rather non-distinct and can be easily confused with other problems. Fortunately K deficiency in canola is rare on the prairies.

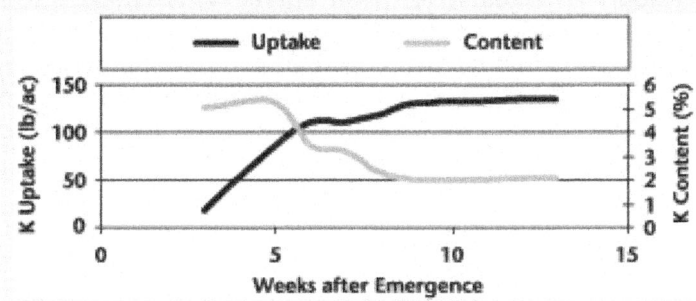

Fig. Potassium Content (per cent) and Uptake by Canola Over the Growing Season.

Canola Response to Potassium Fertilizer

Although canola absorbs large amounts of K, responses to fertilizer K are rare on the prairies. In fact, canola or rapeseed responses to fertilizer K are infrequent around the world testament to the crop strong ability to absorb soil K.

Numerous fertilizer research studies on the prairies and in Ontario have established that canola rarely responds to applied K, even under conditions where cereals normally respond. Although the K soil test is adequate for cereals, the usefulness declines for canola. Critical levels are often stated to be around 280 kg K/ha (250 lb K/ac) or 112 ppm in the top 15 cm (6"), but research indicates that canola will not consistently or economically respond to fertilizer K unless the soil test is very low 78 to 112 kg K/ha (70 to 100 lb K/ac) or 35 to 50 ppm. Very sandy or peaty soils are the most likely soil types to have very low K soil test values.

Other factors that increase the likelihood of K deficiency are :

- Free lime in the rooting zone
- Acid soil
- Poor drainage
- Cool temperatures
- Soil compaction
- Shallow root zone.

Unlike cereal responses, potash applications have not been shown to help with canola disease resistance, lodging or seed quality (oil content or meal protein content).

Potassium Supply from the Soil

Western Canadian soils generally contain ample plant available K due to an abundance of K minerals (such as mica and feldspar) in the parent material

(3 to 4 per cent K). There is often 17,000 to 56,000 kg K/ha (15,000 to 50,000 lb K/ac) in the top 15 cm of prairie mineral (non-peat) soils. The weathering of these minerals slowly releases K+ held in crystal structures typically only about 1 per cent of total soil K is available for plant uptake. This available K is mostly (90 per cent) exchangeable K+ adsorbed to clay surfaces and organic matter, while the other 10 per cent is found dissolved in the soil solution. Approximately 10 to 20 per cent of the total soil K is slowly available from smaller mica particles and certain clays. Figure outlines these K pools. Losses due to leaching or erosion are ignored in this figure, as they are usually small.

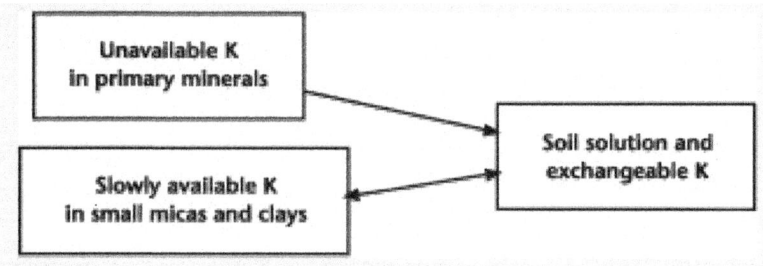

Fig. Potassium Soil Cycle.

Figure shows that the various pools are in dynamic equilibrium. As K+ is removed by plant uptake and through leaching on sandy soils, additional K is released from the mineral soils to become available. Available K moves to plant roots by diffusion through the soil only up to 6 mm (1/4″). Therefore, the equilibrium process that repeatedly moves K from the slowly available to readily available pool is very important for K nutrition. The rate of movement from the slowly available to readily available pool varies among soils due to differences in minerals and clays. This variation in K dynamics creates problems for soil testing. An extractant that measures plant available K in soil solution and exchangeable K over a short time period does not assess the replenishment power. Unfortunately, tests that measure the replenishment power are time-consuming and cost-prohibitive.

Chapter 10

CHEMICAL DYNAMICS OF SOIL

SOME BASIC IDEAS OF SOIL

Until now, our major concern has been whether or not a chemical reaction of interest may or may not occur under specified conditions of temperature, pressure, and concentration. The answer to this question is in the realm of thermodynamics. If thermodynamics tells us that a reaction may occur, then we must be concerned with how rapidly and by what pathway it will occur, and how we may influence this. The speed (rate) at which a reaction takes place is a matter of *chemical dynamics*, or alternately, *chemical kinetics*.

Chemical dynamics (kinetics) is the study of the rates (speeds) and mechanisms (pathways) of chemical reactions. We have briefly encountered the rate concept before in, when we discussed the manner in which vapour pressure changes with time; and in, when we presented dynamic equilibrium as a situation of balanced, opposing rates. Reaction rates will be our chief concern in this chapter. Although chemical kinetics is a well-developed field experimentally, we are still a long way from the theoretical grasp of rate that we have of equilibrium. Thus our predictive ability in the realm of kinetics is very limited.

Thermodynamics may predict that a reaction should proceed essentially completely. Whether it does so depends on the rate with which it occurs. For example, consider the reaction below at 298 K :

$$C(graph) + O_2(g) \rightarrow CO_2(g) \ [\Delta G°_R = -394.4 \text{ kJ}]$$

Thermodynamics predicts that the reaction should go to completion. But carbon (coal) deposits exist in the atmosphere without noticeably reacting; and the "lead" in our pencils does not burst spontaneously into flame. The reaction is favoured, but is very slow at 298 K. We say that C(graph) is thermodynamically unstable in the presence of $O_2(g)$ at 298 K, but is kinetically inert. As a second case, consider the acid-base neutralisation reaction :

$$H_3O^+(aq) + OH^-(aq) \rightarrow 2H_2O(l) \ [\Delta G°_R = -58.28 \ kJ]$$

Thermodynamics predicts that this reaction, too, should go to completion. In fact, it occurs almost instantaneously; equilibrium is attained as rapidly as the reactants are mixed. Thus we say that H_3O^+ is both thermodynamically unstable and kinetically labile in the presence of OH^-. Two sets of descriptive words are necessary to distinguish thermo-dynamics and kinetics :

Thermodynamics : stable $\Delta G > 0$; unstable $\Delta G < 0$

Kinetics : inert (reaction is slow); labile (reaction is fast)

As we have seen above, there is no necessary correlation between "stable" and "inert", or "unstable" and "labile". This lack of correlation reflects the fact that thermodynamics deals with values of state functions in various equilibrium states of a system; kinetics deals with the rate at which a path between states is traversed. Reaction rates are extremely important in many ways in our lives. Sustained muscle action depends on the continuous supply of energy produced by glucose combustion in the mitochondria of cells. Energy production must be fairly rapid, otherwise sustained action is not possible. The extremely rapid (explosive) burning of gasoline in the internal combustion engine provides the motive power for automobiles.

Food cooks (or, as the chemist would say, oxidises) extremely slowly at room temperature, but in a relatively short time at more elevated temperature. Our first concern must be to discuss the manner in which reaction rates are expressed, and to introduce some of the terminology of chemical dynamics.

What is a *rate*? In general, a rate is a number expressing how much a quantity changes in a convenient interval of time. Thus an interest rate is the amount by which a sum of money increases in 1 year; travel rate in an automobile or airplane is the distance covered in 1 hour; heart rate is the number of heartbeats per minute; thus a rate is the change in a quantity per unit time :

Rate = Δquantity/Δt

Note that the units of rate must be "quantity/time, or "quantity–time^{-1}". Thus interest rate has units of dollars/year; travel rate, miles/hour; and heart rate; beats/minute.

How are we to express the rate of a chemical reaction? To develop the answer to this question, we will begin with the simple chemical process in equation 15-1-2. The equation describes quantitatively the conversion of ozone to oxygen that occurs in the earth's upper atmosphere.

$$2O_3(g) \rightarrow 3O_2(g)$$

When this (or any other) reaction takes place, it is the amounts of reactants and products that change. As we know, amount can be expressed in either mass, moles, or molecules. The coefficients in the chemical equation relate both molecules and moles of reactants and products but do not directly relate masses. So we will agree that our measure of amount should be either molecules or moles. Further, since work in the laboratory is carried out on the macroscopic scale, moles is the more suitable unit of amount. So we might describe the rate of process 15-1-2 either in terms of the number of moles of O_3 used or of O_2 produced per unit time. However, there is a major difficulty with this approach. The number of moles of ozone used will depend upon the volume of atmosphere that we choose to examine!

We cannot specify a rate without also specifying the volume examined. Rather than use moles/time, then, we choose to divide by the appropriate volume to give moles/litre-time, or concentration/time. Rates expressed in these units are independent of the actual amount of reaction carried out. At this point then we can express the rate of reaction 15-1-2 in either of two ways :

(15-1-3a) : Rate = Moles O_3 used per litre per time = $-\Delta(\text{conc } O_3)/D\text{time}$

(15-1-3b) : Rate = Moles O_2 produced per litre per time = $\Delta(\text{conc } O_2)/D\text{time}$

The units of the rate are then moles per litre per time, where the time unit should be the one most convenient for the reaction being measured. Chemists agree that the rate of a reaction should be expressed as a positive number. As reaction 15-1-2 proceeds, the concentration of ozone becomes smaller while that of oxygen becomes larger. Thus the change in ozone concentration, Δ (conc O_3), is a negative quantity. To produce a positive rate in 15-1-3a, it is necessary to place the minus sign preceding the Δ (conc) term, as has been done.

In general, whenever the rate of a reaction is expressed in terms of the change in concentration of a reactant, D (conc reactant) is negative and the minus sign is necessary. The minus sign is unnecessary in 15-1-3b because Δ (conc O_2) is positive.

There is an additional difficulty, though. We would like to specify a single number representing the rate of a reaction under given conditions. However, the two expressions above give different numbers for the rate of 15-1-2, because dioxygen appears 1.5 times faster than ozone disappears (*i.e.*, in the same time that 2 moles of ozone react, 3 moles of dioxygen appear). This can be fixed by normalising the two expressions in 15-1-3 with the stoichiometric coefficients. Dividing D(conc O_3) by the coefficient 2 gives the same number as dividing D (conc O_2) by the coefficient 3 :

(15-1-4) : Rate = $-1/2 * \Delta(\text{conc } O_3)/$ time = $1/3 * D(\text{conc } O_2)/D$ time

Replacing "conc O_3" with the brackets to represent molar concentratration gives 15-1-5 :

(15-1-5) : Rate = $-1/2 * \Delta [O_3]/Dt = 1/3 * \Delta [O_2]/D\,t$

The symbol, Δ, signifies a finite (*i.e.*, measurable) change in a quantity. In the limit as the length of the time interval, Δt, approaches zero, we replace $\Delta(\text{conc})$ and Δt with the differential (infinitesimal) quantities d(conc) and dt to obtain the *differential expression of rate* in 15-1-6.

(15-1-6) : Rate = $-1/2\ \Delta[O_3]/dt = 1/3\ \Delta[O_2]/dt$

As we shall see in section 15-3, 15-1-5 is the appropriate formulation for finite difference methods of analysis. For the completely general chemical reaction in 15-1-7, proceeding from left to right from reactants, A and B, to products, D and F, the rate of reaction can be represented using any one of the expressions in 15-1-8 :

(15-1-7) : $aA + bB \rightarrow dD + fF$

(15-1-8) : Rate = $-1/a\,d[A]/dt = -1/b\,d[B]/dt = 1/d\,d[D]/dt = 1/f\,d[F]/dt$

As discussed earlier, negative signs are necessary when rate is expressed in terms of disappearance of a reactant; and division by the stoichiometric coefficient guarantees that the same numerical value of rate is obtained from all four expressions.

Example : In the reaction of ozone to produce oxygen, it is found that under certain conditions of temperature and concentration, 0.0360 moles of ozone per litre react in a 2-hour period. How much dioxygen is produced in this time period? What is the average rate of reaction over this time period?

Solution : The chemical equation, 15-1-2, tells us that 3 moles of O_2 are produced for each 2 moles of ozone reacted. Therefore 0.0360 moles ozone* (3 moles O_2/2 moles O_3) = 0.0540 moles dioxygen are produced.

The average rate of reaction can be expressed in either of two ways :

$$\text{Rate} = -1/2\,\Delta[O_3]/\Delta t = -1/2\,(-0.0360\text{ moles/L})/2\text{ hours}$$

$$= 0.009\text{ moles/L-hour}$$

$$\text{Rate} = -1/3\,\Delta[O_2]/\Delta t = 1/3\,(0.0540\text{ moles/L})/2\text{ hours}$$

$$= 0.009\text{ moles/L-hour}$$

Both expressions give the same positive rate of reaction.

Example : The rate of 15-1-9 was studied at 55°C by measuring the concentration of t-butyl bromide (t-BuBr) as a function of time. The data acquired are given below. Use the data to estimate the rate of reaction 20 seconds after reaction was begun.

(15-1-9) : t-BuBr + H_2O → t-BuOH + HBr

Time, s	[t-BuBr]
0	0.100
10	0.0876
20	0.0768
30	0.0672
40	0.0590
50	0.0517
60	0.0453
80	0.0348
100	0.0267
120	0.0205
180	0.0093
240	0.0042

We begin by plotting the data to obtain a picture of the manner in which the concentration of t-BuBr varies with time. A plot of [t-BuBr] versus time is given in Figure. The shape of the plot is typical of a rate process : the concentration of reactant changes very rapidly at first, but as time goes on, the change in concentration per unit time becomes less and less as the concentration of t–BuBr approaches its final (equilibrium) value. We have seen this characteristic time plot before in our discussions of vapour pressure and approach to chemical equilibrium. Most concentration-time plots for a reaction in progress have this same general appearance.

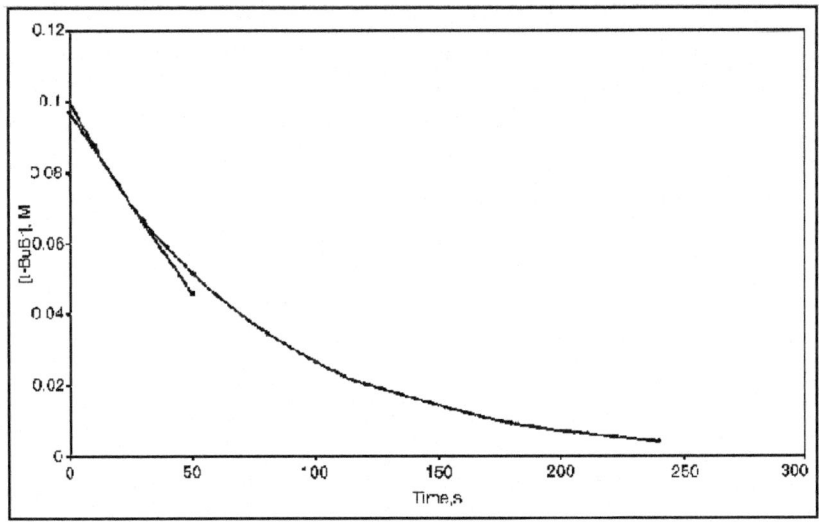

Fig. Time Profile of The Concentration of T-Butyl Bromide.

Two approaches are possible for obtaining the rate at the 20-second mark of the reaction. The finite difference expression of rate in 15–1–5 can be used to calculate the average rate during the time interval preceding or following the 20-second mark. Both average rates are calculated below :

Average rate between 10 and 20 seconds = –(conc at 20 s –conc at 10 s)/(20-10) Ms^{-1} = –(0.0768–0.0876)/(20-10) = 1.08 * 10^{-3} moles $L^{-1} s^{-1}$ Average rate between 20 and 30 seconds = –(0.0672–0.0768) (30-20) = 0.96 * 10^{-3} moles $L^{-1} s^{-1}$ The average rate over the 20–30 second time interval is less than that over the preceding interval because the reaction slows down as reactants are used up. This slowing down is readily apparent in the shape of the plot in Figure. However, the values are not too much different, so either may be used as an estimate of the rate of reaction at the 20 second mark.

The second approach is to use the differential expression of rate in 15-1-6. Just as dy/dx at $x = x_0$ is the slope of a plot of y versus x at $x = x_0$, $d[t\text{-}BuBr]/dt$ at a particular time, t_0, is the slope of a plot of [t-BuBr] versus time at the particular time, t_0. This slope is the same as the slope of the tangent to the curve at t_0. The tangent at t = 20 s is shown in the figure. It's slope is readily calculated to be 1.01×10^{-3} moles $L^{-1} s^{-1}$. As we might have expected, this value gives a rate intermediate between the values obtained above, based on finite time intervals. Because the slope of the tangent at a particular time represents the rate of reaction at that time, the rate obtained in this manner is called the *instantaneous rate* at the particular time. It is analogous to the speed that you measure when you glance at your automobile speedometer at a particular instant. As we have shown in this example, the instantaneous rate can be obtained by constructing tangents; or it can be estimated using the average rate over either the preceding or succeeding interval of time.

We can summarise our introductory discussion of reaction rate as follows :

- From a plot of [reactant] or [product] *versus* time, we can measure the rate at any particular time during the reaction as the slope of the tangent to the curve at that time; the initial rate is the slope of the tangent at t = 0.

- Rate decreases as the reaction proceeds towards equilibrium; this is evident from the shape of the concentration–time plot.

- The units of rate are []/time, or moles/L–time.

Chapter 11

MECHANICS OF SOILS

INTRODUCTION

Loads from foundations and walls apply stresses in the ground. Settlements are caused by strains in the ground. To analyse the conditions within a material under loading, we must consider the stress-strain behaviour.

The relationship between a strain and stress is termed stiffness. The maximum value of stress that may be sustained is termed strength.

Analysis of Stress and Strain

Stresses and strains occur in all directions and to do settlement and stability analyses it is often necessary to relate the stresses in a particular direction to those in other directions.

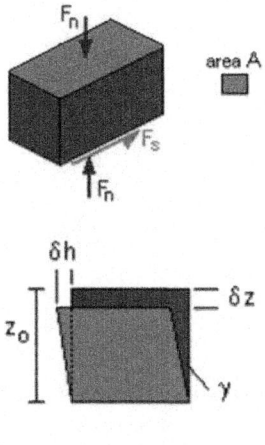

Fig.

- *Normal stress* : $\sigma = F_n / A$
- *Normal strain* : $\varepsilon = \delta z / z_o$
- *Shear stress* : $\tau = F_s / A$
- *Shear strain* : $\gamma = \delta h / z_o$

Note that compressive stresses and strains are positive, counter-clockwise shear stress and strain are positive, and that these are total stresses.

Mohr Circle Construction

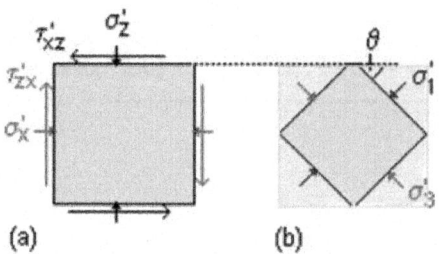

Fig.

Values of normal stress and shear stress must relate to a particular plane within an element of soil. In general, the stresses on another plane will be different. To visualise the stresses on all the possible planes, a graph called the Mohr circle is drawn by plotting a (normal stress, shear stress) point for a plane at every possible angle.

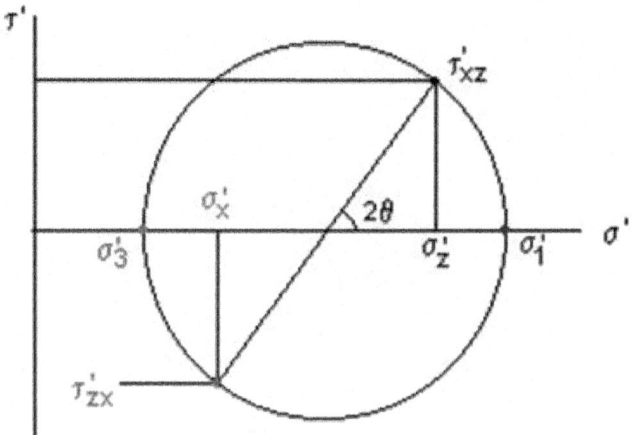

Fig.

There are special planes on which the shear stress is zero (*i.e. the circle crosses the normal stress axis*), and the state of stress (*i.e. the circle*) can be described by the normal stresses acting on these planes; these are called the principal stresses σ'_1 and σ'_3.

Parameters For Stress and Strain

In common soil tests, cylindrical samples are used in which the axial and radial stresses and strains are principal stresses and strains. For analysis of test data, and to develop soil mechanics theories, it is usual to combine these into mean (or normal) components which influence volume changes, and deviator (or shearing) components which influence shape changes.

	Stress	**Strain**
Mean	$p' = (\sigma'_a + 2\sigma'_r)/3$	$e_v = \Delta V/V = (\varepsilon_a + 2\varepsilon_r)$
	$s' = \sigma'_a + \sigma'_r)/2$	$\varepsilon_n = (\varepsilon_a + \varepsilon_r)$
Deviator	$q' = (\sigma'_a - \sigma'_r)$	$e_s = 2(\varepsilon_a - \varepsilon_r)/3$
	$t' = (\sigma'_a - \sigma'_r)/2$	$\varepsilon_\gamma = (\varepsilon_a - \varepsilon_r)$

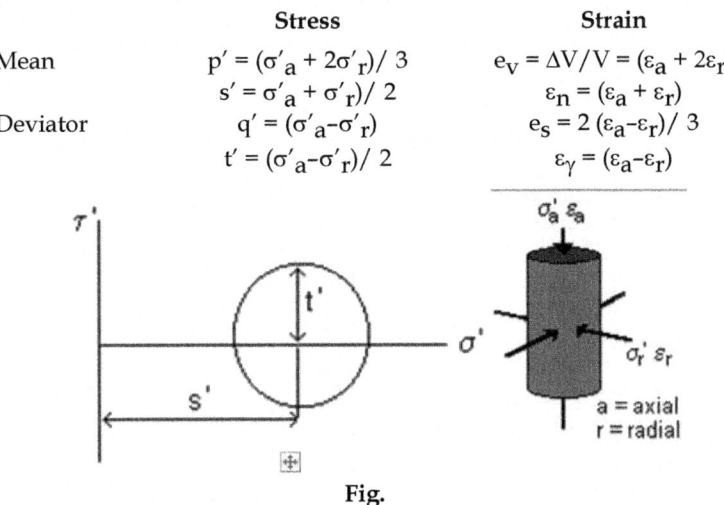

Fig.

In the Mohr circle construction t' is the radius of the circle and s' defines its centre.

Note : Total and effective stresses are related to pore pressure u :

$$p' = p - u$$
$$s' = s - u$$
$$q' = q$$
$$t' = t$$

Strength Back to Basic mechanics of soils :

- Types of failure
- Strength criteria
- Typical values of shear strength.

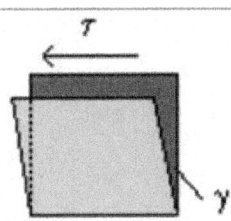

Fig.

The shear strength of a material is most simply described as the maximum shear stress it can sustain : When the shear stress t is increased, the shear strain γ increases; there will be a limiting condition at which the shear strain becomes very large and the material fails; the shear stress t_f is then the shear strength of the material. The simple type of failure shown here is associated with ductile or plastic materials. If the material is brittle (like a piece of chalk), the failure may be sudden and catastrophic with loss of strength after failure.

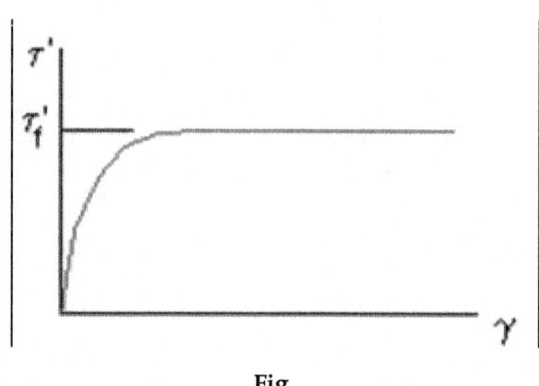

Fig.

Types of Failure

Materials can 'fail' under different loading conditions. In each case, however, failure is associated with the limiting radius of the Mohr circle, *i.e.* the maximum shear stress. *The following common examples are shown in terms of total stresses* :

Shearing

- Shear strength = τ_f
- σ_{nf} = Normal stress at failure

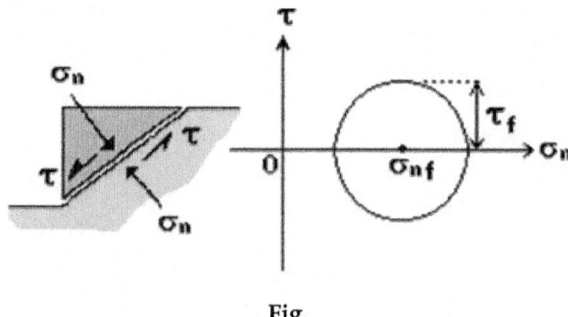

Fig.

Uniaxial Extension

- Tensile strength $\sigma_{tf} = 2\tau_f$

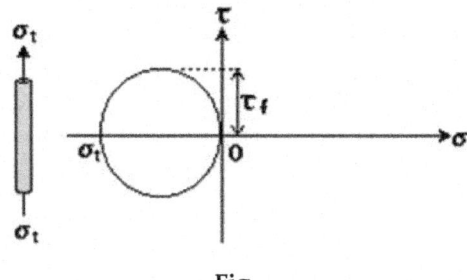

Fig.

Uniaxial Compression

- Compressive strength $\sigma_{cf} = 2\tau_f$

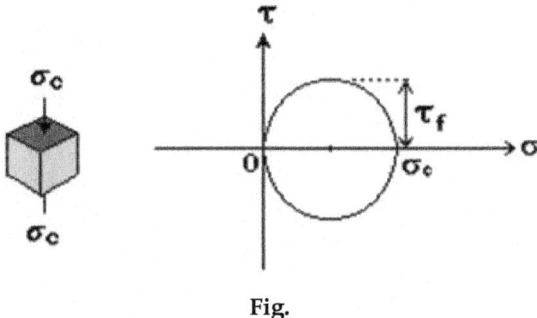

Fig.

Note : Water has no strength $\tau_f = 0$.

Hence vertical and horizontal stresses are equal and the Mohr circle becomes a point.

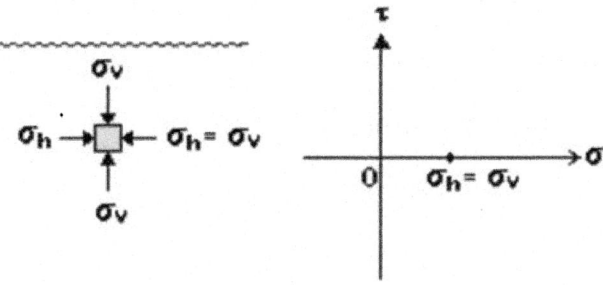

Fig.

Strength criteria

A strength criterion is a formula which relates the strength of a material to some other parameters : these are material parameters and may include other stresses.

For soils there are three important strength criteria : the correct criterion depends on the nature of the soil and on whether the loading is drained or undrained.

In General, course grained soils will "drain" very quickly (in engineering terms) following loading. Thefore development of excess pore pressure will not occur; volume change associated with increments of effective stress will control the behaviour and the Mohr-Coulomb criteria will be valid.

Fine grained saturated soils will respond to loading initially by generating excess pore water pressures and remaining at constant volume. At this stage the Tresca criteria, which uses total stress to represent undrained behaviour, should be used. This is the short-term or immediate loading response. Once the pore pressure has dissapated, after a certain time, the effective stresses have incresed and the Mohr-Coulomb criterion will describe the strength mobilised. This is the long term loading response.

Tresca Criterion

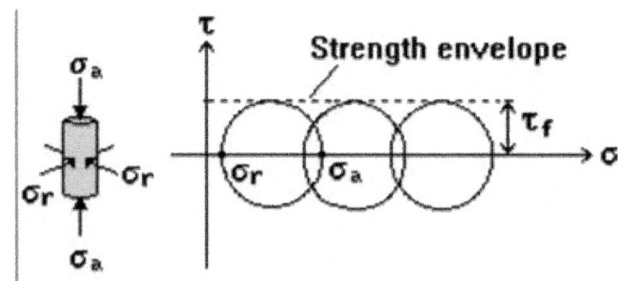

Fig.

The strength is independent of the normal stress since the response to loading simple increases the pore water pressure and not the effective stress.

The shear strength τ_f is a material parameter which is known as the undrained shear strength σ_u.

$$\tau_f = (\sigma_a - \sigma_r) = \text{constant}$$

Mohr-Coulomb (c'=0) Criterion

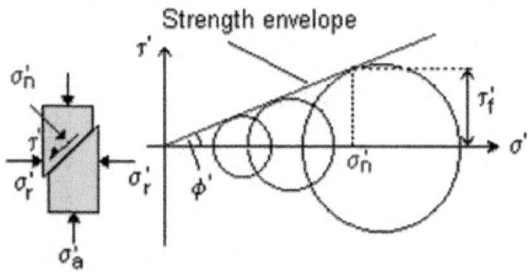

Fig.

The strength increases linearly with increasing normal stress and is zero when the normal stress is zero.

$\tau'_f = \sigma'_n \tan\phi'$

ϕ' is the angle of friction.

In the Mohr-Coulomb criterion the material parameter is the angle of friction f and materials which meet this criterion are known as frictional. In soils, the Mohr-Coulomb criterion applies when the normal stress is an effective normal stress.

Mohr-Coulomb (c'>0) Criterion

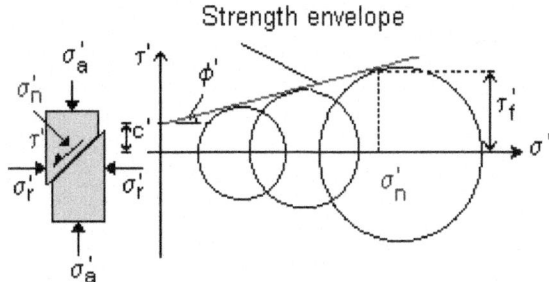

Fig.

The strength increases linearly with increasing normal stress and is positive when the normal stress is zero :

- $\tau'_f = c' + \sigma'_n \tan\phi'$
- ϕ' is the angle of friction
- c' is the 'cohesion' intercept.

In soils, the Mohr-Coulomb criterion applies when the normal stress is an effective normal stress. In soils, the cohesion in the effective stress Mohr-Coulomb criterion is not the same as the cohesion (or undrained strength s_u) in the Tresca criterion.

Typical Values of Shear Strength

Undrained shear strength	s_u (kPa)	
Hard soil	$s_u > 150$ kPa	
Stiff soil	$s_u = 75 \sim 150$ kPa	
Firm soil	$s_u = 40 \sim 75$ kPa	
Soft soil	$s_u = 20 \sim 40$kPa	
Very soft soil	`$s_u < 20$ kPa	
Drained shear strength	c´ (kPa)	$\phi´$ (deg)
Compact sands	0	35°–45°
Loose sands	0	30°–35°

	Unweathered overconsolidated clay	
critical state	0	18° ~ 25°
peak state	10 ~ 25 kPa	20° ~ 28°
residual	0 ~ 5 kPa	8° ~ 15°

Often the value of c′ deduced from laboratory test results (in the shear testing apparatus) may appear to indicate some shar strength at σ′ = 0. *i.e.* the particles 'cohereing' together or are 'cemented' in some way. Often this is due to fitting a c′, ɸ′ line to the experimental data and an 'apparent' cohesion may be deduced due to suction or dilatancy.

Chapter 12

SOILS, WEATHERING, AND NUTRIENTS

SOILS

All our elements, with the exception of hydrogen and helium, came from the dying throes of large stars. Amino acids needed to build animal proteins come from autotrophic plant life.

Biological studies have shown how certain nutrients (*e.g.*, nitrogen, phosphorus, calcium, magnesium, potassium, iron, etc.) are essential for the process of protein formation. All our amino acids and nutrients eventually come to us from plant life (sometimes via the meat of plant-eating animals). Plants synthesise amino acids from the combination of *sunlight, water and soils.*

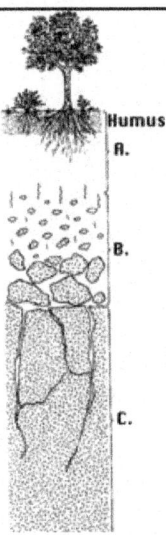

Fig. Layers of Horizons of a Typical Soil Profile.

Soil is therefore of critical importance to life. Simply put: no soil, no life. We first define soil as a *dynamic natural body capable of supporting a vegetative cover.* Where there is no soil, there is no plant life and we have barren rock and/or sand. Soil is composed primarily of *weathered* materials, along with water, oxygen and organic materials. Luckily for us, soil covers most of the land surface with a fragile, thin mantle. Soil and agricultural scientists have identified a huge number of different soil types.

Soil is Layered

Soil is layered into sections called "horizons". Figure shows a typical soil profile developed on granite bedrock in a temperate region. The top horizon is composed of *humus* and contains most of the organic matter. This layer is often the darkest. The "A" horizon consists of tiny particles of decayed leaves, twigs and animal remains. The minerals in the A-horizon are mostly clays and other insoluble minerals. Minerals that dissolve in water are found at greater depths.

The "B" horizon has relatively little organic material, but contains the soluble materials that are *leached* downwards from above. The "C" horizon is slightly broken-up bedrock, typically found 1-10 meters below the surface. While this is a typical soil profile, many other types exist, depending on climate, local rock conditions and the community of organisms living nearby. The U.S. Department of Agriculture has classified 10 orders and 47 suborders of soils. If you include other subsets, there are over 60,000 types of soil. The lunar surface, which has been produced by meteoroid impacts, is not classified as a soil, but is rather given the name "regolith" (derived from the Greek words meaning cover and stone). The layered nature of soil indicates its long evolution under the effects of atmospheric and biological processes.

The process that creates soil from bare rock is called "weathering". In the weathering process, the atmosphere and water interact with bare rock to slowly break it down into smaller and smaller particles. Rock climbers who encounter talus slopes (regions of pebble-life rocks that form in great conical piles at the feet of mountains) experience an intermediate step in the inexorable transition from solid granite to sand and soil. We next discuss the process whereby bare volcanic rock can be slowly turned into soils that can support life.

Soil Development

Soil forms from a complex interaction between earth materials, climate, and organisms acting over time. The brightly coloured soils of the humid tropics reflect the intense chemical reactions occurring in warm climates. The fertile prairie soils of the American Midwest evolved from the nutrient-rich organic matter left by decaying grasses. Regardless of soil characteristics, the whole process starts with the breakdown of earth material.

Fig.

Weathering

Weathering refers to processes that physically breakdown and chemically alter earth material. *Physical weathering*, also known as *mechanical weathering*, is the breakdown of large pieces of earth material into smaller ones. Think of physical weathering as the *disintegration* of rock without changing its chemical composition. There are many ways earth material can be physically weathered. When water freezes in rock crevices it expands creating stress in the crevice. As the stress increases, the crevice widens ultimately breaking the rock. Plant roots wedge rocks apart as they grow into rock crevices too. The shrinking and swelling by alternating heating and cooling weakens mineral bonds causing the rock to disintegrate. A very important result of physical weathering is its impact on the surface area of weathered material. When a block of earth material is broken into several smaller pieces, the amount of exposed surface increases. Examine the diagram below. A block with a width, depth, and height of 1 cm has a total surface area of 6 square centimeters. If we break the block in half in all directions it yields eight smaller pieces all with width, height, and depth of .5 cm. Breaking the block apart creates additional exposed surfaces such that the total surface area is now 12 square centimeters. Having more total exposed surface provides more area upon which chemical reactions can take place to further weather the material. The shape of the pieces also affects the the amount of exposed surface area. Plate-like pieces have more exposed surface area than do block-like pieces.

Chemical weathering breaks down earth material by chemical alteration. This usually means adding a substance like water or air to the material. For instance, when oxygen is added to iron bearing minerals, *oxidation* takes places and a loose mantle of iron oxide is created (rust). *Hydrolysis* is an exchange reaction involving minerals and water. Free hydrogen (H^+) and hydroxide (OH)⁻ ions in water replace mineral ions and drive them into solution. As a result, the mineral's structure is changed into a new form. Hydrolysis is a common process whereby silicate minerals are weathered into a clay mineral. Think of chemical weathering as the decomposition of earth material.

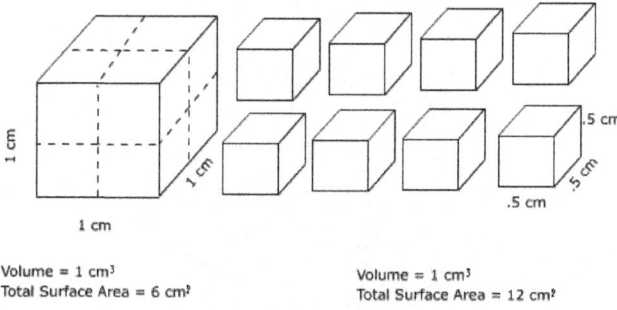

Volume = 1 cm³
Total Surface Area = 6 cm²

Volume = 1 cm³
Total Surface Area = 12 cm²

Fig. Effect of Physical Weathering on Surface Area.

The spatial variation of climate and organisms play a significant role in the weathering of earth materials. Dry locations tend to be dominated by physical weathering and moist places by chemical weathering. The type of earth material available also determines the amount of weathering that might take place. Limestone is easily broken down where abundant rainfall and high temperatures prevail. However, limestone will remain intact in dry locations. The end result of the weathering process is the creation of a *weathered mantle*. The weathered mantle is not yet a soil until it undergoes further change. This involves the addition, transformation, translocation and removal of materials from the weathered mantle to form distinctive soil layers.

Since early geologic time, the atmosphere has interacted with the Earth's exposed crust though a process known as *weathering*. Weathering takes place through a combination of both mechanical and chemical means. We have all experienced the results of weathering first hand. Any visit to an old cemetery find us peering at the blurred inscriptions on old marble tombstones. These inscriptions were once perfectly legible, but with the passage of time, the small fractures and cracks in the rock have made it vulnerable to attack by aqueous solutions. A dramatic example of weathering can be seen in these two images of the same 3000 years old Egyptian obelisk just before relocation to damp New York and 100 years after accelerated weathering in New York Weathering rates are obviously a strong function of climate!

Fig.

Many of the original volcanic gases (*e.g.*, carbon dioxide, sulphur-bearing gases, etc.) were able to dissolve in water and produce acids. The acids, in turn, reacted with surface minerals. Later, oxygen in the atmosphere reacted with the exposed reduced materials, making the red beds discussed in an earlier lecture. Since the advent of land plants, soil and surface minerals have been exposed to relatively high concentrations of carbon dioxide maintained in soil pores as a result of decomposition and the metabolic activities of roots. The reaction of carbon dioxide with water in the soil produces carbonic acid (H_2CO_3) which determines the rate of rock weathering in most ecosystems.

$$CO_2 \text{ (gas)} + H_2O \text{ (liquid)} - H_2CO_3 \text{ (solution)}$$

Acid rain, produced by human effluents of nitrogen and sulphur-bearing gases will increase the rate of rock weathering in downwind areas. To understand why weathering occurs and why the rate of rock weathering is so dependent on climate, we need to discuss the chemistry of the process in a little more detail.

Igneous Rock Weathering

There is a well known expression that captures much of the story of rock weathering and illustrates the important role it has played in Earth history:

"Igneous Rocks + Acid Volatiles = Sedimentary Rocks + Salty Oceans"

What we mean by this will become clearer if we look at the details of the weathering processes. Consider a boulder or rock containing *Feldspar* minerals. Feldspar is a general term for a group of aluminosilicate minerals containing sodium, calcium, or potassium and having a lattice framework structure that makes for rigidity. Feldspars turn out to be one of the most common minerals in the Earth's crust. Feldspars are weathered through the chemical process of *hydration:*

$$K\,Al\,Si_3O_8 + H_2O \rightarrow Al_2SiO_5(OH)_4$$

In this chemical formula feldspar reacts with water to produce a kaolinite (clay). Notice that the chemical equation does not exactly balance, that is, not all the elements on the left hand side appear on the right hand side. This is because soluble elements, such as potassium (K) are *leached* out during the chemical reaction and carried away as dissolved salts. The process of leaching can perhaps best be understood by analogy with the making of coffee. When hot water is passed over crushed coffee beans, the soluble components (making the coffee) are leached away, leaving the insoluble crushed coffee bean remnants behind.

In this way, rocks containing feldspars are weakened though the conversion of rigid feldspar to more plastic clays which do not have anything like the same structural rigidity. The process occurs at exposed surfaces of the minerals making up the rock. Through geologic time, large amounts of sedimentary r0ocks have been deposited as part of this process. In fact about 75 per cent of all exposed rocks on the Earth's surface today are of sedimentary origin and have been brought to the surface by geologic uplift.

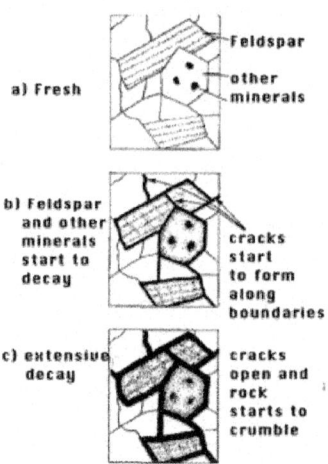

Fig.

Of course, geological processes return some of the sedimentary rocks to the mantle of the Earth, where they are converted back to the primary materials under conditions of great temperature and pressure. Rock weathering is also critical for the release of bio-chemical elements that have no gaseous form-examples are calcium, Ca, Potassium, K, Iron, Fe, and Phosphorus, P. The latter element plays a key role in cell metabolism. Thus, we can say that weathering provides key nutrients for life through the process of leaching.

The soluble nutrients are transferred to soils layers below the immediate surface (*e.g.*, the "B" layer). This is a major reason that plants have evolved root systems-to search out these critically needed nutrients below ground.

Igneous rock weathering proceeds in stages. It is useful to picture the inside of a rock-made up of interlocking minerals of irregular shapes, each being rigid. Figure shows a microscopic view of such an interior rock composition, with some of the grains being feldspars.

As weathering proceeds, the boundaries of the vulnerable feldspar (and other) mineral grains start to decay. As the decay proceeds, water can reach more and more feldspar surfaces and the process accelerates. This process can also be accelerated by melt-freeze cycles that force the grains apart due to the difference between the volume occupied by water and ice. We can see that weathering is due to the combined effects of chemical and mechanical decay.

The chemical and physical processes of weathering transform the igneous rock into sand and clay particles and dissolved salts. Chemical weathering can add carbon dioxide, water, and oxygen. The link provided courtesy of the National Park Service shows an example of a weathering rock. It is interesting to note that, since most of the Earth's exposed rock is of sedimentary origin and since sedimentary rocks are a by-product of weathering, most of the rocks that are weathering away in today's world are second or perhaps even third generation rocks.

That is, they originated as igneous material, became sedimentary through weathering and transport to the bottom of shallow waters, were then subject to geologic uplift to become again exposed and, finally, began to undergo weathering yet again. The natural world is full of such endless cycles.

How Fast Does a Rock Decay

The best answer to this question is... it depends. It depends on the local environment and the type of rock. For example, an iron nail buried in the ground in Michigan will only take a year or so to decay to the point that it is easily snapped in two. Iron nails rust much more slowly in drier environments. Aluminum cans decay very slowly, even in humid climates. Glass decays even more slowly, while plastic is considered essentially non-biodegradable. It is somewhat ironic that a plastic tombstone will endure much longer than one made in marble!

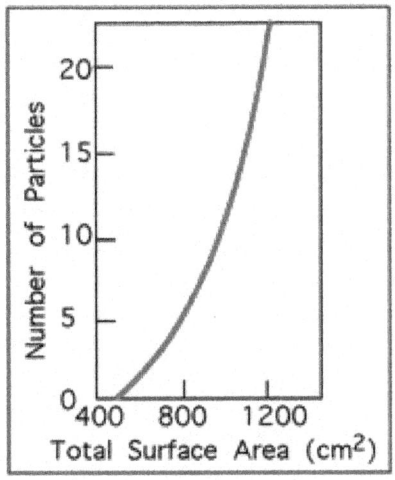

Fig.

From what we have already discussed, soils themselves aid in rock decay, as do melt-freeze cycles and bacterial action. Thus, soils are a consequence of weathering, but also a factor in accelerating weathering. The production of soil is a *positive feedback* process. The following table illustrates rates of weathering for three rock types as a function of climate. As more of a rock becomes amenable to weathering, the speed of weathering increases. This can be understood if we plot the rate at which the available exposed surface area of a rock increases as the rock fragments. Figure 3.6 shows this relationship.

Nutrients

Although living tissue is composed of carbon, hydrogen, and oxygen in the approximate proportion of CH_2O, as many as 23 other elements are necessary for biochemical reactions and for the growth of structural biomass.

Examples of important nutrients are:

Table. Rates of Weathering of Clean Rock Surfaces (Micro-Meters/1000Years).

Rock Type	Cold Climate	Warm, Humid Climate
Basalt10	100	
Granite	1	10
Marble	20	200
Nutrient		Role

Nitrogen The proteins found in plants and animals contain about 16 per cent by weight of nitrogen

Phosphorus   Pis part of the important ribulose
    biphosphate carboxylase molecule and is part of ATP-adenosine triphosphate, the
    universal molecule for energy   transformations

Calcium   Ca is a major structural component of the proteins forming plants and animals

Other important nutrients include magnesium, potassium, iron, sulphur, etc. We should note that, although carbon, nitrogen and sulphur can be obtained from the atmosphere, calcium, magnesium, potassium, iron, and phosphorus all come from rock weathering processes. The atmosphere has no store of these essential nutrients.

As mentioned earlier, one of the main purposes of plant root systems is to get access to nutrients stored in the soil. Some plants go to enormous lengths to do this.

For example, Emiliani quotes that "A single plant of winter rye, 50 cm high, was found to have a root system consisting of 143 main roots, 35,600 secondary roots, 2.3 million tertiary roots, and 11.5 million quaternary roots ! The root system was found to have a total length of 600 km and a total surface of about 250 square meters". Delivery of nutrients into plant root systems can occur by several pathways. In some cases, direct uptake in water solution occurs. Sometimes, plants actually have to protect themselves against too much nutrient intake.

Too much of a good thing can prove poisonous. An example of this can be seen in the accumulations of calcium carbonate deposits that surround the roots of some desert shrubs.

Table. Nature's Vitamin Requirements

Amount	Nutrient
100 parts	N
15 parts	P
50 parts	K
5 parts	Ca
5 parts	Mg
10 parts	S

Some nutrients, such as nitrogen, phosphorus, and potassium are often harder for roots to find and specialised (incredibly efficient) enzymes have evolved located in root membranes to seek out these scarce and needed resources. If some nutrients are not readily available, plants will grow more slowly and/or increase their root/shoot ratio.

In general, the availability of nutrients (deficit or surplus availabil ity) often controls the form of the ecosystem, determining its overall productivity, and influencing which particular set of plants come to predominate. Excess nitrogen can lead to the loss of fine root biomass and deficiencies in other nutrients.

The pool of nutrients held in the soil and vegetation is many times larger than the annual receipt of nutrients from the atmosphere and rock weathering. Thus, life *husbands* its needed nutrients on land, storing much of the total in the humus. Recycling of nutrients is critical to the productivity of natural ecosystems, although less critical to crop production, due to the availability of commercial fertilizers.

**Table. Percentage of Annual Nutrient Requirement
for Growth of Hardwood Forest**

	Process (kg/ha/yr)						N	P	K	Ca	Mg
Total Growth	115	12	67	62	10		Atmospheric Inputs	18	0	1	4
Requirement	6	Rock Weathering Inputs			0	13	11	34	37	Reabsorptions	
(Intra-system)	31	28	4	0	2	Detritus Turnover	69	81		86	85
						87					

The data of the table came from a study of the famous Hubbard Brook ecosystem in New Hampshire. It shows how effective the soils are in storing the needed nutrients and how limited are the rates of supply from atmosphere and rock weathering. Acid rain caused by human emissions of nitrogen and sulphur oxides leads to enhanced weathering and changes in nutrient ratios. For example, recent studies suggest that forest growth has declined in areas down wind of air pollution. Acid rain appears to increase the movement of aluminum ions, which may, in turn, reduce the intake rates of calcium and other nutrients.

In the oceans, life is also limited by the availability of nutrients. The productivity is highest on continental shelves and in regions of upwelling. Nutrients are removed from surface waters by downward sinking and are regenerated in deep waters. A paucity of nutrients limits production in the open oceans. On the other hand, marine productivity is threatened by excessive human inputs of nitrogen and phosphorus in coastal regions, where it leads to excessive algal blooms, whose subsequent decay robs deeper water of oxygen, creating "dead zones".

Soil Erosion

We define soil erosion as the movement of surface litter and topsoil. The forces responsible are wind and water flow. It is important to recognise at the outset that soil erosion is a natural process. It is slowed, however, by plant roots that serve to stabilise the soil. In any undisturbed ecosystem, the rate of soil loss is matched by its rate of production. In a disturbed ecosystem, however, major changes in the rate of erosion can occur. In fact, almost every activity that can be characterised as a "development" causes enhanced soil erosion. This includes farming, logging, building, grasing, off-road travel, etc. Soil erosion, discussed later in this lecture, is responsible for the loss of a great deal of our topsoil.

The problem of accelerated soil erosion is a major one. On a global scale, topsoil is eroding faster than it can be replenished in over one third of the world's croplands. For example, in China and India combined, more than 12 million square kilometers have been severely eroded since 1945. The causes of this have been deforestation (30 per cent), overgrasing (35 per cent), and farming (28 per cent). The soil erosion problem is particularly severe if one considers that each year we must feed an extra 90 million more people, with about 25 billion tons less topsoil !

In the U.S., the situation is also of concern. The Dust Bowl of the 1930's was caused by plowing of the prairies. Before the pioneers, the soil was held in place by the long root systems of prairie grasses. In response to the dust bowl, the Soil Conservation Service (SCS) was established in 1935. The Dust Bowl was brought on by a long drought and lasted about a decade. Much was learned about soil conservation and later droughts have not caused as much damage.

Fig.

The Great Plains has lost about one-third of its original topsoil in the past 150 years. Iowa has been particularly hard hit, already having lost half of its topsoil since the arrival of the first European settlers. In California, the current topsoil is being eroded at a rate that exceeds its replenishment rate by a factor of more than 70. The amount of topsoil lost in a day is staggering: Emiliani estimates that the lost topsoil would fill a line of dump trucks 3,500 miles long! The cost of this loss of natural resource is incalculable. The map below summarises points brought up in lecture on geographic distribution of soil degradation :

Soil is important for the support of life processes. Real soil only exists on Earth. Soil is the place where atmosphere, biosphere, hydrosphere, and lithosphere meet. Over 60,000 types of soils have been catalogued. Without soil, there is no food. Soils take 10's of thousands of years to form in rock weathering processes. Weathering occurs due to a combination of chemical and mechanical processes that are subject to strong positive (reinforcing) feedback mechanisms. Over geologic time, rock weathering processes have led to sedimentary rocks and salty oceans !

Plants obtain inorganic minerals (nutrients) from the soil and incorporate their elements into bio-chemical materials. Animals may eat plants and each other, and synthesise new proteins, but the building blocks are the amino acids originally synthesised in plants. Soil effectively husbands the needed nutrients in the upper horizons.

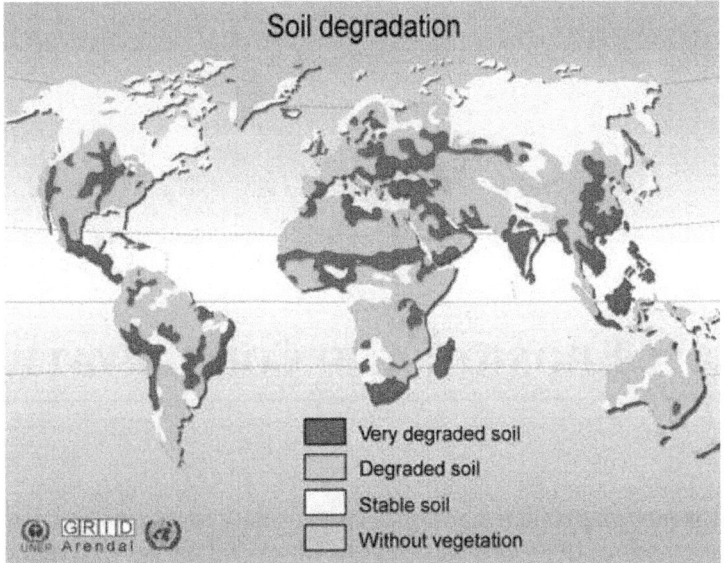

Fig.

The availability (or lack of availability) of important nutrients places critical constrains on the type of life that can survive in particular ecosystems. Soil erosion is a very serious problem globally because of its inextricable link to development activities. Soil conservation is therefore a very serious business and an important part of future global change studies.

Chapter 13

SOIL EROSION AND CONSERVATION

Environmental problems are by no means new to American agriculture. The Dustbowl episode of the 1930s was but one of several historical instances in which concern has been raised about the environmental performance of U.S. farmers. Nonetheless, current controversies about the management of agricultural resources in the United States date largely from the rise of the environmental movement in the 1960s and have included issues relating to soil erosion, the use of agricultural pesticides, energy and water consumption, and the like. The soil conservation implications of American agriculture have remained among the most longstanding agricultural-environmental issues of concern to rural sociologists and are given principal emphasis in the section that follows.

The majority of U.S. farms are family proprietorships in which there is household or family ownership and management of the land and water resources. Accordingly, the logical point of departure in understanding the soil erosion problem in agriculture has been to conduct micro-level research on the individual, farm firm, and ecological factors associated with soil erosion rates or the utilisation of soil conservation technologies. This has been a productive line of research, but it has also become a controversial one, with the major controversies centering around theoretical approach, the relationships between micro and macro levels of analysis, and the public policy inferences that should be drawn from social science research.

SOIL CONSERVATION

Soil conservation is a set of management strategies for prevention of soil being eroded from the Earth's surface or becoming chemically altered by overuse, acidification, salinisation or other chemical soil contamination.

It is a component of environmental soil science. Decisions regarding appropriate crop rotation, cover crops, and planted windbreaks are central to the ability of surface soils to retain their integrity, both with respect to erosive forces and chemical change from nutrient depletion.

Crop rotation is simply the conventional alternation of crops on a given field, so that nutrient depletion is avoided from repetitive chemical uptake/deposition of single crop growth.

Erosion Prevention

Practices

There are also conventional practices that farmers have invoked for centuries. These fall into two main categories : contour farming and terracing, standard methods recommended by the U.S. Natural Resources Conservation Service, whose Code 330 is the common standard.

Contour farming was practiced by the ancient Phoenicians, and is known to be effective for slopes between two and ten per cent. Contour plowing can increase crop yields from 10 to 50 per cent, partially as a result from greater soil retention.

There are many erosion control methods that can be used such as conservation tillage systems and crop rotation. Keyline design is an enhancement of contour farming, where the total watershed properties are taken into account in forming the contour lines. Terracing is the practice of creating benches or nearly level layers on a hillside setting. Terraced farming is more common on small farms and in underdeveloped countries, since mechanised equipment is difficult to deploy in this setting. Human overpopulation is leading to destruction of tropical forests due to widening practices of slash-and-burn and other methods of subsistence farming necessitated by famines in lesser developed countries. A sequel to the deforestation is typically large scale erosion, loss of soil nutrients and sometimes total desertification.

Perimeter Run-off Control

Trees, shrubs and groundcovers are also effective perimeter treatment for soil erosion prevention, by insuring any surface flows are impeded. A special form of this perimeter or inter-row treatment is the use of a "grassway" that both channels and dissipates run-off through surface friction, impeding surface run-off, and encouraging infiltration of the slowed surface water.

RESEARCH ON SOIL EROSION AND CONSERVATION

One of the most crucial aspects of soil erosion research is the very conceptualisation of why soil erosion is a socioeconomic problem. Traditionally this has not been an issue, since sustained high levels of soil loss were presumed to irreversibly degrade the productivity of agricultural resources. Accordingly, if the soil erosion problem is conceptualised as medium to long-term land productivity decline detrimental to the interests of farmers, it would follow that conservation should be achieved through voluntary farmer compliance (rather than through mandatory regulation) based on the logic of the long-term interests of farmers.

Historically, federal and state government policy thus has emphasised education of farmers and modest levels of financial inducements (such as Agricultural

Stabilisation and Conservation Service cost-sharing programmes) to lever farmer decision making. But it has been increasingly recognised that the proportion of prime U.S. farmland that stands to be irrevocably degraded by soil erosion is relatively small and that the "off-site" costs of soil erosion — the impacts of soil erosion and run-off on water and land resources of a farmer's parcel — are very significant and may very well be in excess of the on-site costs.

These alternative conceptualisations of the soil erosion problem — productivity loss versus destruction of off-site water and land resources — have very profound implications for appropriate types of social science research and for public policy. To the degree that soil erosion primarily results in land productivity losses, research should focus on individual- and farm-level factors that influence resource management, and public policy should incorporate these research findings in order to develop educational and incentive programmes to achieve soil conservation through farmer self-interest. To the degree that soil erosion is primarily a problem because of off-site costs, farmers cannot be expected to conserve because the long-term productivity benefits of conservation are small, and research at a more macro level (*e.g.*, to determine the degree to which farmers' incentive structures and resource management behaviours are congruent with the public's interests in clean, navigable waters) would be most appropriate.

Debate over the appropriate kind of analysis in soil conservation research has been closely related to, and in many respects was stimulated by, debate over the adoption-diffusion approach. Pampel and van Es, reported that the correlates of adoption of conservation technologies tended to be different from those of commercial non-conservation technologies. They also suggested that effective conservation practices will tend not to be profitable for farmers, so that to achieve the public interest in soil conservation voluntary compliance among farmers may be insufficient. The issues raised by Pampel and van Es have continued to pervade the erosion and conservation literature. Many researchers have conducted their research by seeking to revise the micro-sociological diffusion of innovations approach, while others have argued that the constraints to soil conservation behaviour tend to be more macro in nature and that the diffusion of innovations approach will be inadequate.

Prior to 1970 soil and water conservation technologies and practices were generally understood to be a diverse collection of cropping patterns (crop rotations, contour planting, strip cropping, cover cropping, sod waterways, filter strips) and physical and biomass structures (terraces, sediment retention basins, diversion channels, animal waste structures, and hedge rows). Since that time, however, soil and water conservation has, for the majority of researchers and policy-makers, come to be largely conterminous with the adoption and use of no-till and related conservation and reduced tillage equipment, which enables farmers to leave a mulch layer of residue on the soil to reduce run-off and sediment losses.

Two observations can be made with regard to traditional and reduced-tillage equipment approaches to soil and water conservation in agriculture. First, these two approaches, while by no means incompatible, are nonetheless quite different and will tend to apply most appropriately to different types of farm operations. Traditional cropping practices and structures are often most attractive to smaller, more diversified farmers, while reduced tillage tends to be most attractive to larger, more highly mechanised producers of row crops.

Second, reduced tillage, although acknowledged to be effective in reducing erosion in row-crop mono-cultures, has become somewhat controversial. One controversial aspect of the reduced-tillage approach has been that it requires increased use of herbicides, which may lead to contamination of water and thereby negate some of the run-off reduction benefits of reduced tillage. Many studies have indicated, moreover, that reduced tillage practices have been adopted by farmers largely because of labour savings rather than because of the ability of the technology to reduce soil erosion.

Indeed, the farmers who have been most likely to adopt no-till and other reduced tillage technologies have been found to be large operators with highly specialised operations involving practices such as continuous cropping and row-crop mono-cultures. Thus farmer-adopters of such conservation technologies ironically represent the major trends in U.S. farm structure that have been demonstrated to lead to environmental degradation in agriculture. However, farmer perceptions of a threat to soil productivity may lead to adoption of soil-conserving innovations, whereas stewardship related motivations may have little effect on motivations to prevent other kinds of consequences such as off-site pollution.

IMPORTANT OF SOIL EROSION AND ITS CONSERVATION

Soil is one of the most important natural resources of man. Soils are essential for man for growing crops, fodder and limber. Once the fertile portion of the earth's surface is lost, it is very difficult to replace it. In India, the destruction of the top-soil has already reached an alarming proportion. Land degradation problems have resulted in increasing depletion of the productivity of the basic land stock through nutrient deficiencies. In addition to the direct loss of crop producing capacity, soil erosion increases the destructiveness of floods and decreases the storage capacity of water in reservoirs. It is therefore essential that the soils should not be allowed to wash or blow-away more rapidly than they can be regenerated, their fertility should not be exhausted and their physical structure should remain suited to continued production of desired plant materials. Protection of land from further degradation, adoption of various conservation measures, including reclamation and scientific management of available land stock is very important for a country like India to achieve higher productivity of food, fodder, fuel and industrial raw materials on a substantial basis. Besides, demand for land for providing social priorities such as shelter, roads, industrial activities is increasing at a very fast rate with the rise in population and very often good agricultural and forest lands are being diverted to such use. It is, therefore, necessary to keep soil in place and in a state favourable to its highest productive capacity.

Soil Erosion

The process of destruction of soil and the removal of the destroyed soil material constitute soil erosion. According to Dr. Bennett "the vastly accelerated process of soil removal brought about by the human interference, with the normal disequilibrium between soil building and soil removal is designated as soil erosion".

Types of Soil-Erosion

Erosion of soil by water is quite significant and takes place chiefly in two ways (a) Sheet erosion, (b) Gully erosion.

Sheet Erosion

Sheet movement of water causes sheet erosion and depends on the velocity and quantity of pronounced surface run-off and the erodability of the soil itself. In such cases, the soil is eroded as layers from the hill slopes, sometimes slowly and in-sidiously and sometimes more rapidly. Sheet erosion is more or less universal on :-

- all bare follow land,
- all uncultivated land whose plant cover has been thinned out by over grasing, fire or other misuse, and
- all sloping cultivated fields and on sloping forest, scrub jungles where natural porosity of soil has been removed by heavy grasing, felling of trees or burning etc.

The particles loosened and shifted by the rain drops are carried down slope by a very thin sheet of water which moves along the surface. The impacts of the raindrops increases the turbulance and transporting capacity of this unchannelised sheetwash which results in the uniform skimming of the top soil. Sheet erosion is considered as dangerous as it may continue for years but may or may not leave any trace of the damage. Sheet erosion is common in the Himalayan foot-hills, in Assam, Western ghats and Eastern ghats. When sheet erosion continues unchecked, the silt laden run-off forms well-defined minute finger shaped grooves over the entire field. Such thin channeling is known as 'rill-erosion', which is active over wide areas in Bihar, Uttar Pradesh, Madhva Pradesh and in semiarid areas of Maharashtra, Karnataka, Andhra Pradesh and Tamil Nadu.

Gully Erosion

On a gentle slope, adequately covered by vegetation, clay soil will resist erosion to a great extent and the water forms small rivulets which can then erode deeper. The rivulets in turn join together to form larger channels until gullies are formed gradually deep gullies cut into the soil and then spread and grow until all the soil is removal from the sloping ground. This phenomenon once started and if not checked, goes on extending and ultimately the whole land is converted into a bad-land topography. Gully erosion is more common in areas where the river system has cut down into elevated plateaus so that feeders and branches carve out an intricate pattern of gullies.

Apart from this, it also takes place in relatively level country whenever large blocks of cultivation give rise to con-centration of field run-off.

Wind Erosion

It occurs in dry climatic areas having a sparse and low vegetation cover on mechanically weathered, loosened surficial material. Dust storms are the principal agents of wind erosion. The top soil is often blown off from the surface rendering it infertile. Besides, with the decrease in the wind velocity coarse sand particles get deposited in some areas covering the existing soil and rendering it unproductive.

SOIL BIOLOGY AND THE BIOLOGICAL MICRO-ENVIRONMENT

Organic matter, microscopic and macroscopic organisms (*e.g.*, fungal hyphae and invertebrates), detritus from fungi and animals, and bacteria, and biological exudates, all assist in stabilising soil structure. The role of each part of the bio-mass differs according to its size. Broadly, large aggregates greater than 250 m m diameter (macro-aggregates), are stabilised by their inherent physical structure, wetting and drying cycles, and organic matter. Micro-aggregates (< 250 m m) are stabilised by live or dead roots, fungi, invertebrates and micro-organisms.

The populations of soil organisms of all sizes are linked functionally through their roles in the degradation of various forms of organic material. The latter in-cludes live and dead plant material and other live or dead organisms. This shows that animals such as nematodes and some fungi feed directly on live plants while other fungi and bacteria feed predominantly on litter. Earthworms and other large invertebrates create, and inhabit, burrows and pores, and are very mobile. The most notable of these are termites, which are divided into three groups according to the structure of their nests : those that build mounds (a) above ground, (b) on the soil surface, and (c) below ground. Small arthropods, microfauna and fungi live mostly in larger voids and in association with roots. Foster (1988) reviewed the location of the various types of soil-dwelling organisms and found that fungi, which constitute about 80 per cent of the biomass in many soils, tend to be restricted to the rhizosphere of roots, to larger pores between aggregates and to the surface of aggregates. Bacteria, by contrast, are found on roots in the rhizosphere, in small colonies in the larger micropores, within aggregates and on and within cell debris.

Organic Matter

Both plants and animals provide inputs of organic matter to soils. Once within the soil organic residues can be distinguished on the basis of their chemi-cal structure (*e.g.*, old lignified humic substances that degrade slowly), by their source (plant or animal) or by location.

The standing crop of litter in semi-arid grasslands is usually more than 3 t/ha and in temperate dry steppe may exceed 11 t/ha. There has been much debate about the relative contents of organic matter in tropical and temperate soils. Within those wet-and-dry climates that have hot summers assisting rapid decomposition, there is no evidence of inherently lower levels of organic matter in the tropics than in comparable temperate regions. Kowal and Kassam (1978) and Juo and Payne (1993) review the role of organic matter in tropical soils. Here, it is sufficient simply to state that organic matter has various interrelated effects on soil fertility. In particular it should be noted that both chemical and physical effects are of relatively great importance in the soils of the semi-arid tropics because these generally have low cation exchange capacity. The relative importance of litter (crop residue) and manure as inputs of organic matter, varies between cropping systems and spatially within a system. Here, most of the above-ground crop residue is fed to animals, but an equal amount of below-ground crop material enters the soil organic matter pool.

Where alley cropping and agroforestry are practised, values are more variable, but possible inputs could be very significant where the trees, from which the litter is taken, are grown away from the annual crops. If, say, two-thirds of the leaves from leguminous trees are harvested annually, litter values will be substantially higher and the material of better quality than the leaf and stem residue from an annual crop likely to be recycled in the field. Some qualifications, however, should be made. Tree root material is not available for decomposition in the crop field unless it is spatially overlapping (*e.g.* as an intercrop), in which case the trees will compete with the crop for soil nutrients, water, light and space.

The proportion of animal and human manure used on cropland is more variable. Some farmers have developed stable systems which strongly emphasise the use of animal manure on crops. For example, Norman *et. al.* (1982) describe how farmers in northern Nigeria managed to apply 4 t/ha of manure to their heavily-cropped crop land though they had only 3 cows each. Many other farmers do not ensure adequate recycling, either because they are more concerned with livestock management or they do not know the importance of maintaining a 'zero nutrient budget' to replace nutrients removed by the crop. For example, Norman *et. al.* (1982) also describe farmers with 10 cows each, who applied only 1.9 t manure/ha to their crops.

Within a cropping system, manuring practice varies with location. There is transference towards the centre of the system. On traditional farms, the area near the household or village is highly fertilized with human and animal manure while more distant fields receive little or no organic matter. Fussell (1992) describes such a traditional 'ring' farming system in semi-arid west Africa. Here, if houses are thatched, the village needs rebuilding or moving every 2 to 4 years. Moving takes advantage of the fertility gradient. Where the huts are not moved, the fertility gradient becomes steeper with time. Rather than trying to even out fertility by labour-intensive transport of manure, farmers vary the cropping of the fields. Continuous cropping of millet is sustainable close to the hut or village where there is plenty of human and animal manure but crop rotations are essential at the periphery.

MANAGEMENT FOR MAINTENANCE OF SOIL BIOLOGY AND NUTRITION

The aim of management should be to create balanced organic matter and mineral budgets. It should ensure that, over several years (a complete crop rotation), soil organic matter is not depleted and that nutrients added equal or exceed those removed by cropping or lost in various ways.

When managing organic matter farmers should recognise that the effects of animal and human manure, sewage sludge and plant residues last longer than those of green manure crops. Green manures, though valuable, usually last only one or two seasons because they are incorporated before they are mature and lignified.

Juo and Payne (1993) advise : 'In spite of the many proven benefits of soil organic matter, its management and recycling in an intensified, modem agroecosystem must necessarily revolve around two fundamental characteristics of the system, namely, the availability of organic material at the farm level, and the economic incentive for conserving and recycling organic matter'. To this may be added consideration of the benefits and costs of treating and transporting human wastes from cities. On-site organic matter, organic material brought in from outside and industrially-produced fertilizers will be used according to their benefit cost ratios and the attitudes of governments and crop managers. Tillage reduces the frequency of VAM, at least in invertebrates. Tillage reduces earthworm populations 10-30-fold, both killing them directly and destroying their burrows. Haines and Uren (1990) show that differences in earthworm populations between tillage treatments and stubble burning are reflected by 50 per cent differences in numbers of soil pores.

Tillage also effects the location, numbers and activity of micro-organisms within the soil, depending on the type of tillage. For example a mouldboard plough that inverts topsoil has more effect than minimal tillage. Doran (1987) found microbial biomass and potentially-mineralisable nitrogen to be 54 per cent and 37 per cent higher respectively in the top 7.5 cm of no-till than ploughed soil.

These would be critical differences if translated to the semi-arid tropics. Microbial and fungal biomasses may be 20-70 per cent higher after 1 and 2 years no-till. However, conventional tillage, by burying topsoil, may cause microbial populations and their activities to be higher than under no-till at depths of say 7.5 to 15 cm. Differences in soil organisms (*e.g.*, microbial biomass) under differing tillage treatments are seasonal, generally being greatest in the period when the weather is most favourable for soil organism activity and plant growth. Rasmussen and Collins (1991) summarise many data on the impact of tillage on soil properties and conclude that stubble-mulching with zero tillage conserves up to 2 per cent more organic matter per year than ploughing.

SOIL BULK DENSITY DETERMINATION

Soil weight is referred to as soil bulk density. Density is the mass of material contained within a given volume. This is the idea that a given size of box may be heavy or light, depending upon what kind of material it contains. If the box were filled with wood, it would be light when compared to having it filled with lead. The weight of water is the reference for density measurements : 1 gram of water=1 cubic centimeter (cc), & 1 cc water=1 ml. & 1 cubic foot of water=62.4 lbs.

The bulk density of a soil is the mass of dry soil per unit of "bulk" (total volume of soil or soil particles & pore space).

B.D. = mass of oven dry soil (grams) ÷ total volume of soil (cm³)=grams/ cc

The bulk density takes into account the total soil volume (the space occupied by the solid particles plus the space occupied by the air of the pores or pore space). To determine the mass of the soil – since air does not have any significant weight – we can just weigh the oven dry soil on a balance. The volume of the soil can be determined by pouring the soil into a graduated cylinder and measuring the volume that it occupies.

The problem with this procedure is that structural aggregates may be crushed or compacted once they are removed from the soil and placed into the cylinder. A better procedure is one in which the aggregates could be removed from the soil and frozen exactly the way they were in the soil. Plastic fixatives enable us to do this.

A large aggregate has been removed from the soil and is being fixed by dipping it into a saran solution. This will make the clod impervious to water. The weight of our clump of soil can be obtained by hanging it on the balance or placing it on the metal weighing tray.

The volume of the clod also needs to be determined. This can be obtained by weighing the sample again, only this time the sample will be in water. The weight of the sample now will be less, since the sample will be buoyed up by the amount of water it is displacing.

Archimedes most famous theorem gives the weight of a body immersed in a liquid, called Archimedes' principal. If you want to know more about the mathematician who determined that this procedure would work, go to Archimedes.

Thus, by subtracting the weight of the clod in water from its weight in air, we obtain the weight of the water displaced by the clod, which equals the volume of water displaced by the clod (because 1 cm³ of water=1 gm. of water) and is equal to the volume of the soil clod.

Clod Method of BD Determination =Mass of Clod ÷ Volume of water displaced.

Volume of water displaced=weight of water or the (gms clod in air)-(gms clod in water)=(grams or volume of water displaced).

Cylinder method of BD Determination=Mass of oven dry soil (gms) ÷ total volume of soil (cm³).

Another method of determining bulk density is by using soil cores. These are obtained with a probe that forces a cylinder into the soil. The soil can be removed from this cylinder and cut into sections and placed in another cylinder for transportation back to the laboratory. Care needs to be taken to avoid compacting the soil during and after obtaining the core.

BD of soil (using core method)={(mass of soil + mass of core)-(mass of core} ÷ (volume of core)

volume of core=Pi (3.1416) r^2 × h

r=radius of core; h=the height of core.

For the cores in the lab, r=2.45cm (½ of Diameter); h=5 cm

V=3.1416 × 2.45^2 × 5=94.3 cm^3;

Bulk density is a fairly easy determination that will yield information about the soil that will be significant for determining the soil's potential for plant growth or a building foundation. Remember it is the oven dry mass/volume soil. or g/cc.

Significance of Bulk Density

The bulk density of the soil will play an important role in determining if the soil has the physical characteristics necessary for plant growth, building foundations or other uses. From the laboratory investigations you will obtain very high bulk densities for the Bt and sandy soil. The weight of the soil is important if you are going to be lifting it or hauling it long distances. One of the main reason for sod farms to be on organic soils is to reduce the cost of transportation because the organic soil is so much lighter than mineral soils. The organic sod is easier to handle. Most sod farms in Minnesota are on the peat soils north of the Twin Cities.

Sometimes we are interested in the weight of an acre of soil for erosion comparison purposes. Erosion of 5 tons per acre sounds like a lot. However the weight of an acre furrow slice on average is 2,000,000 lbs., so 5 tons per acre seems like a small amount. Five tons per acre is only about 34 thousandths of an inch thick (.034 inches). However, many areas have lost 10 to 100 times that amount. (Note : an acre furrow slice, or AFS, is a three-dimensional volume of soil which is one acre in area and 7 inches deep.)

Weight per AFS=BD × Volume

Example : Volume of soil=Soil depth × 43560 ft.^{2} (area of 1 acre)

For conversion of g/cc to lbs/ft^3, multiply BD × 62.4 lbs/ft^3 (weight of ft^3 water)

How many tons are lost if a soil with a B.D. of 1.2 g./ cm^3 erodes 5 inches per acre?

Answer : (1.2 × 62.4)=(weight in lbs/ft^3) X Volume : (where Volume=(5 in. ÷12 in/ft) × 43560 ft)^2 or (74.88 lbs/ft^3) × 18,150 ft^3=1.3 million lbs or divide by 2000 lbs/ ton or weight=679 tons.

Soil Factors Impacting Tillage

Another factor similar to carrying soil is its potential to be moved short distances, such as in plowingor rototilling. The soil that is heavier will be more difficult to move. However, another factor is the ability of the soil to stick together, or the soil's consistence. Clay soil is noted for its stickiness and large energy requirements in tillage. Farmers refer to it as "heavy," but they really mean it is difficult to plow, not that it has a high bulk density; clay soils generally will have a lower bulk density than sandy soils (clay=1.3 g/cc vs sand=1.6g/cc). Sandy soils have a higher bulk density, but are easier to plow since they have weaker consistence. Thus they are often referred to as "light soils." The lower B.D. of clays is due to their better aggregation. Rototilling the soil reduces the bulk density by "fluffing" the soil. For a look at tillage implements go to Tillage Implements

Soil Consistence

Soil consistence is the soil's ability to cohere or stick together. The soil's consistence may be evaluated at three moisture conditions : air dry, moist, and wet. Moist consistence is evaluated by placing the soil between the thumb and forefinger and gently applying pressure. The ease with which a ped can be crushed determines the consistency.

Terms commonly used to describe moist consistence are :

- *Loose*-Non-coherent when dry or moist; does not hold together in a mass.
- *Friable*-When moist, crushes easily under gentle pressure between thumb and forefinger and can be pressed together into a lump.
- *Firm*-When moist, crushes under moderate pressure between thumb and forefinger, but resistance is distinctly noticeable.
- *Plastic*-When wet, readily deformed by moderate pressure but can be pressed into a lump; will form a "wire" when rolled between thumb and forefinger.
- *Sticky*-When wet, adheres to other material and tends to stretch somewhat and pull apart rather than to pull free from other material.
- *Hard*-When dry, moderately resistant to pressure; can be broken with difficulty between thumb and forefinger.
- *Soft*-When dry, breaks into powder or individual grains under very slight pressure.
- *Cemented*-Hard; little affected by moistening.

In air dry conditions, the resistance to rupturing when rubbed is measured. At intermediate moisture content, the soil's resistance to shearing forces by thumb and finger is noted. In the wet condition, its plasticity — ability to be molded and stickiness — are measured. One significance of the B.D. of a soil is the amount of surface area a soil has. The more surface area, the more ability to retain water and nutrients. Notice that the two soils pictured below weigh the same, but are significantly different in surface area. Which soil has the higher B.D.?

Compaction, Porosity and Soil Temperature

Soil Compaction

Soil compaction works something like the soil particles seen below. When the soil is randomly fluffed, as on the left, the amount of pore space will be 45 to 55 per cent. When we push the soil particles closer together, we remove the pore space, and the bulk density will increase. Compaction increases the bulk density by reducing the pore space.

Compaction can also result in the changing of the proportion of pore sizes. Notice in the compaction diagram that compacted soils not only have a lower total pore space (resulting in higher bulk density), but also less macro pores and more micro pores.

This will often result in excess water retention.

When we walk on the soil we compact it. Eventually the compaction prevents the growth of plants. This trail TRAIL behind the Soils Building shows the effect of compaction and samples from the soil will be used in the lab to determine the amount of compaction.

Compaction changes the amount and size of pores.

Deep plowing can reduce compaction below the plow zone.

C.O.L.E.

One of the important engineering properties of soils is related to the clay content and consistency and is called the coefficient of linear extensibility or C.O.L.E. It is a measure of the shrink-swell potential, or the volume change of the soil with changes in moisture content. If the C.O.L.E. exceeds .09, significant shrink-swell potential can be expected.

Soils with high C.O.L.E. values can cave in basements, displace walls, and break gas or water pipes. Knowing which soils have high C.O.L.E. values is an important step in selecting sites for construction activities. Compacting soils reduces the ability of soils to take on water and reduces their shrink/swell potential. C.O.L.E.=(Length Moist ÷ Length Dry)-1.

SOIL COMPACTION AND BUILDING

Soil compaction is defined as the method of mechanically increasing the density of soil. In construction, this is a significant part of the building process. If performed improperly, settlement of the soil could occur and result in unnecessary maintenance costs or structure failure. Almost all types of building sites and construction projects utilise mechanical compaction techniques.

Soil density is used almost exclusively by the transportation industry to specify, estimate, measure, and control soil compaction. Soil density can be easily determined via weight and volume measurements. The objective of compaction is to stabilise soils and improve their engineering properties. Source : US Dept. of Transportation.

There are five principle reasons to compact soil :-

*   Increases load-bearing capacity-
*   Prevents soil settlement and frost damage-
*   Provides stability-
*   Reduces water seepage, swelling and contraction-
*   Reduces settling of soil-source Soil Compaction Handbook.

Porosity

Bulk Density is an indirect measure of soil pore space. In fact, there is a formula : 100-((B.D. ÷ Particle Density) X 100)= per cent pore space or POROSITY. Under field conditions, pore spaces are occupied at all times by air and water. "Tortuous pathways" best describes soil pores. Porosity, when expressed as a per cent, is the same thing as per cent pore space. Soil particles have irregular shapes, and thus the spaces or pores between them vary irregularly in size, shape and direction. Sandy soils have large continuous pores, while clays have small pores which transmit water slowly. Clays, however, contain more pore space than sandy soils, because of the pores inside of the soil peds. To growing plants, pore sizes are of more importance than total pore space.

Particle Density

In the laboratory you will determine porosity and particle density for the sandy soil used previously. Particle density will also be determined for this soil. Particle density only takes into account the volume occupied by the solid particles. It excludes the volume occupied by air and water. Since a large portion of most soils is composed of particles derived from quartz minerals, the particle density of most soils is near 2.6 g/cc, which is the density of quartz. Variations in the particle density are due to the presence of heavier minerals like iron oxides or lighter organic components.

Significance of Porosity

Above, a weakly developed soil is on the left and a well developed soil is on the right. Where clay accumulation is minimal, the macro pore space is about the same with increasing depth. However, where clay has moved into pores, the amount of macro pores and the rate of water movement both decrease. Soils with clay films will have slow water flow through the Bt horizon.

Some soils are difficult to aerate by plowing due to the turf cover. For lawns, playgrounds, or football fields, aeration is accomplished by making small holes in the turf with solid tines or using hollow tines to pull out a small core. Later the holes can filled with sand to promote air exchange. This is called *aerifying* and *topdressing*.

Many home lawns are compacted because of the construction activities when the home was built or due to people walking or riding on the turf when the soil is wet. Aerification will improve the turf growth on these lawns.

Aeration Equipment=Tines & John Deere

Soil Temperature Regimes

The temperature of a given soil at a given time is dependent upon the gains and losses of heat energy. Generally, dark surfaces will absorb more heat than light surfaces. However, the amount of water in the soil is an important factor. Losses of the absorbed heat are by radiation back into the atmosphere as long-wave radiation, heating the air and cooling the soil. These soils warm quickly in the spring due to their dark surface.

Soil temperature regimes are used to classify soils. They are defined according to the average annual soil temperature in the root zone. The use of soils for agriculture and forestry is closely related to soil temperature, due to the specific requirements of plants. Over most of the earth, daily soil temperatures below 50 cm deep seldom change. To approximate the mean annual soil temperature, 2 ° F is added to the mean annual air temperature. If we do this for the annual temperature map of Minnesota and follow the 45°F line, which would be 47° soil temperature, the state is divided along a line from the Twin Cities to the South Dakota border, and soils are frigid to the north and mesic to the south.

SOIL ANALYSIS AND SOIL NUTRIENT MANGEMENT

High yields of top-quality crops require an abundant supply of 16 essential nutrient elements. In addition to providing a place for crops to grow, soil is the source for most of the essential nutrients required by the crop. Our soil resource can be compared to a bank where continued withdrawal without repayment cannot continue indefinitely. As nutrients are removed by one crop and not replaced for subsequent crop production, yields will decrease accordingly. Accurate accounting of nutrient removal and replacement, crop production statistics, and soil analysis results will help the producer manage fertilizer applications.

A soil analysis is used to determine the level of nutrients found in a soil sample. As such, it can only be as accurate as the sample taken in a particular field.

The results of a soil analysis provide the agricultural producer with an estimate of the amount of fertilizer nutrients needed to supplement those in the soil. Applying the appropriate type and amount of needed fertilizer will give the agricultural a more reasonable chance to obtain the desired crop yield.

Objectives of Soil Analysis

- To provide an index of nutrient availability or supply in a given soil. The soil extract is designed to evaluate a portion of the nutrients from the same "pool" used by the plant.

- To predict the probability of obtaining a profitable response to fertilizer application. Low analysis soils may not always respond to fertilizer applications due to other limiting factors. However, the probability of a response is greater than on a high analysis soil.
- To provide a basis for fertilizer recommendations for a given crop.
- To evaluate the fertility status of the soil and plan a nutrient management programme.

Chemical analysis of plant composition indicates chemicals or elements present in a crop at maturity or when it is harvested. For example, 1,250 lb of lint cotton contains approximately 125 lb of nitrogen (N), 20 lb of phosphorus (P), and 75 lb of potassium (K).

The essential question in fertilization is, "How much nutrient must be added to the soil as fertilizer for a given amount to be taken up by the growing plant?" The crop utilises only a portion of the available nutrients in the soil. This means that more nutrients must be present than are removed by the crop. The amount added varies according to the level already present in the soil and the crop's need for the nutrient involved. The soil analysis is the starting point, since it measures the level or content presently in the soil.

The soil analysis along with the information provided in the information sheet, is interpreted and reported in terms of the nutrients needed to supplement those in the soil. With this information, producers can add sufficient nutrients for the correct balance to obtain high yields.

Limiting Factors

Crop yields are determined by a variety of factors including crop variety selection, available moisture, soil fertility, crop adaptation to the area, and the presence of diseases, insects, and weeds. The soil analysis and its interpretation deal only with the fertility level (plant nutrients) of the soil. Recommended fertilizer will provide sufficient nutrients for the best possible yields. Other factors of production or management may still cause low yields, even though nutrients are adequate.

Carryover

If yields are only partial in relation to a large amount of fertilizer applied, many of the nutrients are carried over for use by the next crop. It is this carryover, or residual effect, from one year to the next that makes heavy fertilizer applications practical in the face of other limits to yield.

Yields to Expect

A certain fertilizer application cannot be expected to produce a specific yield such as two bales of cotton or nine tons of hay. It is more realistic to assume that a balanced fertilizer programme assures that nutrients are not the limiting factor in yields obtained. Research has shown that producers who use a balanced fertilizer programme obtain consistently better yields than those who don't.

The Soil Analysis Report

After the soil is analysed, fertility recommendations are made based on amounts of actual nutrients in the soil, not on the amount of any particular fertilizer or mixture. For example, if 100 lb of N were recommended, that amount could be supplied by approximately 300 lb of ammonium nitrate (33 per cent N), 220 lb of urea (45 per cent N), or 120 lb of anhydrous ammonia (82 per cent N). Likewise, a recommendation of 60 lb of P205 per acre could be added as 133 lb of 45 per cent triple superphosphate.

Fertilizer Labelling

Nitrogen is expressed on the elemental basis as "total nitrogen" (N). Phosphorus is expressed on the oxide basis as "available phosphoric acid" (P205). Potassium is expressed as "soluble potash" or potassium oxide (K20).

In reality, there is no P205 or K20 in fertilizers. Phosphorus exists most commonly as mono-calcium phosphate, but also occurs as other calcium or ammonium phosphates. Potassium is ordinarily in the form of potassium chloride or sulfate. Furthermore, P205 and K20 are not absorbed by plants. Plant roots absorb most of their phosphorus in the form of orthophosphate ions, H2P04-, and most of their potassium as potassium ions, K+. For these reasons, the elemental expression (N-P-K) is used in all of the recent research publications. Conversions from one form of P and K to another can be made using the following formulas.

per cent P= per cent P205 x0.437　　　　per cent K= per cent K20x0.826

per cent P205= per cent Px2.29　　　　per cent K20= per cent Kx1.21

INTERPRETATION OF THE SOIL ANALYSIS

The soil analysis report contains two parts : characterisation and fertility status of the soil, and fertility recommendations. Soil characterisation (pH, texture, per cent exchangeable sodium, per cent organic matter, and salinity expressed as electrical conductivity) is explained in the report. The fertility status is reported as nutrients available to the plant. The second part, fertility recommendation, contains the suggested amounts of fertilizer to apply. These amounts are based on the crop requirements, management practices affecting the crop (as shown in the information sheet), the present fertility level of the soil, and the yield goal desired by the producer. Special notification is given if the tests indicate that a salt or sodium hazard exists or if the information provided shows any other specific problems.

Soil amendments or treatments to reduce a sodium or salt hazard will be recommended if requested. In general, application of gypsum is suggested for reducing a sodium hazard, and leaching is recommended in most cases to lower salt content in the soil. Gypsum or leaching requirements are calculated and reported if requested.

Soil Analysed

There are many soil testing laboratories in New Mexico, Texas, Colorado, and Arizona. Basic soil testing packages vary in price and number of analyses. Many labs are participating in the Western Region Soil Testing Proficiency Programme. Programme participants share identical soils and compare results quarterly. This process assures the clients that the lab is striving for consistency and accuracy in lab analyses. Recommendations will undoubtedly vary from lab to lab. Often the best recommendation will come from the local Extension service. The choice of labs is at the client's discretion but should be based on report readability, result accuracy, turn-around time, and cost factors. New Mexico specialists can assist with many questions regarding plant health. Remember, a soil analysis is only as good as the soil sample taken.

Nutrient Management Planning

Soil provides a reservoir of nutrients required by crops and also therefore for animals but not necessarily at optimum levels of immediate availability to plants. The purpose of soil analysis is to assess the adequacy, surplus or deficiency of available nutrients for crop growth and to monitor change brought about by farming practices.

This information is needed for optimum production, to avoid transferring undesirable levels of some nutrients into the environment and to ensure a suitable nutrient content in crop products. Farm assurance schemes, buyer's protocols and codes of practice are increasingly demanding more accurate fertilizer recommendations which must depend on the nutrient-supplying capacity of the soil. Regular soil analysis should be undertaken as a vital part of good management practice.

Chapter 14

Soil Temperatures and Conditions

TEMPERATURE

Soil temperatures are affected by a forest cover in the same way as air temperatures but to a much greater degree. Mean daily maximum temperatures at the soil surface may be reduced in summer by as much as 50° F. and absolute maxima even more. The influence of the forest varies inversely as the distance from the surface, but under some conditions may be recognisable to a depth of 30 feet.

Minimum temperatures show a similar but less pronounced increase. This influence is particularly noticeable in the winter, when soils usually freeze later and less deeply in the forest than in the open. Sometimes the soil in the forest remains unfrozen while that in adjacent open fields is frozen to a considerable depth. On the other hand, in situations where there is a heavy blanket of snow in the open and a very light blanket under a dense forest canopy, this relationship may be reversed.

These influences are due partly to the reduction of insolation and radiation by the overhead forest cover and partly to the insulating effect of the litter and humus of the forest floor. These reduce maximum temperatures more than they increase minimum temperatures and hence the net effect is a cooling one. In both cases the effect is beneficial. During the summer young plants are protected from high temperatures which in exposed situations in the open may be so extreme as to kill the living tissue; while during the winter reduction in the depth and period of freesing diminishes surface run-off by permitting the penetration of more water into forest soils.

MOISTURE CONTENT

Forests tend to decrease the amount of moisture in the soil by intercepting precipitation, by retaining water in the forest floor and by transpiration. They tend to increase the amount by reducing surface run-off, by increasing the permeability of the soil and by decreasing evaporation from it.

Much precipitation in a forest never reaches the ground because it is first intercepted by and then evaporated from, the leaves, twigs, branches and trunks. This loss may run as high as one tenth of an inch per shower. The amount of water that actually reaches the ground may vary from none in light showers to the great bulk of the total precipitation in heavy rains. Interception naturally increases with the density of the stand and of the foliage and is much greater with deciduous trees in summer than in winter.

When the precipitation reaches the ground it is first absorbed by the forest floor, which has a field moisture capacity (that is, an ability to retain water against the pull of gravity) of from one to five times its own dry weight and when saturated contains much larger quantities. Its absorptive and storage ability is much greater than that of mineral soil, the field capacity of which usually ranges from 10 to 50 per cent of its dry weight. The thicker and the more decomposed the litter and humus of the forest floor, the more water it will hold.

The amount of water that actually gets through to the mineral soil may vary from zero to 100 per cent, depending on the amount of precipitation and on the condition of the forest floor. With light showers in times of drought all of the precipitation may be retained by the forest floor; while with heavy showers and in wet periods little or none may be so retained.

On land without vegetative cover and particularly with heavy soils, the surface layer tends to become hard and impermeable, with marked reduction in the amount of precipitation that sinks into the main body of the soil. In the forest, on the other hand, the litter slows down surface run-off and thus gives more time for the water to be absorbed by the humus, which acts as a huge sponge. More important still, the humus prevents muddying of the underlying mineral soil, which remains porous and thus absorbs and retains much more water than does exposed soil in the open.

Enormous quantities of water are removed from the soil by transpiration. A well-stocked forest of mature trees uses more water than does any other form of vegetation. It may require from 100 to 1,400 pounds of water for every pound of dry matter produced. Transpiration is far greater in summer than in winter and during the growing season is somewhat greater with broadleaf deciduous trees than with conifers. Its tendency is obviously to reduce the moisture content of soil in the forest as compared to that in the open. On the other hand, the reduction of evaporation in the forest, which was mentioned in the discussion of atmospheric conditions, tends to maintain the moisture content at a higher level than in the open.

The net effect of these conflicting influences varies widely with the character of the topography, the soil, the climate and the forest. So far as any generalisation is possible, it may be concluded that soils under a forest cover tend to be somewhat drier than similar exposed soils in the open on fairly level ground and somewhat moister on steep slopes where more water is lost by surface run-off in the open than by interception and transpiration in the forest.

Erosion and Stream Flow

Of all the influences of the forest on its environment, the most obvious and the least controversial is the reduction of surface run-off of water. This prevents or substantially checks soil erosion and modifies stream flow.

COMPOSITION AND STRUCTURE

Forests have a marked influence on the composition and structure of the underlying soil. Every year they add to the forest floor large quantities of leaves, twigs and branches, which under normal conditions are constantly decomposing to form humus. Part of this humus gradually mixes with the mineral soil beneath and also supplies it with soluble compounds carried downward by percolating water.

The amount of organic material in the forest floor varies widely with differences in climate, soil and vegetation. Recorded data run from less than 2 tons per acre in old-growth longleaf pine in Florida to nearly 120 tons in a forest of birch, sugar maple and spruce in New Hampshire. In general, high temperatures result in its rapid decomposition and disappearance. Ordinarily there is a gradual transition from undecomposed litter through partially decomposed duff and humus to a mixture of organic and inorganic material and eventually to mineral soil. An exception occurs in coniferous forests in the north where slow decomposition of the needles results in acid "raw humus." This raw humus does not mix well with the underlying mineral soil and in extreme cases may be peeled off like a blanket.

The usual effect of a forest cover is to improve both the chemical and the physical characteristics of the soil. Nitrogen, calcium, phosphorus, potassium and other elements are made available in larger quantities. Even more important is the action of organic materials in making the soil more friable and crumbly. They tend to make light soils, such as sands, heavier; and heavy soils, such as clays, lighter. These effects are strengthened by the constant growth and death of the tree roots. While growing they keep pushing into and loosening up new areas of soil; when dead they add organic material to the soil and leave channels through which water may percolate more readily. Porosity of frozen soil is also increased in the forest, where it is more permeable than in the open.

Accelerated Erosion

Soil is constantly being moved by water and gravity from higher to lower elevations. Under normal conditions, with an undisturbed vegetative cover, this is a slow process known as "geologic" erosion. Over the centuries it has built up fertile alluvial soils in the valley bottoms. With the removal or disturbance of the vegetative cover, a striking change takes place. Larger and larger particles of soil and even huge boulders, can now be moved and the rate of removal is greatly increased.

The beneficial process of geologic erosion is replaced by the destructive process of "accelerated" erosion. Among the harmful results are deterioration or ruin of the lands where the erosion takes place and where the coarse detritus is deposited, siltation of reservoirs, impairment of the quality of water for municipal and industrial uses, destruction of fish habitats, clogging of river channels and increase in the volume of floods.

All kinds of vegetation serve to check erosion, but by far the most effective are well-stocked forests and well-sodded grasslands. Their influence is primarily due to their ability to reduce the amount and velocity of surface run-off.

The amount of material that water can carry at a given velocity is directly proportional to its volume. As the velocity increases, however, there is a much more than proportional increase in the cutting and carrying power of a stream. Thus, if the velocity of water is doubled, its cutting power is increased fourfold, its carrying power thirty-two-fold and the size of the material it can carry sixty-four-fold. These facts explain the occurrence of "mud-rock" flows, the transportation by small streams of boulders weighing many tons and up to a six-thousand-fold increase in the rate of erosion following destruction of the forest and the forest floor.

During ordinary storms much, if not all, of the precipitation is absorbed by the forest litter and humus, retained a short time and then passed on gradually to the mineral soil beneath. The humus layer is not comparable, as has sometimes been alleged, to a blotter on a slate roof, the efficacy of which in storing water ceases as soon as it becomes saturated. Except in "raw humus," there is no sharp line between the organic surface and the inorganic subsurface material in the forest.

On the contrary, there is a gradual transition from the top layer of newly fallen, undecomposed litter to partially and completely decomposed humus, to a decreasing mixture of organic with inorganic material and finally to mineral soil. Thus there is a constant and uninterrupted downward movement of water from the temporary storage provided by the upper layers of the forest floor into the larger and more permanent soil reservoir beneath. From there the water moves slowly as a subsurface flow to springs, streams and lakes, from which it is in time evaporated and again precipitated in the course of the hydrologic cycle. Tree roots also help to keep the soil from being washed away in gullies and along the banks of streams and tend to check the occurrence of landslides. Erosion by wind is virtually non-existent in the interior of a forest and is greatly reduced for some distance to its lee. In many ways forests are influential in preventing the development of accelerated erosion — the greatest threat to the world's soil resources. Fortunately their influence is most effective where the danger is most acute, as on steep slopes, in clayey soils and with heavy rainfall. Wind erosion can also be prevented or greatly reduced by properly placed windbreaks in open country, as in the Great Plains of the United States.

Floods

What effect these influences of the forest have on floods is a hotly debated question. Certainly forests cannot prevent an abnormally high flow of streams when heavy precipitation falls on saturated or frozen soils. If the storage basin is already full, there is nothing for any additional supply of water to do but to flow off by way of the nearest stream. On the other hand, when the soil reservoir is well below the saturation point, there may be far less surface run-off from forested areas than from those not covered by vegetation, even in severe storms.

By and large there can be no doubt that forests tend to reduce flood flow in small watersheds in hilly country. One would expect them to have a similar influence on large rivers made up of the flow of many small streams. Here, however, the situation is complicated by many factors, the combined effect of which it is difficult to evaluate. For example, the way in which precipitation is distributed over a river basin as a whole and the rate of flow of the tributary streams largely determine whether their peak discharges reach the main river simultaneously or in successive waves.

Obviously the first combination will result in a much higher flood than the second, even with the same or perhaps even a smaller total run-off. Therefore, the clear-cut influence of the forest on the flow of streams from small watersheds at the headwaters of large rivers may not always be reflected in the flow of the main river.

Nevertheless, the general tendency will be for the forest to decrease both maximum and total run-off and to increase minimum run-off, throughout the river basin. The action of the forest in reducing erosion also has an important influence on flood flow and flood damage. Small streams with steep slopes issuing from deforested or burned-over watersheds may carry several times as much solid material as water. One instance is on record in southern California where solids comprised 88 per cent of the total flow. As the stream flattens out, the larger and heavier materials are, of course, deposited; but even rivers with small gradients may carry heavy loads of sediment. This material adds significantly to the volume of the stream and increases its erosive power, particularly in time of flood. It may, in fact, be even more important than the volume of water as a source of damage.

Distribution of Run-Off

For all practical purposes the total stream flow from any watershed consists of the precipitation less the sum of the losses from interception, transpiration and evaporation. To the extent that the forest increases interception and transpiration, it tends to reduce stream flow; and to the extent that it decreases evaporation, it tends to increase stream flow. The net effect depends on the relative weight of these influences. It will obviously not be the same under all environments or at all seasons of the year.

Measurements from forested and from deforested or partially forested areas in Switzerland, Colorado and North Carolina have shown a strong tendency towards smaller annual run-off from the forested areas. The difference was particularly marked in the spring, when melting snow and heavy rains caused the maximum seasonal run-off. The greater discharge of water from the non-forested areas also carried with it much more eroded material.

The total flow of a stream is made up of two main parts: the surface run-off and the sub-surface or ground-water run-off. The forest has a profound effect in reducing surface run-off and thereby increasing the amount of water available for sub-surface run-off. The evidence is conclusive that in most regions surface run-off is very small or negligible from areas of undisturbed vegetation, while it may amount to half of the precipitation where the vegetation has been destroyed or seriously disturbed. However, the outstanding influence of forests in this respect is greatly weakened when the forest floor is destroyed by repeated fires, even though the trees themselves are not killed.

Reduction of surface run-off results in more uniform flow of a stream throughout the year. Maximum flow is nearly always less and minimum flow usually more from forested than from comparable non-forested areas. The tendency is to avoid sharp peaks of flow in the spring and deep troughs in the summer. The total amount of flow may be less, but its greater uniformity is highly desirable.

Watershed Management

From the point of view of water supply, the ideal management of the forest, as a part of the broader field of watershed management, is that which will produce the maximum total run-off, well distributed throughout the year, with minimum erosion.

Such management may not always be identical with that which will produce the largest supply of wood. Serious conflicts will, however, be the exception rather than the rule. When they do occur, the method of management to be selected must be based on relative values, among which water for domestic and industrial use, irrigation, power, navigation and recreation may rank high. Striking evidence of the basic importance of conserving both soil and water resources is afforded by the greatly increased attention being paid by such agencies as the Soil Conservation Service, the Tennessee Valley Authority and the Forest Service in the United States to the promotion of more effective watershed management, in which forest planting and improved forest practices play a prominent part.

Wood, the Universal Raw Material

No less prominent than the influence of the forest on climate and run-off is its influence on industry. In fact, its tangible products affect our economic activities even more clearly and more directly than do the intangible services that it renders. They provide opportunity for the profitable employment of labour and capital in many important industries and furnish a wide variety of indispensable consumers' goods.

Wood is sometimes known as "the universal raw material" because it is so widely distributed and can be used for so many purposes. Although for some uses it has been largely displaced by other materials, there are few articles into the composition, manufacture, or transportation of which it does not enter. While perhaps no more "indispensable" in modern civilization than many other materials, such as iron, cement and glass, its range of use is certainly much broader than most of them.

For thousands of years, until the advent of coal, oil and gas, wood constituted man's chief source of fuel. Even today more than half of the total consumption of wood in the world goes for this purpose. Another third in the form of lumber and structural timbers is used for the construction of buildings of all kinds. Except in urban centers and where forests are scarce or lacking, most people continue to live in wooden houses.

Lumber and dimension stock are in turn remanufactured into innumerable articles such as toothpicks, toys, sporting goods, musical instruments, handles, boats, caskets and furniture. In the United States approximately an equal amount of lumber (about one seventh of the total cut) goes into boxes, crates and dunnage used in the shipment of commodities. Large additional amounts of wood are used for railroad ties, poles, piling, fence posts, mine timbers, excelsior, shingles, cooperage, veneer and plywood. Modern adhesives are greatly helping to expand the use of wood in the fields of ply and laminated construction and are making practicable the salvage of much low-grade material. Advances in engineering techniques now make possible the use of wood in members requiring high strength and long spans, where metal was formerly regarded as indispensable. Even more spectacular advances have taken place in the chemical utilisation of wood. The demand for paper and paper products seems unlimited. Each year larger and larger quantities of wood go into the manufacture of newspaper, book and magazine paper, writing paper, wrapping paper, wallpaper, building paper, pasteboard cartons and paper board. From "dissolving pulp" come cellophane, rayon, artificial wool and many plastics. Today one could, if one chose, be clothed wholly in textiles that originated in the forest.

A lively imagination is required to visualise the many other chemical substances that can be derived from wood. Among these are wood gas, potash, charcoal, acetone, methyl alcohol, sugar, ethyl alcohol, yeast, salicylic acid and vanillin. Potentially the forest is the source of large quantities of motor fuel and food.

During World War II, when gasoline was scarce, many automobiles in Europe were operated entirely on wood gas, which is still used to a considerable extent in trucking. Another potential substitute for gasoline is ethyl (grain) alcohol produced by the fermentation of sugar, which in turn has been produced by the hydrolysis of wood cellulose. The alcohol can also be converted into lubricating oil and synthetic rubber.

Wood sugars have already been used extensively in Europe as feed for livestock and are a possible source of food for human beings. They can also be turned into yeast of high nutritive value by inoculating them with the appropriate organisms.

Edible carbohydrates and proteins are among the innumerable products obtainable from wood cellulose, which in its original state is a highly indigestible product. Lignin, the other main constituent of wood, is a comparatively unknown substance, of which little use is now made but which is likely in time to provide a wide variety of important chemical products. Many substances other than wood and its derivatives come from trees and are ready for use with relatively little further processing. Among these are nuts, maple syrup, chewing gum, tannin, dyes, rubber and resin. The last yields turpentine, universally used as a solvent in paints and varnishes and in the production of synthetic camphor; rosin, used by violin players and baseball pitchers and in the manufacture of soap, varnish and paper; pine oil, used in the flotation process of separating metals from their ores and in the textile industry in the fixation of dyes; and pitch, for which chemists are just beginning to find profitable uses.

Many pages would be required merely to list the useful substances that come from trees. Wood, once regarded by some as obsolescent, is now used more widely and for more purposes than ever before. Its inherently desirable properties of high strength in relation to weight, its workability, elasticity, non-conductivity of heat and electricity and beauty can be enhanced and its undesirable properties of shrinking and swelling, inflammability and susceptibility to attack by insects and fungi can be mitigated by methods developed by modern technology. Wood as wood, properly treated and handled, is one of the most useful materials known to man; while cellulose, lignin and resin are a veritable treasure house of chemical products already in use and yet to be discovered. The significance of forest products is emphasised further by the facts that forests are both renewable and more efficient as producers of solid substance (ligno-cellulose) than is any other crop. Here is at least one case where we can have our cake and a rich cake at that and eat it too. Industrially the age of the forest is ahead of us as well as behind us.

Forest and Wood-Using Industries

This situation makes it possible for man, through intelligent handling of the forest resources, to raise his standard of living by the continued and increasing production of a host of valuable, often indispensable, raw materials and finished commodities. It also provides widespread opportunities for the profitable employment of land, labour and capital.

SOIL RESOURCES

The soil resources in this table fall into three broad groups, reflecting their underlying inter-relationships: those related to soil structure (water use, structure, erosion) are dealt within this Chapter, those related to nutrition or biological factors.

As this *Bulletin* deals with dryland crops, water availability is the major constraint on crop performance. Table suggests that water use efficiency (WUE) is a key measurement of performance. WUE reflects water availability (seasonably of rainfall) but also, importantly for management, it integrates the influences of factors such as the volume of soil exploitable by roots.

Long-term trends with time in WUE and trends between areas provide estimates of the relative sustainability of specific cropping systems. The theoretical WUE line are less efficient or effective than they should be. One reason for such ineffectiveness is poor management during the growing season, for example weediness and nutrient deficiency. This is correctable, as shown by the vertical lines, where WUE was increased during experimental treatments. Other reasons for poor WUE are related more to fundamental issues associated with soils.

SOIL RESOURCES AND CROPPING SYSTEMS

A traditional view of the influence of soil is that it provides an opportunity, or a constraint on the type of cropping system that can be implemented and its productivity. A more responsible view is that 'the soil' combines various properties which interrelate and are directly influenced by the procedures of cropping. A non-sustainable cropping system, in which soil resources are declining through changes in surface properties, sub-soil compaction, loss of organic matter and reduced biological activity. It shows that each of these properties may affect the others and that all can directly and indirectly reduce plant performance, as well as affecting other aspects including erosion, salinisation and acidification. Some of the mechanisms underlying these inter-connections.

Soil Pores and Water Characteristics

Soils are made up of three parts: mineral materials, organic matter and space (called pore-or void-volume). The relative importance of these varies with soil type but pore space can occupy about half the volume of a medium-textured soil. At optimum water content for plant growth, approximately half the pore space is filled with water and half with air. The proportions of water and air can change rapidly depending on weather, evapotranspiration and other factors. The dimensions (size, shape and arrangement) and number of pore spaces are most important in determining soil water and soil structure.

Porosity is the volume of soil voids (*pore* space). It is expressed in relation to the bulk volume of the soil. The water holding capacity of a soil depends on its porosity, and the size distribution of its pores. Small pores retain water at greater suctions than larger pores. The moisture (or water) potential is the amount of energy required to remove water from a soil; field capacity is the water-holding capacity after a free-draining soil has been allowed to drain.

Available soil water (ASW) is the amount of water which is available for uptake by plants, namely that held at suctions between wilting point and field capacity. It varies with soil type and can be correlated with the clay content and structural arrangement of the soil. It varies also with soil treatment because the size and distribution of pores in the topsoil reflects surface exposure, normal seasonal wetting and drying, and management. Williams *et. al.* (1983), studying the water content of 244 soil samples, found that the ASW of well-structured soils was one-third to twice as large as that in comparable (similarly-textured) poorly structured or degraded soils. Bearing in mind that ASW varies with natural weathering and management.

Hydraulic conductivity (K) of a soil is its conductivity to movement of water down a pressure gradient. High values of K are associated with well-structured soil and contiguous pores; they allow high infiltration rates and rapid drainage. Earthworm channels, which can have populations of 500 m^{-2} in Mediterranean climates, and continuous deep voids left by dead roots (5-10 000 m^{-2}) contribute greatly to hydraulic conductivity. Hydraulic conductivity varies with soil type and management. K values below 10 mm/h are low and likely to cause run-off following rainfall or problems with irrigation, given that steady rain falls at about 10 mm/h. K values of 10 to 20 mm/h can give intermittent run-off (a downpour falls at about 50 mm/h) while values up to 120 mm/h are associated with occasional, increasingly rare run-off. Values above 120 mm/h may facilitate regular drainage to the groundwater, causing potential problems for heavily-fertilized soils, and those treated with effluent, herbicides or pesticides.

Both soil water content and saturated hydraulic conductivity generally relate to the number and continuity of pores, particularly the larger macro-pores. It is, however, difficult to measure these soil attributes and they are highly location-specific, so that variability is great and they sometimes have little interpretive value. Moran *et. al.*, (1988), however, in a study of a soil in a wet-and-dry environment, show that a soil treated with minimum tillage had more pores, identified directly by image analysis, and higher hydraulic conductivity, measured in the field, than did a similar soil traditionally cultivated. The impact of management on soil pore soil water characteristics.

Surface sealing and crusting are common in wet-and-dry climates. Sealing increases run-off and seriously reduces the amount of water infiltrating into the soil thus reducing the water held in the soil. Sealing can also increase ponding at the surface and thereby evaporation.

Infiltration rates are often reduced 1000-fold by crusting. The crust can have a skin with a conductivity of only about 0.1 mm/h, able only to accommodate the lightest rate of precipitation (fine mist) and commonly overlies a layer of poorly-aggregated material which also has a conductivity substantially lower than that of the underlying soil. Chase and Boudouresque (1989) and Chase *et. al.*, (1989) illustrate the impact of crusting in the Sahel on increased run-off from soils. They show how run-off may be reduced, and the depth of wetting increased, by covering the soil surface with mulch.

These solid fractions contribute to the consistence and strength of the soil, and their packing determines bulk density.

Bulk density is a measure of the packing or compression of the three constituents of soil. Just as the inherent bulk density of a soil will vary by 30 per cent according to its constituents, so the limiting values of bulk density for root penetration will range from about 1.4 g cm^3 in a soil of clay texture to 1.8 g cm^3 in a sandy one.

Soil strength is the resistance of soil to shearing or structural failure. This reflects the friction which is built up between the soil and an implement, and depends on the density, and the roughness and shape of the soil particles. The shear strength of an individual clod decreases with wetting but, more importantly, the strength of the bulk soil increases with increasing moisture to about the lower plastic limit (known to field operators as the 'sticky point'), at which each particle is surrounded by a film of water which acts as a lubricant. Soil strength drops sharply from that point to the upper plastic limit, where the soil becomes viscous. The difference between the moisture content at the upper and lower plastic limits, termed the plasticity index, is an index of the workability of the soil. A large range or high plasticity index implies a need for large amounts of energy to work the soil to a desired tilth.

Chapter 15

WATER USE EFFICIENCY IN SALINE SOILS UNDER COTTON CULTIVATION IN THE TARIM RIVER BASIN

Xiaoning Zhao[1,*], Hussein Othmanli[1], Theresa Schiller[1], Chengyi Zhao[2], Yu Sheng[2], Shamaila Zia[3], Joachim Müller[3] and Karl Stahr[1]

[1] Institute of Soil Science and Land Evaluation, University of Hohenheim, Emil-Wolff-Str. 27, Stuttgart 70593, Germany; E-Mails: husseinothmanli@hotmail.com (H.O.); theresa. schiller@gmx.net (T.S.); karl.stahr@uni-hohenheim.de (K.S.)

[2] Key Laboratory of Oasis Ecology and Desert Environment, Xinjiang Institute of Ecology and Geography, Chinese Academic of Science, Urumqi 830011, China; E-Mails: zcy@ms.xjb. ac.cn (C.Z.); shengyu@ms.xjb.ac.cn (Y.S.)

[3] Institute of Agricultural Engineering, Hohenheim University, Stuttgart 70593, Germany; E-Mails: shamailazia@googlemail.com (S.Z.); joachim.mueller@uni-hohenheim.de (J.M.)

[*] Author to whom correspondence should be addressed; E-Mail: xiaoningzhao2012 @gmail.com; Tel.: +49-711-459-239-80; Fax: +49-711-459-231-17.

Academic Editor: Markus Disse

ABSTRACT

The Tarim River Basin, the largest area of Chinese cotton production, is receiving increased attention because of serious environmental problems. At two experimental stations (Korla and Aksu), we studied the influence of salinity on cotton yield. Soil chemical and physical properties, soil water content, soil total suction and matric suction, cotton yield and water use efficiency under plastic mulched drip irrigation in different saline soils was measured during cotton growth season. The salinity (mS cm^{-1}) were 17–25 (low) at Aksu and Korla, 29–50 (middle) at Aksu and 52–62 (high) at Aksu for ECe (Electrical conductivity measured in saturation-paste extract of soil) over the 100 cm soil profile. The soil water characteristic curves in different saline soils showed that the soil water content (15%–23%) at top 40 cm soil, lower total suction power (below 3500 kPa) and lower matric suction (below 30 kPa) in low saline soil at Korla had the highest water use efficiency (10 kg·ha^{-1}·mm^{-1}) and highest irrigation water use efficiency (12 kg·ha^{-1}·mm^{-1}) and highest yield (6.64 t·ha^{-1}). Higher water content below 30 cm in high saline soil increased the salinity risk and led to lower yield (2.39 t·ha^{-1}). Compared to low saline soils at Aksu, the low saline soil at Korla saved 110 mm irrigation and 103 mm total water to reach 1 t·ha^{-1} yield and increased water use efficiency by 5 kg·ha^{-1}·mm^{-1} and 7 kg·ha^{-1}·mm^{-1} for water use efficiency (WUE) and irrigation water use efficiency (IWUE) respectively.

Keywords

Salinity; soil matric suction; soil osmotic suction; water use efficiency; Tarim River Basin.

1. INTRODUCTION

The Tarim River Basin, the most important location for Chinese cotton production (corresponding to 3.7% of the world cotton production [1]), as a result of exploitation, gained attention because of serious environmental problems developing over the last 50 years: serious degradation of soil (more than 12×10^3 km^2 of land desertification; approximately 112 T_g of organic carbon was released into the atmosphere from 1970 to 2000 in the Tarim River basin [2]); increased water salinity (maximum salt concentration of the irrigation water increased between 1960 and 1998 from 1.3 to 7.8 g·L^{-1} [3]); water resource degradation (a 4–6 m drop in ground water levels from the 1960s to 1980s [2]; approximately 300 km of the Tarim River's lower reaches ran dry between the 1950s and 1970s, including the previous terminal lake Lop Nor [3]; arsenic concentration in the Tarim River was 4.2 times higher than international limits due to the use of pesticides [4]); and plant coverage reduction (*Populus euphratica* (Salicaceae family) forest acreage and biomass declined by 67% and 50% respectively from 1958 to 1978 and 3820 km^2 of *P. euphratica* forest, and 200 km^2 of shrub- and grassland were lost in the lower reaches between the 1950s and 1990s [2]). Accumulative salt and gypsum in the Tertiary sediments induced the big amount of saline and alkaline soil in Xinjiang (71.61 $\times$ 10^4 ha), which occupied 33.26% of all agricultural fields [5]. The salinity in surface water moved to the basin. Soil salinity increased with increasing irrigation water salinity levels also in plastic mulch drip irrigation [6].

"The main causes of Tarim River desiccation were the increase in the irrigated area of the headstream section in the upstream region, the rise in water consumption in the upper and middle reaches, and the construction of reservoirs in the mountain area" [7].

A cotton irrigation experiment demonstrated that drip irrigation under a cover of plastic mulch is an effective way to protect from unproductive soil evaporation and that a mild water deficit during the budding stage could significantly enhance cotton fiber yield and improve water use efficiency [8]. Plastic mulching significantly increased the harvesting of rainwater and significantly increased yield [9], An experiment in Shihezi University under varying soil water content, with 90%, 75%, 60% of field water capacity, showed that a higher soil water saturation is unfavorable for the growth of the cotton root system and the yield of cotton under mulched drip irrigation in Xinjiang [10]. Research on water use efficiency of cotton in the Tarim River Basin showed that the lower limits of optimum soil water indices for high yields, water-saving, and good quality of seeding, squaring, flowering, boll-opening stage of cotton are 65%, 65%, 72%, 63% of soil water capacity (at 100 cm depth), respectively [11]. In Xinjiang, the experimental fields with different soil matric potentials at 20 cm soil depth showed that percolation and the ratio of deep percolation with irrigation water all increased with increasing soil matric potential [12]. Irrigation type, irrigation amount, and irrigation time are the factors in agricultural production which most affect water use efficiency (WUE). Irrigation of cotton in Xinjiang indicated that the flowering and budding stages were the most suitable times to supply limited irrigation water, thus significantly improving WUE by 57% [11]. At Aksu station (Xinjiang), a study of different limited irrigation (80%, 70%, 60%, 50% and 40% of field capacity) impact on winter wheat growth was conducted and showed that periods of mild soil water depletion in the early vegetative growth period together with severe soil water depletion in the maturity stage of winter wheat is an optimal limited irrigation regime in this oasis [13]. The effects of soil moisture on cotton root length density and yield under drip irrigation with plastic mulch in the same station showed that the water stress caused root length density increase in lower soil layers [14].

Germination, emergence, and early seedling growth are considered to be more sensitive to salinity than later stages of cotton growth [15]. "The key to salinity control and to irrigation sustainability is leaching and it interacts closely with crop growth, irrigation methods and soil-physical properties. Whereas most soils in the saline wasteland of Xinjiang have low permeability, which is considered critical in reclamation, their infiltration capacity tends to decrease greatly due to corruption of soil structure as soils are saturated" [16]. To control soil secondary salinization, one should mainly establish irrigation-drainage systems and reduce irrigation amount [17]. The volume of irrigation water is a key factor in controlling salt accumulation; insufficient irrigation cannot guarantee enough leaching of soil salt because of a low infiltration volume [18]. The rate of irrigation also affects the salt accumulation: The lower the drip rate (1.24 L·h^{-1}, 3 h per time), the less the salt content along the soil depth; the higher the drip rate (2.55 L·h^{-1}, 3 h per time), the greater the tendency of salt content to increase with horizontal distance [19].

Matric potential had a greater effect on organic matter decomposition than clay content [20]. The different soil matric potentials for the drip agricultural systems were studied at 20 cm soil depth in China and provided the best estimates for increasing crop yield, which included, for example, matric potentials higher than -20 kPa for cotton in Xinjiang province [12] and for oleic sunflower in Tianjing [21], -35 kPa for Radish field in the North China Plain [22], and -10 kPa for corn in Northwest China [23]. Soil water retention is influenced by soil texture [24,25] and structure [25–27], organic matter content [28,29], and bulk density [30].

The calcium carbonate content of soils in arid and semi-arid areas should also be taken into account, when available water values are estimated from textural considerations [31,32].

As research mentioned above indicates, those studies focused mainly on the effect of soil matric potential on water use, the osmotic potential on plant growth, and water use on the osmotic potential, respectively, but seldom mentioned their combination on water use under field plastic mulched drip irrigation in soils. The aims of this study were: (i) to quantify cotton agricultural hydrological features; (ii) to combine soil matric and osmotic suction on water use; (iii) to investigate water use efficiency in different saline soils under cotton cultivation as affected by plastic mulch and drip irrigation in the Tarim River Basin in China.

2. MATERIALS AND METHODS

2.1. Site Description

The experiment was conducted at the Aksu National Experimental Station of Oasis Farmland Ecosystems (40°37′ N, 80°45′ E, altitude 1028 m) and at Xinier Township, Korla City (41°35′ N, 86°09′ E, altitude 903 m), Xinjiang, located in the Tarim River Basin (Table 1, Figure 1). It is a typical temperate arid climate, with mean minimum and maximum temperature during the study period (April–November) ranging between 16.6 and 34.8 °C. There are different degrees of soil salinity and alkalinity (Table 2).

Table 1. The basic information of two experimental stations in Tarim River Basin.

Site	Location	Temp (°C)	Prec (mm)	Ele (m) a.s.l.	GWD (m)	Relative Humidity [a] (%)	Wind Speed [a] (km·h⁻¹)	Soil Type
Aksu	40°37′ N 80°45′ E	11.0	71.6	1028	2.0	50.5	5.3	Solonchak
Korla	41°35′ N 86°09′ E	12.2	100.8	903	1.4	42.8	7.7	Solonchak

Notes: Temp, annual average temperature from 1982 to 2012 [33]; Prec, annual total precipitation from 1982 to 2012 [33]; Ele, elevation; GWD, groundwater depth; a.s.l, above sea level; [a] the annual average data from 1982 to 2012 [33].

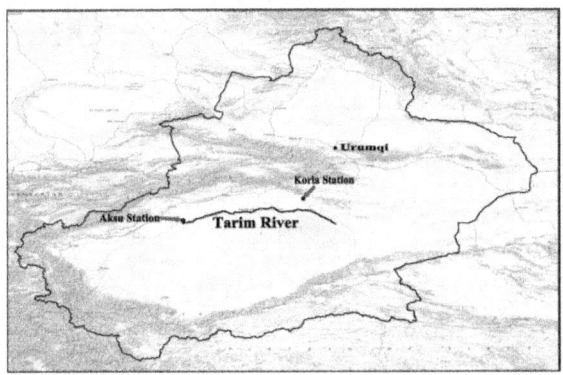

Figure 1. The location of the Aksu (left arrow) and Korla (right arrow) experimental stations in Tarim River Basin.

Table 2. The soil chemical and physical properties of different saline soils in two experimental stations in Tarim River Basin.

Soil Salinity Level	Sample Depth	CEC	BD	pH$_{H2O}$ (1:5)	EC (1:5)	ECe	C$_{org}$	N$_{tot}$	CaCO$_3$	CO$_3^{2-}$	HCO$_3^-$	Cl$^-$	SO$_4^{2-}$	Ca^{2+}	Mg^{2+}	Na$^+$	K$^+$	Clay < 2 μm	Silt 2-63 μm	Sand 63-2000 μm	Soil Texture
	(cm)	(cmol/kg)	(g/cm³)			(mS/cm)						(g/kg)$^{-1}$							(%)		
	27	2.9	1.57	7.8	1.7	23.8	4.8	1.1	116.1	0.00	0.2	0.2	2.1	0.8	0.2	0.2	0.1	2.6	38.1	59.3	Sandy loam
	52	2.0	1.55	8.1	1.5	21.0	1.7	0.9	123.5	0.00	0.3	0.4	0.6	0.3	0.1	0.3	0.1	2.6	36.0	61.4	Sandy loam
Low (Korla)	63	1.5	1.50	8.2	1.5	21.0	1.6	0.9	120.7	0.00	0.2	0.2	0.9	0.4	0.1	0.2	0.1	2.0	27.2	70.8	Loamy sand
	85	2.9	1.56	8.2	1.8	25.2	2.4	0.9	115.9	0.01	0.3	0.6	1.2	0.4	0.2	0.5	0.1	2.5	41.3	56.1	Sandy loam
Low (17–25 mS cm^{-1})	120	1.2	1.50	8.5	1.2	16.8	1.5	0.9	111.1	0.01	0.3	0.2	0.4	0.2	0.1	0.2	0.1	1.7	21.3	77.0	Loamy sand
	140	1.9	1.57	8.4	1.3	18.2	2.1	0.9	116.5	0.01	0.2	0.2	0.4	0.2	0.1	0.2	0.1	3.2	43.8	53.1	Sandy loam
	27	5.0	1.37	8.0	1.8	25.2	6.8	1.3	161.4	0.01	0.4	0.3	1.6	0.5	0.3	0.3	0.1	8.8	82.2	9.0	Silt
Low (Aksu)	38	7.4	1.54	8.2	1.4	19.6	8.7	1.4	157.1	0.00	0.4	0.2	0.7	0.3	0.1	0.2	0.1	6.8	75.2	18.0	Silt loam
	64	6.1	1.51	8.1	1.5	21.0	8.2	1.4	159.8	0.00	0.4	0.2	0.8	0.3	0.2	0.2	0.1	8.2	77.0	14.7	Silt loam
	130	1.7	1.33	8.3	1.2	16.8	2.1	0.9	67.1	0.00	0.3	0.2	0.3	0.2	0.1	0.1	0.1	2.1	71.1	26.8	Silt loam

(Contd...)

Soil Salinity Level	Sample Depth	CEC	BD	pH$_{H2O}$ (1:5)	EC (1:5)	ECe	C$_{org}$	N$_{tot}$	CaCO$_3$	CO$_3^{2-}$	HCO$_3^-$	Cl$^-$	SO$_4^{2-}$	Ca^{2+}	Mg^{2+}	Na$^+$	K$^+$	Particle Size Distribution			Soil Texture
																		Clay < 2 µm	Silt 2–63 µm	Sand 63–2000 µm	
	(cm)	(cmol/kg)	(g/cm³)		(mS/cm)				(g/kg)^1									(%)			
Middle (29–50 mS cm⁻¹) (Aksu)	35	5.6	1.52	7.5	3.5	49.0	4.4	0.3	138.7	0.00	0.1	0.5	8.2	2.5	0.3	0.8	0.1	5.4	74.7	19.9	Silt loam
	67	1.8	1.42	7.5	3.6	50.4	1.5	0.1	94.8	0.00	0.1	1.0	8.1	2.9	0.1	0.9	0.0	2.6	51.1	46.3	Silt loam
	104	5.5	1.40	7.9	2.1	29.4	2.2	0.2	161.7	0.00	0.2	0.8	1.4	0.3	0.1	0.7	0.0	3.6	70.8	25.5	Silt loam
	130	4.8	1.48	7.9	1.6	22.4	2.1	0.1	170.6	0.00	0.2	0.2	1.2	0.2	0.1	0.3	0.0	4.5	75.1	20.4	Silt loam
High (52–62 mS cm⁻¹) (Aksu)	32	2.8	1.70	7.5	3.7	51.8	2.1	0.1	100.4	0.00	0.1	1.0	8.3	2.7	0.3	1.0	0.1	4.0	57.5	38.5	Silt loam
	57	2.8	1.71	7.6	4.1	57.4	1.5	0.1	108.5	0.00	0.1	1.7	8.9	2.9	0.3	1.5	0.1	4.7	68.5	26.8	Silt loam
	85	3.8	1.39	7.6	4.4	61.6	1.8	0.1	107.9	0.00	0.1	2.1	8.8	2.8	0.3	1.9	0.0	5.1	74.6	20.3	Silt loam
	110	3.9	1.49	7.5	4.3	60.2	1.7	0.1	138.1	0.00	0.1	1.9	8.5	2.8	0.2	1.8	0.0	6.7	81.7	11.6	Silt loam
	115	4.1	n.d.	7.4	4.2	58.8	1.7	0.1	121.9	0.00	0.1	1.7	8.7	2.9	0.1	1.8	0.0	5.6	78.7	15.7	Silt loam

Notes: ^{1} the eight ions content was all total content; n.d., not identified.

2.2. Experimental Design

The cotton planting design was double rows with irrigation tubes (two tubes at Korla and one tube at Aksu) and one bare soil row (Figure 2). The salinity (mS·cm⁻¹) was 17–25 (low) at Aksu and Korla, 29–50 (middle) at Aksu and 52–62 (high) at Aksu for ECe value over the 100 cm soil profile and two replicates per treatment, in which the soil matric potential at a depth of 25, 45, 65 cm was recorded. Every treatment had three replicates for TDR (Time Domain Reflectometer) and tensiometer.

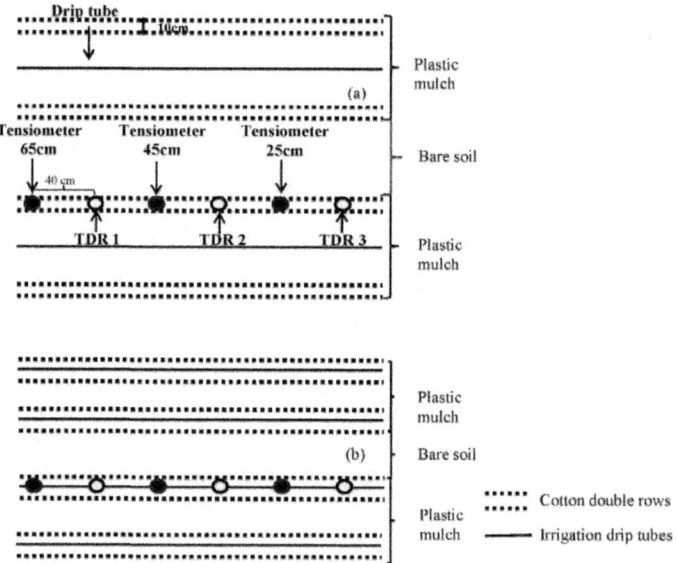

Figure 2. Field experimental design in different saline soils during cotton season from May to September 2012 in Tarim River Basin (**a**) at Aksu station and (**b**) at Korla station.

2.3. Field Sampling and Laboratory Analysis

Soil water content was measured by TDR-Time Domain Reflectometer (Time domain Reflectometry with Intelligent MicroElements-TRIME-PICO IPH) up to a depth of 85 cm. Soil water potential was monitored in the morning hours using tensiometer P-80 (Mechanical ecoTech Tensiometer, ecoTech Umwelt-Meßsystem GmbH, Bonn, Germany) at 25, 45 and 65 cm depths. The TDRs and tension meters were set up between the cotton double rows and around 20 cm from the irrigation tube at Aksu and near the irrigation tube at Korla (Figure 2). The tensiometers were filled with distilled water and readings were taken every three days at the same time as TDR measurement in the fields. Soil samples were taken during harvesting season at Aksu and Korla. All soil samples were air-dried and sieved. Soil texture was determined using a granulometer. Bulk density was determined by the cylinder method. Organic carbon was measured by potassium bichromate titrimetric outside heating method. Total N was measured by perchloric acid-sulfate digestion using LWY84B an aluminum body digestion furnace and determination of nitrogen distiller. $CaCO_3$ was measured by Gas Method. CO_3^{2-} and HCO_3^- were measured by double indicator neutralization method. Cl^- was measured by $AgNO_3$ titration. Ca^{2+}, Mg^{2+} and SO_4^2 were measured by EDTA complexometry. K^+ and Na^+ were measured by flame photometry. All above methods used the analytical methods from Soil Agricultural Chemical Analysis [34]. At harvest, numbers of cotton bolls were recorded and cotton was collected from each replicate, oven-dried at 70 °C for 24 h and weighed to estimate yield. The cotton yield calculation was from the methods of the ministry of agriculture of the People's Republic of China [35].

Cotton seed yield = plant density × average boll number per cotton plant × weight per boll × 0.85 (1)

Average boll number per cotton plant = boll with cotton + boll without cotton + 1/3 × small boll (smaller than 2 cm) (2)

2.4. Calculations and Statistical Analysis

2.4.1. Soil Water Retention

The soil water potential energy is the sum of matric potential (ψ_M), osmotic potential (ψ_O), gas pressure potential (ψ_P) and gravitational potential (ψ_Z) [36,37].

$$\psi_T = \psi_M + \psi_O + \psi_P + \psi_Z \quad (3)$$

ψ_T: the total soil water potential.

In unsaturated soils, gas pressure potential is zero and gravitational potential is a relative value from an arbitrary reference level [37], so the equation is changed to Equation (4). We selected 25 cm below the soil surface as the reference level.

$$\psi_T = \psi_M + \psi_O + \psi_Z \quad (4)$$

"The osmotic potential results from the reduction in energy of the water (relative to that of pure, free water) resulting from mixing the water with a solute" [36]. Osmotic potential is due to the solute in soil water. The $EC_{1:5}$ was converted to ECe using the following equation [38].

$$ECe = (14.0 - 0.13 \times clay\ \%) \times EC_{1:5} \quad (5)$$

The osmotic potential of soil water was determined using the following equation from the United States Salinity Laboratory [39].

$$\psi_O = -0.036 EC_{meas}\ \theta_{ref}/\theta_{act} \quad (6)$$

ψ_O: the osmotic potential (MPa) at the actual moisture content; EC_{meas}: the measured electrical conductivity (mS.cm^{-1}) of the extract at the reference water content (1:5 soil/water); θ_{ref}: the reference water content (g.g^{-1}) at 1:5 soil/water; θ_{act}: the actual moisture content (g.g^{-1}).

2.4.2. Water Use Efficiency (WUE)

The total cotton evapotranspiration (ETc) for different salinity (low at Korla, low, middle, high at Aksu) soils during cotton season from May to September

2012 in Tarim Basin was estimated using the water balance method as follows [12]:

$$ETc = I + P \pm \Delta S - R - D \tag{7}$$

I: irrigation amount; P: precipitation; ΔS: change of soil water storage in 1 m; R: surface runoff; D: downward flux below the crop root zone.

The soil water content of the soil profile (down to 80 cm) was measured by TDR during the cotton growing season in 2012, which was used for ΔS estimation, because 85% of the cotton roots were distributed in the top 30 cm of soil under mulched drip irrigation [10]. Surface runoff (R) was ignored because precipitation was not high and no gradient of movement was observed.

Then water use efficiency (WUE t $\cdot$ ha^1 $\cdot$ mm^1) and irrigation water use efficiency (IWUE t $\cdot$ ha-1 $\cdot$ mm^1) is defined by the following equations [12]:

$$WUE = Y/ETc \tag{8}$$

$$IWUE = Y / I \tag{9}$$

Y is the seed cotton yield (t ha^1) and I is the irrigation water applied (mm).

2.4.3. Statistical Analysis

Using SAS 9.1 software, one way ANOVA was used to evaluate the effects of treatments on water use efficiency. Student t test ($p \le 0.05$) was used to compare and rank the treatment means. To count the average data, two replicates were randomly located in the field except for the edge of the field. Statistic 10.0 software was used for soil water model lineal parameter estimation with Quasi-Newton estimation method.

3. RESULTS

3.1. Soil Chemical and Physical Properties of Different Saline Soils

The ECe value increased from 17 to 62 mS $\cdot$ cm^1 throughout the soil profiles. Sodium increased with the increase of EC. Using the linear relationship between EC (dS $\cdot$ m^1) and soil salt content (g $\cdot$ kg^{-1}) with equation $y = 4.6x$ (EC was the variable) in Aksu water balance station [40]. The cotton critical soil salt content, cotton threshold soil salt content, the soil salt content at the fastest rate of cotton relative yield reduction, and the soil salt content at the 50% cotton relative yield reduction were 0.302% (0.66 mS $\cdot$ cm^{-1}), 1.119% (2.43 mS $\cdot$ cm^{-1}), 0.558% (1.21 mS $\cdot$ cm^{-1}), 0.581% (1.26 mS $\cdot$ cm^{-1}) at 0-20 cm soil layer at Aksu river irrigation district respectively [41]. In the experiment, low salinity at Korla and Aksu was under the cotton soil salt content threshold; however the middle and high salinity level at Aksu were higher than cotton soil salt content threshold. Here the ECe data were much higher than the international accepted limit of 15 dS $\cdot$ m^1 for high salinity [42], but the general high level of salt in Xinjiang was also documented in the locally used limits. There were higher yield data (3.0 to 5.9 t $\cdot$ha^{-1}) with top soil EC (3-11 dS $\cdot$ m^1) in the south

Xinjiang documented [43]. The Na^+ content increased as the soil salinity level increases from 0.1 to 1.8 g·kg⁻¹. Na^+ content was also used to define the different soil salinity levels because the ions that lead to salinization increase in importance in the following order: $Mg^{2+} << Ca^{2+} < SO_4^{2-} < Cl^- = Na^+$ [44]. Soil texture in the top 30 cm soils were: sandy loam in low saline soil at Korla, silt in the low saline soil at Aksu, and silt loam in the middle and high saline soils at Aksu (Table 2).

The high saline 30 cm topsoil at Aksu had the lowest CEC (2.8 cmol·kg⁻¹), highest bulk density (1.70 g·cm⁻³), lowest organic carbon content (2.1 g·kg⁻¹), highest SO_4^{2-} content (8.3 g·kg⁻¹) and the highest calcium content (2.7 g·kg⁻¹). The low saline 30 cm topsoil at Aksu had the lowest bulk density (1.37 g·cm⁻³), the highest organic carbon content (6.8 g·kg⁻¹), the highest total nitrogen content (1.3 g·kg⁻¹), lowest SO_4^{2-} content (1.6 g·kg⁻¹), the lowest calcium content (0.5 g·kg⁻¹) (Table 2). Within the data, bulk density of 1.7 g·cm⁻³ is already high, but 1.37 g·cm⁻³ is medium.

3.2. Soil Water Retention in Different Saline Soils at Different Soil Depths

The water content was higher in low saline soil (20%–29%) than in middle saline soil (18%–24%) at Aksu and in low saline soil (15%–23%) at Korla in 0 to 40 cm depth (Figure 3), where the most cotton roots were distributed. Soil water content was highest in August in all soils. Soil water content changed strongly in the high saline soil (Figure 3d). The highest soil water content (57%) was observed in July at 70 cm depth soil of the high saline soil, while the lowest soil water content (15%) was observed in July at 30 cm depth soil of the low saline soil at Korla.

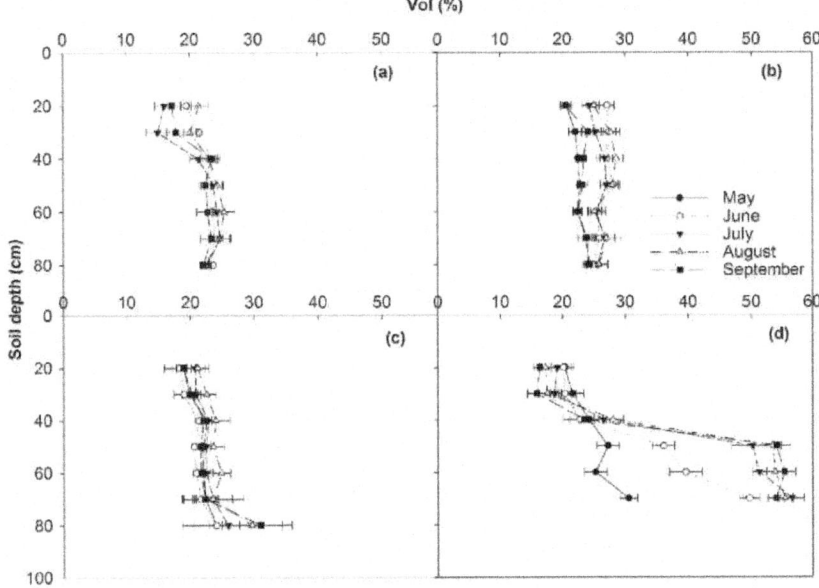

Figure 3. The soil water content in different soil depths (0–80 cm) during cotton season from May to September 2012 in Tarim River Basin (a) in low saline soil at Korla; (b) in low saline soil at Aksu; (c) in middle saline soil at Aksu; and (d) in high saline soil at Aksu.

The soil in 25 cm depth had the highest matric suction in all saline soils from May to October in 2012, which was followed by 45 cm depth and 65 cm depth because of surface evapotranspiration, new soil organic carbon input from root growth and increasing moisture in deeper soil profile (Figure 4). The soil matric suction fluctuated from 2 to 72 kPa in low saline soil and 12–52 kPa in middle saline soil at Aksu from May to October in 2012 compared to that in high saline soil (2–12 kPa) and low saline soil at Korla (5–31 kPa) (Figure 4). In the 25 cm soil depth, the matric suction was highest (72 kPa) and lowest (2 kPa) in low saline soil at Aksu compared to the other soils (Figure 4). The matric suction power of the high saline soil remained constant in all three soil depths because of the high soil water saturation problems (Figure 4).

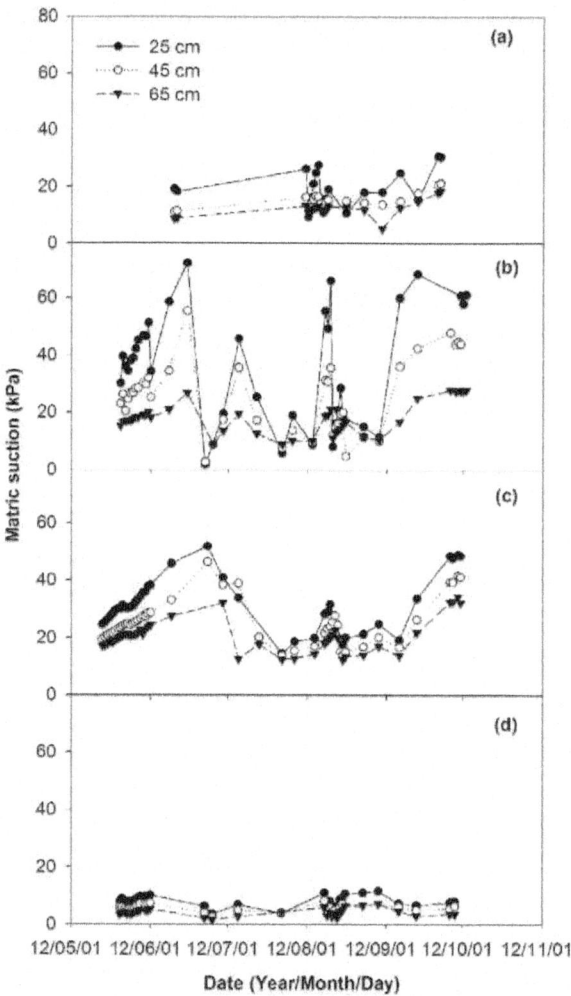

Figure 4. The soil matric suction in different soil depths (25 cm, 45, 65 cm) during cotton season from May to October 2012 in Tarim River Basin (**a**) in low saline soil at Korla; (**b**) in low saline soil at Aksu; (**c**) in middle saline soil at Aksu; and (**d**) in high saline soil at Aksu.

The soil salinity level mainly effects osmotic suction and the total suction. The highest total suction power was 5400 kPa at 25 cm soil depth in higher saline soil and the lowest (1100 kPa) was in the low saline soil at 45 cm soil depth at Aksu (Figure 5). At all soil depths, the suction power of the high and middle saline soils was higher than that in the low saline soils (at Aksu and Korla), the order was: the total suction power of the high saline soil (2200–5400 kPa) > middle saline soil (3800–5200 kPa) > low saline soil at Aksu (1100–2200 kPa) and Korla (1600–3500 kPa). From the water content and the matric suction, all soil would not have strong water stress at any time. The irrigation obviously was insufficient to bring the total suction above the wilting point (15,000 kPa), therefore the cotton could withdraw water at any time, but the medium and high saline soil already had these problems.

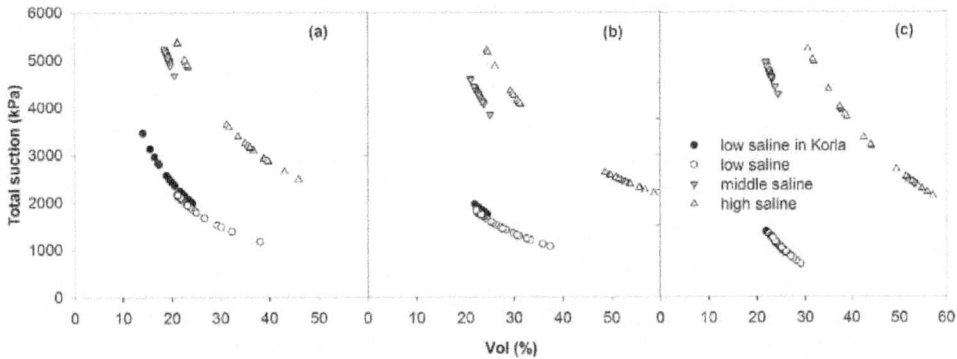

Figure 5. The soil water characteristic curves (matric and osmotic suction) in different saline (low at Korla, low, middle and high at Aksu) soils during cotton season from May to September 2012 in Tarim Basin (a) in 25 cm soil depth; (b) in 45 cm soil depth; and (c) in 65 cm soil depth.

3.3. Water Use Efficiency in Different Saline Soils

Seed cotton yield was higher in low saline soil than that in higher saline soil. The seed cotton yield was highest (6.64 t·ha⁻¹) in the low saline soil at Korla and lowest (2.39 t·ha⁻¹) in the high saline soil at Aksu (Table 3). The yield measured in this paper was covered within the regime from the reported cotton yield data in different treatments and different locations in Xinjiang (1.8–3.6 t·ha⁻¹ on a saline wasteland [45]; 3.0 to 5.9 t·ha⁻¹ in the southern Xinjiang [43]; 5.5 to 6.5 t·ha⁻¹ at Aksu [46]; 5.3–6.5 t·ha⁻¹ at Shihezi [47]; 7.0 t·ha⁻¹ in the south of Xinjiang [48]). The Chinese average cotton yield data for 2012/13 was 1.44 kg·ha⁻¹, which was higher than the world average data (0.77 kg·ha⁻¹) and is bigger than that in USA (0.99 kg·ha⁻¹), India (0.49 kg·ha⁻¹), Pakistan (0.68 kg·ha⁻¹) and Brazil (1.43 kg·ha⁻¹) [49].

Table 3. The field management data and water use efficiency of different saline soils in two experimental stations in Tarim River Basin in 2012.

Location	Soil Salinity Level	Sowing Date	Harvest Date	Fert N	Fert P	Fert K	Irrigation [1]	Precipitation [1]	Yield [2]	IWUE	WUE
				(kg·ha⁻¹)			(mm)	(mm)	(t·ha⁻¹)	(kg·ha⁻¹·mm⁻¹)	
Korla	Low	04.05	04.09	331	124	108	571	128	6.64	11.6 [a]	9.5 [a]
Aksu	Low	08.04	15.09	306	294	55	878	49	4.48	5.1 [b]	4.8 [b]
Aksu	Middle	25.04	10.09	317	88	135	878	49	4.68	5.3 [b]	5.0 [b]
Aksu	High	08.04	05.09	327	215	70	804	49	2.39	3.0 [c]	2.8 [c]

Notes: [1] the amount was within the growing season; [2] seed cotton yield; Fert, fertilizer; IWUE, irrigation water use efficiency; WUE, water use efficiency; values in column IWUE or WUE followed by the different letters ([a, b, c]) indicate significant differences among treatments at 0.05 levels.

Precipitation at Korla (128 mm) was much higher than that at Aksu (49 mm) during the cotton growing season in 2012. More irrigation water (804–878 mm) was used at Aksu, compared to 571 mm irrigation water at Korla. A significant difference was determined in water use efficiency for different salinity levels between low saline soil (WUE 10 kg·ha⁻¹·mm⁻¹, IWUE 12 kg·ha⁻¹·mm⁻¹) at Korla, low and middle saline soil (5 kg·ha⁻¹·mm⁻¹), and high saline soil (3 kg·ha⁻¹·mm⁻¹) at Aksu by the student t test at the 0.05 level. EC had affected water use efficiency mainly through the cotton yield. EC was negatively correlated with cotton yield ($p < 0.01$) at Maigaiti county in Xinjiang [50]. Two years of different salinity and fertilization treatment under cotton showed the IWUE changed from 0.7 to 1.5 kg·m⁻³ at Shihezi in Xinjiang [47]. Southern Xinjiang had an average cotton yield over three years from 3.6 to 5.1 t·ha⁻¹, and irrigation water productivity between 0.91 and 1.16 kg·m⁻³ with a low EC in the top 30 cm soil from 3 to 11 dS·m⁻¹ [43]. As research mentioned above, our WUE data were lower (0.53 kg·m⁻³), except for WUE and IWUE data at Korla.

4. DISCUSSION

4.1. Soil Water Retention in Relation to Different Soil Properties

Soil texture, organic matter content, and bulk density all together can influence soil water retention. The order of sand content in 25 cm soil depth was low salinity soil at Korla (59.3%) > high salinity soil at Aksu (38.5%) > middle salinity soil at Aksu (19.9%) > low salinity soil at Aksu (9%) (Table 2). The soil textures have effect on soil water retention through soil physical process by increasing or decreasing of the field capacity. The texture mainly influenced the matric suction power. At 25 cm depth soil, the low saline soil at Aksu had the lowest sand and highest clay and silt content (Table 2), which had the highest matric suction power compared to the others (Figure 4). Research on the effect of clay content on well-graded sands due to infiltration indicated an increase in matric suction with an increase in the clay content in the mixture of sand and clay [51]. Soil organic carbon increases soil water retention mainly through increasing aggregation, increasing the biological activity and reducing bulk density. The order of soil organic carbon content in 25 cm soil depth was low salinity (6.8 g·kg⁻¹) soil at Aksu > low salinity soil at Korla (4.8 g·kg⁻¹) > middle salinity soil (4.4 g·kg⁻¹) at Aksu> high salinity soil (2.1 g·kg⁻¹) at Aksu (Table 2). The high saline soil at Aksu with the lowest SOC content had the highest bulk density and the low saline soil at Aksu with the highest soil organic carbon content had the lowest bulk density (Table 2). The soil water retention increased, when the bulk density was reduced [31,52]. At high organic carbon values all soils from the U.S. National Soil Characterization database showed an increase in water retention and the largest increase was in sandy and silt soils [29]. The low saline soil at Aksu had the lowest total suction power, lowest bulk density (1.37 g·cm⁻³), sand content (11.8%, more fine-textured soil), highest SOC content (6.8 g·kg⁻¹) in topsoil, which indicated that the low saline soil at Aksu had better macro-aggregate structure, compared to the high salinity soil, which had the highest total suction power, the highest bulk density (1.70 g·cm⁻³), and lowest soil organic carbon content (2.1 g·kg⁻¹) (Figure 5, Table 2). The water logging problem in deeper soil of higher salinity soil induced low suction power but the higher total suction power because of the higher osmotic potential (Figures 3–5). This hindered a deeper rooting in the saturated subsoil.

4.2. Soil Water Retention Curves

The different soil matric potentials play an important role in the salt concentration in the soil. Many experiments have shown a good relationship between plant growth and soil matric potential. Average ECe value in the root zone, after the growing increased and as the control target of soil matric potential decreased had a linear relationship between these factors [53]. A three year experiment on salt distributions and the growth of cotton under different irrigation regimes in Xinjiang in an extremely dry and saline wasteland with drip irrigation showed a favorable low salinity zone existed in the root zone throughout the growing season, when the soil matric threshold was controlled below -25 kPa [53]. Highest irrigation water use efficiency values were recorded when the soil matric potential was around -20 kPa [53]. Matric potential plays an important role in salt accumulation in soil. The amount of salt removed from 0 to 80 cm depth decreased with decreasing soil matric potential [16]. In our research, there is a little different from the earlier studies. Only the low saline soil at Korla and the high saline soil at Aksu kept the soil matric suction below 30 kPa in 65 cm soil profile during whole cotton growing season (Figure 4) and the yield at Korla showed the highest value (6.64 t·ha⁻¹) (Table 3). The water logging in high saline soil at Aksu (Figure 3) induced oxygen deficiency. The low soil matric potential could not offset the high osmotic disadvantages, which suggested that the irrigation volume should not be greatly reduced in high saline soil at Aksu. To reduce the matric suction, the irrigation frequency should be increased in June, July, and August to loose rewetting effects of the low saline soil at Aksu, which induced the higher soil suction power fluctuation (Figure 4). When we consider the higher soil water content in low saline soil (20%–29%) than in middle saline soil (18%–24%) at 40 cm soil profile at Aksu with the same irrigation and precipitation amount and the value of the yield (4.48 t·ha⁻¹ in low salinity, 4.68 t·ha⁻¹ in middle salinity), the irrigation water was overused in low saline soil at Aksu.

The difference in SWCCs (Soil Water Characteristic Curves) was mainly related to the soil salinity level (Figure 5). Salt stress reduced the growth of plants [15,21,53–56]. The high saline soil had the highest suction power (5400 kPa) (Figure 5). Salinity had a pronounced negative effect on microorganism activity, mainly through the metabolic burden imposed by the need for stress tolerance mechanisms [57]. Salt stress reduces the growth of plants [21,54–56] and also effects the soil microbial activity. Research on the different salinity effects on soil microbial activity in soils of varying texture showed cumulative CO_2 in soil decreasing significantly with increasing osmotic potential [58,59].

4.3. Water Use Efficiency in Different Saline Soils

In Xinjiang province drip irrigated cotton fields, the highest seed cotton yield was obtained, when the matric potential threshold was controlled above -30 kPa in 2008 and -0 kPa in 2009 and 2010 and water use value tended to increase as the soil matric potential threshold from -30 kPa increased to -20 kPa in 2009 and 2010 under plastic mulched drip irrigation in Xinjiang [12]. The irrigation type [60], irrigation rate [46] and the irrigation amount [10] have effected on the cotton yield. Especially, Soil salinity and sodicity can be maintained at acceptable low levels by appropriate preplant irrigation[61].

The cotton seed yield increased as the soil matric potential control target increased [53]. The low saline soil at Korla kept the soil matric suction below 30 kPa, the soil suction below 3500 kPa and had the highest yield (6.64 kg·ha^{-1}), which prevented high salinity stress in the cotton during the growing season, producing the highest yields and highest water use efficiency (12 kg·ha^{-1}·mm^{-1}) (Table 3, Figures 4 and 5). Although the soil matric suction of high saline soil was below 30 kPa (Figure 4), considering the highest osmotic suction (Figure 5), and the waterlogging problem below 40cm soil profile (Figure 3), it did not bring much profit to the yield (2.8 kg·ha^{-1}) (Table 3). The main reason for decreased root length in cotton under drip irrigation with mulch film was localized accumulation of salinity [62] and the cotton yield increased with the root biomass increase [63]. For the high saline soil one would need a better drainage. This will deepen the root zone, improve the leaching, reduce the salinity and finally increase the yield. Thereby it could save water and increase WUE as well. On the other hand, one need water more irrigation to alleviate the salinity problem. Compared to low saline soils at Aksu, the low saline soil at Korla saved 110 mm irrigation and 103 mm total water to reach 1 t·ha^{-1} yield and increased water use efficiency by 5 kg·ha^{-1}·mm^{-1} and 7 kg·ha^{-1}·mm^{-1} for WUE and IWUE respectively (Table 3).

With the relationship of soil water and soil matric suction, the osmotic suction, the soil texture and soil organic carbon content and total nitrogen, there were the models with strong closed relationship ($R^2 = 1$) to modelling the soil water content (Figure 6). The soil organic carbon, and soil total nitrogen content, soil texture, which affected the soil matric suction, also affected the soil water content and reduced the salt effect to cotton. The laboratory experiment showed the SOC could restrict the soil water evapotranspiration, salt accumulation and increased the salt leaching [64]. A field experiment showed that farmyard manure could reduce the soil salinity and sodicity and increased cotton yield [65]. In the cotton field in Xinjiang, the soil fertility amelioration is a way to increase the leaching and soil water content, water use efficiency and resist the harm of salinity.

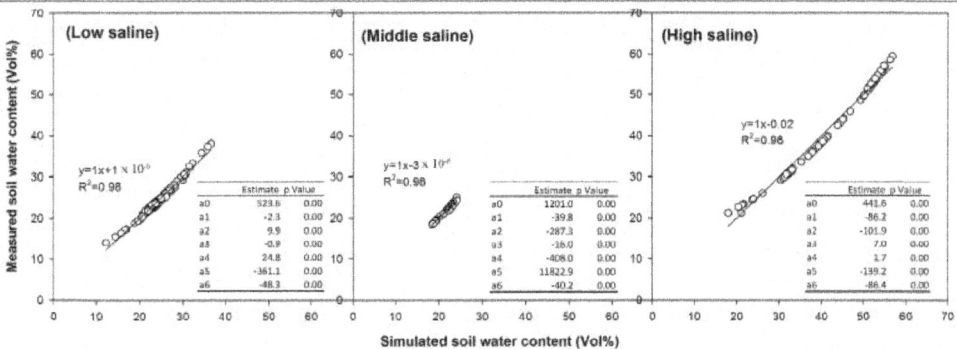

Figure 6. The relationship between the simulated and the measured soil water content with modelling (Vol% = a0 + a1 × pF1 + a2 × clay% + a3 × silt% + a4 × Corg + a5 × Ntot + a6 × pF2) (pF1: pF matric, pF2: pF osmetic, Corg: (g/kg), Ntot: (g/kg)) (**a**) in low saline soil; (**b**) in middle saline soil; and (**c**) in high saline soil.

5. CONCLUSIONS

Based on the experimental data, the relationships between soil water retention, water consumption, water use efficiency and yield were systematically analyzed. The mechanisms of soil moisture and salinity distribution and transport together with the relationship of soil water retention in different saline soils showed that the soil water content (15%–23%) at the top 40 cm soil, lower suction power (below 3500 kPa) and lower matric suction soil (below 30 kPa) in low saline soil had the highest water use efficiency and higher yield. The water resource limitation and increasing salinization danger, the physical and chemical properties, matric and osmotic suction, and water logging problem should all be considered for water use efficiency in field management. For example: draining the water logging fields, increasing irrigation frequency, reducing the irrigation amount depending on the soil texture, increasing the manure fertilization. A feasible irrigation with reduction of the salinity harm combined with increasing the soil fertility should be the way to increase water use efficiency in the Tarim River Basin.

ACKNOWLEDGMENTS

The study was funded by the Project SuMaRiO of BMBF (German Federal Ministry of Science and Technology)-Sustainable Management of River Oases along the Tarim River, Institute of Soil Science and Land Evaluation of Hohenheim University, Germany and the National Natural Science Foundation (41201042), Xinjiang Province Youth Science and Technology Innovation Talents Project (2013731029), the Chinese academy of sciences "Western Light" of personnel plan

the western doctor special (XBBS201208) and National Scientific and Technological support Program of China (Grant No. 2013BAC10B01). The delivery of data
from the research stations Aksu and Korla, especially, the data cooperation with
the Pengnian Yang of College of Hydraulic and Civil Engineering of Xinjiang
Agricultural University are greatly acknowledged. We are also grateful to Kath-
leen Regan, Institute of Soil Science and Land Evaluation and Wolfram Spreer,
Institute of Agricultural Engineering, Hohenheim University for providing help
in revision of this manuscript.

Author Contributions

Theresa Schiller and Karl Stahr designed the field experiment. Theresa Schiller
performed the experimental work and data collection. Xiaoning Zhao performed
statistical analysis and prepared the manuscript. Hussein Othmanli performed
the soil data collection and analysis. Chengyi Zhao, Yu Sheng, Shamaila Zia
and Joachim Müller gave their support through the field experiments and data
exchange.

Conflicts of Interest

The authors declare no conflict of interest.

REFERENCES

1. Hao, X.M.; Chen, Y.N.; Li, W.H. Impact of anthropogenic activities on the hydrologic characters
    of the mainstream of the Tarim River in Xinjiang during the past 50 years. *Environ. Geol.* **2009**,
    57, 435–445.

2. Qi, F.; Wei, L.; Jianhua, S.; Yonghong, S.; Yewu, Z.; Zongqiang, C.; Haiyang, X. Environmental
    effects of water resource development and use in the tarim river basin of Northwestern China.
    Environ. Geol. **2005**, *48*, 202–210.

3. Feng, Q.; Liu, W. Environmental conditions leading to the formation of holocene soil layers in
    the Northern Taklimakan Desert, Tarim Basin, Northwest China. *Geol. J.* **2005**, *40*, 23–34.

4. Feng, Q.; Cheng, G.D.; Masao, M.K. Trends of water resource development and in arid North-
    West China. *Environ. Geol.* **2000**, *39*, 831–838.

5. Chen, X.B.; Yang, J.S.; Liu, C.Q.; Hu, S.J. Soil salinization under integrated agriculture and its
    countermeasures in Xinjing (in Chinese with English abstract). *Soils* **2007**, *39*, 347–353.

6. Hou, Z.A.; Wang, H.Y.; Gong, J.; Xiao, L.; Ma, L.; Qi, T. Salt distribution and accumulation
    as affected by under-film drip irrigation with saline water in an arid region (in Chinese with
    English abstract). *Chin. J. Soil Sci.* **2008**, *39*, 16–24.

7. Chen, Y.; Ye, Z.; Shen, Y. Desiccation of the tarim river, xinjiang, china, and mitigation strategy.
    Quatern. Int. **2011**, *244*, 264–271.

8. Zhang, X.Y.; Cai, H.J. Effects of regulatd deficit irrigation on plastic effect-mulched cotton
    (in Chinese with English abstract). *J. Northwest Sci. Tech. Univ. Agric. For.* **2001**, *29*, 9–12.

9. Wang, C.R.; Tian, X.H.; Li, S.H. Effects of plastic sheet-mulching on ridge for water-harvesting
    cultivation on wue and yield of winter heat. *Sci. Agric. Sin.* **2004**, *37*, 208–214.

10. Hu, X.T.; Chen, H.; Wang, J.; Meng, X.B.; Chen, F.H. Effects of soil water content on cotton root
    growth and distribution under mulched drip irrigation (in Chinese with English abstract).
    Agric. Sci. China **2009**, *8*, 709–716.

11. Hu, S.; Song, Y.; Zhou, H.; Tian, C. Experimental study on water use efficiency of cotton in the tarim river basin. *Agric. Res. Arid Areas* **2002**, *20*, 66–70.

12. Kang, Y.; Wang, R.; Wan, S.; Hu, W.; Jiang, S.; Liu, S. Effects of different water levels on cotton growth and water use through drip irrigation in an arid region with saline ground water of Northwest China. *Agric. Water Manag.* **2012**, *109*, 117–126.

13. Zhao, C.Y.; Sheng, Y.; Yimam, Y. Quantifying the impacts of soil water stress on the winter wheat growth in an arid region, Xinjiang. *J. Arid Land* **2009**, *1*, 34–42.

14. Zhao, C.Y.; Yan, Y.Y.; Yimamu, Y.; Li, J.Y.; Zhao, Z.M.; Wu, L.S. Effects of soil moisture on cotton root length density and yield under drip irrigation with plastic mulch in aksu oasis farmland. *J. Arid Land* **2010**, *2*, 243–249.

15. Dong, H.; Li, W.; Tang, W.; Zhang, D. Early plastic mulching increases stand establishment and lint yield of cotton in saline fields. *Field Crop Res.* **2009**, *111*, 269–275.

16. Wang, R.; Kang, Y.; Wan, S.; Hu, W.; Liu, S.; Jiang, S.; Liu, S. Influence of different amounts of irrigation water on salt leaching and cotton growth under drip irrigation in an arid and saline area. *Agric. Water Manag.* **2012**, *110*, 109–117.

17. Chen, X.B.; Yang, J.S.; Zhang, F.D.; Hu, S.J.; Li, H. Control of soil water salinity variatons based on crop-salt-water production function in Tarim irrigation area (in Chinese with English abstract). *J. Irrig. Drain.* **2007**, *26*, 75–78.

18. Liu, M.X.; Yang, J.S.; Li, X.M.; Yu, M.; Wang, J. Effects of irrigation water quality and drip tape arrangement on soil salinity, soil moisture distribution, and cotton yield (gossypium hirsutum l.) under mulched drip irrigation in Xinjiang, China. *J. Integr. Agric.* **2012**, *11*, 502–511.

19. Ma, D.; Wang, Q.; Lai, J. Field experimental studies on the effects of water quality and drip rate on soil salt distribution in drip irrigation under film. *Trans. Chin. Soc. Agric. Eng.* **2005**, *21*, 42–46. (In Chinese)

20. Thomsen, I.K.; Schjønning, P.; Jensen, B.; Kristensen, K.; Christensen, B.T. Turnover of organic matter in differently textured soils. II. Microbial activity as influenced by soil water regimes. *Geoderma* **1999**, *89*, 199–218.

21. Chen, M.; Kang, Y.; Wan, S.; Liu, S.P. Drip irrigation with saline water for oleic sunflower (helianthus annuus l.). *Agric. Water Manag.* **2009**, *96*, 1766–1772.

22. Kang, Y.; Wan, S. Effect of soil water potential on radish (raphanus sativus l.) growth and water use under drip irrigation. *Sci. Hortic.* **2005**, *106*, 275–292.

23. Jiao, Y.P.; Kang, Y.H.; Wan, S.Q.; Liu, W. Effect of soil matric potential on waxy corn (zea mays l. Sinesis kulesh) growth and water use under drip irrigation in saline soils of arid areas. In Proceedings of the 8th International Dryland Development Conference, Beijing, China, 25–28 February 2006.

24. Nimmo, J.R. Modeling structural influences on soil water retention. *Soil Sci. Soc. Am. J.* **1997**, *61*, 712–719.

25. Kironchi, G.; Kinyali, S.M.; Mbuvi, J.P. Environmental influence on water characteristics of soils in two semi-arid catchments in Laikipia, Kenia. *Afr. Crop Sci. J.* **1995**, *3*, 479–486.

26. Pachepsky, Y.A.; Rawls, W.J. Soil structure and pedotransfer functions. *Eur. J. Soil Sci.* **2003**, *54*, 443–451.

27. Juhász, C.E.P.; Cooper, M.; Cursi, P.R.; Ketzer, A.O.; Toma, R.S. Savanna woodland soil micromorphology related to water retention. *Sci. Agric.* **2007**, *64*, 344–354.

28. Hollis, J.M.; Jones, R.J.A.; Palmer, R.C. The effects of organic matter and particle size on the water-retention properties of some soils in the west midlands of England. *Geoderma* **1977**, *17*, 225–238.

29. Rawls, W.J.; Pachepsky, Y.A.; Ritchie, J.C.; Sobecki, T.M.; Bloodworth, H. Effect of soil organic carbon on soil water retention. *Geoderma* **2003**, *116*, 61–76.

30. Reeve, M.J.; Smith, P.D.; Thomasson, A.J. The effect of density on water retention properties of field soils. *J. Soil Sci.* **1973**, *24*, 355–367.

31. Walczak, R.T.; Moreno, F.; Sławihski, C.; Fernandez, E.; Arrue, J.L. Modeling of soil water retention curve using soil solid phase parameters. *J. Hydrol* **2006**, *329*, 527-533.

32. Abrol, I.P.; Khosla, B.K.; Bhumbla, D.R. Relationship of texture to some important soil moisture constants. *Geoderma* **1968**, *2*, 33-39.

33. Tutiempo. Global Climate Data. Available online: http://www.tutiempo.net/en/Climate/ (accessed on 1 January 2014).

34. Lu, R.K. *Soil Agricultural Chemical Analysis;* Beijing Agricultural Technology press: Beijing, China, 1983. (In Chinese)

35. The Ministry of Agriculture of People's Republic of China. The Notification of Cotton Yield Calculation, 2010. Available online: http://www.moa.gov.cn/govpublic/ZZYGLS/201008/t20100816_1619412.htm (accessed on 1 January 2013). (In Chinese)

36. Campbell, G.S. Soil water potential measurement: An overview. *Irrig. Sci.* **1988**, *9*, 265-273.

37. Or, D.; Wraith, J.M. Soil water content and water potential relationships. In *Handbook of Soil Science;* CRC press: Boca Raton, FL, USA, 2000; pp. A53-A85.

38. Rengasamy, P. Soil salinity and sodicity. In *Growing Crops with Reclaimed Wastewater;* Inkata press: Melbourne, Australia, 2006; pp. 125-138.

39. Regional Salinity, L. *Diagnosis and Improvement of Saline and Alkali Soils;* U.S. Department of Agriculture: Washington, DC, USA, 1954.

40. Hu, S.; Shen, Y.; Chen, X.; Gan, Y.; Wang, X. Effects of saline water drip irrigation on soil salinity and cotton growth in an oasis field. *Ecohydrology* **2013**, *6*, 1021-1030.

41. Zhang, Y.; Wang, L.H.; Sun, S.M.; Chen, X.L.; Liang, Y.J.; Hu, S.J. Indexes of salt tolerance of cotton in akesu river irrigation destrict (in Chinese with English abstract). *Sci. Agric. Sin.* **2011**, *44*, 2051-2059.

42. Yadav, S.; Irfan, M.; Ahmad, A.; Hayat, S. Causes of salinity and plant manifestations to salt stress: A review. *J. Environ. Biol.* **2011**, *32*, 667-685.

43. Wang, Z.; Jin, M.; Šimůnek, J.; van Genuchten, M.T. Evaluation of mulched drip irrigation for cotton in arid Northwest China. *Irrig. Sci.* **2014**, *32*, 15-27.

44. Zhang, N.; Zhao, Y.S.; Yu, G.R. Simulated annual carbon fluxes of grassland ecosystems in extremely arid conditions. *Ecol. Res.* **2009**, *24*, 185-206.

45. Wang, R.; Kang, Y.; Wan, S. Effects of different drip irrigation regimes on saline-sodic soil nutrients and cotton yield in an arid region of Northwest China. *Agric. Water Manag* **2015**, *153*, 1-8.

46. Danierhan, S.; Shalamu, A.; Tumaerbai, H.; Guan, D. Effects of emitter discharge rates on soil salinity distribution and cotton (gossypium hirsutum l.) yield under drip irrigation with plastic mulch in an arid region of Northwest China. *J. Arid Land* **2013**, *5*, 51-59.

47. Min, W.; Guo, H.; Zhou, G; Zhang, W.; Ma, L.; Ye, J.; Hou, Z. Root distribution and growth of cotton as affected by drip irrigation with saline water. *Field Crop Res.* **2014**, *169*, 1-10.

48. Deng, Z.; Bei, D.; Zou, G.L.; Cong, J.; Li, Y.; Cai, J.M.; Feng, J.J. Effects of water and nitrogen regulation on the yield and water and nitrogen use efficiency of cotton in South Xinjiang, Northwest China under plastic mulched drip irrigation. *Chin. J. Appl. Ecol.* **2013**, *24*, 2525-2532.

49. Dai, J.; Dong, H. Intensive cotton farming technologies in China: Achievements, challenges and countermeasures. *Field Crop Res.* **2014**, *155*, 99–110.

50. Zheng, Z.; Zhang, F.; Ma, F.; Chai, X.; Zhu, Z.; Shi, J.; Zhang, S. Spatiotemporal changes in soil salinity in a drip-irrigated field. *Geoderma* **2009**, *149*, 243–248.

51. Jeong, S.; Kim, J.; Lee, K. Effect of clay content on well-graded sands due to infiltration. *Eng. Geol.* **2008**, *102*, 74–81.

52. Horn, R.; Taubner, H.; Wuttke, M.; Baumgartl, T. Soil physical properties related to soil structure. *Soil Tillage Res.* **1994**, *30*, 187–216.

53. Wang, R.; Kang, Y.; Wan, S.; Hu, W.; Liu, S.; Liu, S. Salt distribution and the growth of cotton under different drip irrigation regimes in a saline area. *Agric. Water Manag.* **2011**, *100*, 58–69.

54. Bassil, E.S.; Kaffka, S.R. Response of safflower (carthamus tinctorius l.) to saline soils and irrigation i. Consumptive water use. *Agric. Water Manag.* **2002**, *54*, 67–80.

55. Bassil, E.S.; Kaffka, S.R. Response of safflower (carthamus tinctorius l.) to saline soils and irrigation ii. Crop response to salinity. *Agric. Water Manag.* **2002**, *54*, 81–92.

56. Kang, Y.; Wang, F.X.; Liu, H.J.; Yuan, B.Z. Potato evapotranspiration and yield under different drip irrigation regimes. *Irrig. Sci.* **2004**, *23*, 133–143.

57. Schimel, J.; Balser, T.C.; Wallenstein, M. Microbial stress-response physiology and its implications for ecosystem function. *Ecology* **2007**, *88*, 1386–1394.

58. Setia, R.; Marschner, P.; Baldock, J.; Chittleborough, D.; Smith, P.; Smith, J. Salinity effects on carbon mineralization in soils of varying texture. *Soil Biol. Biochem.* **2011**, *43*, 1908–1916.

59. Chowdhury, N.; Nakatani, A.S.; Setia, R.; Marschner, P. Microbial activity and community composition in saline and non-saline soils exposed to multiple drying and rewetting events. *Plant Soil* **2011**, *348*, 103–113.

60. Tang, L.-S.; Li, Y.; Zhang, J. Biomass allocation and yield formation of cotton under partial rootzone irrigation in arid zone. *Plant Soil* **2010**, *337*, 413–423.

61. Nightingale, H.I.; Davis, K.R.; Phene, C.J. Trickle irrigation of cotton: Effect on soil chemical properties. *Agric. Water Manag.* **1986**, *11*, 159–168.

62. Mai, W.X.; Tian, C.Y.; Li, C.J. Soil salinity dynamics under drip irrigation and mulch film and their effects on cotton root length. *Commun. Soil Sci. Plan* **2013**, *44*, 1489–1502.

63. Luo, H.H.; Tao, X.P.; Hu, Y.Y.; Zhang, Y.L.; Zhang, W.F. Response of cotton root growth and yield to root restriction under various water and nitrogen regimes. *J. Plant Nutr. Soil Sci.* **2015**, doi:10.1002/jpln.201400264.

64. Shan, X.Z.; Wei, Y.Q.; Yang, H.J.; Liu, J.F.; Zhang, R. The modelling research on the effect of soil organic carbon content on the soil water and salt movement (in Chinese with English abstract). *Soil and Fertilizer.* **1996**, *5*, 1–5.

65. Kahlown, M.A.; Azam, M. Effect of saline drainage effluent on soil health and crop yield. *Agric. Water Manag.* **2003**, *62*, 127–138.

This page left intentionally blank.

INDEX

A

Abiotic Soil Components, 139
Accelerated Erosion, 287
Adaptations of Behaviour, 113
Adaptations of Body Form, 110
Agronomic Measures for Soil and Water
    Conservation, 297
Analysis of Stress and Strain, 217
Application of Inorganic Fertilizers, 80
Applications, 264
Arthropod Fauna in Arable Crops, 161
Arthropod Fauna in Orchards, 161

B

Bacteria, 67
Band Placement, 190
Basic Plant Nutrition, 167
Basin-Listing, 303
Bench-Terracing, 305
Beneficial Arthropods, 153
Biodegradation of Pesticides in Soil, 116
Body Form and Lifestyle, 110
Broadcast Incorporation, 192
Burning, 81

C

Cage Tests Using Selected Arthropod
    Species, 160

Carbon, 147
Carryover, 282
Characteristics of Phosphorus, 197
Checking the Growth of Gullies, 309
Chemical Composition of the Soil, 27
Chemical Dynamics of Soil, 211
Clasification of Soil Biota, 59
Class I (Green Colour), 314
Class II (Yellow Colour), 315
Class III (Red Colour), 315
Class IV (Blue Colour), 316
Class V (Dark Green or Uncoloured),316
Class VI (Orange Colour), 317
Class VII (Brown Colour), 317
Class VIII (Purple Colour), 317
Classification of Gullies, 307
Classification of Soils, 43
Close Packing, 239
Components of Climate, 2
Compound Structure, 15
Conservation of Water and Soil 296
Constitution of the Soil, 319
Contour and Peripheral Bunding, 309
Contour Strip-cropping, 301
Contour-Bunding, 303
Contour-Farming, 297
Coping with Living in Soil, 110
Correct Sampling Depth, 176

Crop Losses from Soil Organisms, 81
Crop Rotations, 107
Cultivation, 101

D

Date of Fall Banding, 193
Decomposition, 135
Denitrification, 186
Design of Composite Check Dams, 311
Direct Methods, 72
Diseases from Soil Bacteria, 92
Disease-Suppressive Soils, 130
Distribution of Run-Off, 289
Diversity, 57
Diversity of Biota, 48
Dry Deposition, 143

E

Earthworm Field Tests, 162
Earthworms, 70
Ecological Cycling, 137
Ecosystem Processes, 59
Ecosystem Services, 138
Effects of Pesticides, 115
Enchytraeids, 151
Enhancement of Soil Structure, 96
Enhancing Beneficial Organisms, 73
Erosion, 188
Erosion and Stream Flow, 287
Erosion Prevention, 269
Essential Plant Nutrients, 168
Estimation of Peak Rate of Run-off, 306
Evolutionary History, 47
Extent of Soil Erosion, 296
Extra-Structural Cracks, 15

F

Fallow and Topsoil Storage, 109
Fertilizer Labelling, 283

Fertilizers, 102
Field Strip-Cropping, 302
Field Studies (Biomonitoring), 164
Field Tests, 159
Fire, 108
Fire and Pest Control, 64
Floods, 288
Foliar Application, 192
Forest and Wood-Using Industries, 292
Function of Soil Organisms, 96
Fungal Diversity, 50
Fungi, 68
Fungicides, 106

G

General Nutrient Uptake, 169
Grasing, 100
Gully Erosion, 272
Gully Plugging, 309

H

Herbicides, 106
Higher Plants, 149
Honey Bee Field Test, 160
Host Plant Resistance, 77
How Fast Does a Rock Decay, 231
How to Practice Contour-Farming, 298

I

Igneous Rock Weathering, 229
Immobilisation, 187
Importance of Soil Biota, 62
Increasing Plant Diversity, 75
Indirect Methods, 73
Inorganic Fertilizers, 102
Insecticides, 106
Insects, 69
Interpretation of the Soil Analysis, 283
Ionic Compounds, 244

L

Laboratory and Field Bioassays, 162
Land-capability Classes, 314
Land-Capability Classification, 314
Land-capability Sub-class, 318
Landslide Control, 312
Large Soil Animals, 97
Lattice Imperfections, 252
Leaching, 187
Lime, 104
Limiting Factors, 282
Lsopods and Millipedes, 149
Lumbricids, 152

M

Macro-fauna, 54
Macroscopic Properties of Solids, 236
Major Elements, 143
Management Considerations, 64
Management Effects on Soil Biota, 99
Managing the Soil Biota, 71
Mechanics of Soils, 217
Mesofauna, 52
Microbes, 98
Microcosm Tests, 154
Micro-fauna, 51
Mineral Weathering, 144
Mites (Acari), 53
Mixed Cropping, 302
Model of the Liquid Phase, 254
Molluscs, 153
Mulching, 298
Mutagenicity Tests, 164
Mycorrhizal Fungi, 82

N

N Fertilizers, 104
Nematodes, 68

Nematodes, 52
Nitrification Inhibitors, 194
Nitrogen, 145
Nitrogen Cycle : Major Processes, 46
Nitrogen Effects on Canola Growth, 180
Nitrogen Fertilizer, 188
Nitrogen Fixation, 144
Nutrient Management Planning, 284
Nutrient Storage and Release, 63
Nutrients, 231

O

Objectives of Soil Analysis, 281
Organic Activity of Soil, 16
Organic Chemicals and Metals, 144
Organic Fertilizers, 104
Organic Matter, 134
Organisms Affect of Soil Formation, 3
Oribatid Mites, 150
Origin of Soils, 21

P

P and S Fertilizers, 103
Parameters For Stress and Strain, 219
Parent Material, 5
Particle Density, 280
Pasture Food Web, 94
Patterns, 141
Perimeter Run-off Control, 269
Persistence of Pesticides in Soil, 116
Pesticide Use, 80
Pesticides, 105
Phase Transformations, 261
Phosphorus, 147
Phosphorus Fertilizer Placement, 202
Plant and Tissue Testing, 177
Plant Residue Retention, 106
Planting for Pest Management, 84
Poisonous Fungi, 91

Porosity, 280
Potassium (K), 207
Potassium Supply from the Soil, 209
Practices, 269
Principals of Soil Organisms, 65
Principles and Strategies, 122
Proper Sample Handling, 176
Proper Sampling Date, 175
Proper Soil Sampling Equipment, 174
Properties of Liquids, 253
Properties of Soil, 20
Protozoans and Nematodes, 149

Q

Quantification of Input of Chemicals in
   the Soil, 143
Quirky Bit, 91

R

Reclamation of Medium Gullies, 311
Reclamation of Small Gullies, 310
Reduced Tillage, 78
Relief and Topography of Soil, 5
Research Priorities and Approaches, 85
Residue Decomposition, 63
Ridging and Terracing, 79
Rock Phosphate, 204
Role of Invertebrate Animals, 98
Role of Microbes, 99
Role of Potassium, 207
Roles of soil organisms, 96

S

Sanitation, 79
Seed Row Placement, 188
Seed Saving, 79
Shearing, 220
Sheet Erosion, 272
Significance of Bulk Density, 277

Significance of Porosity, 280
Soil Analysed, 284
Soil and Available Water, 12
Soil Biodiversity, 44
Soil Bulk Density Determination, 276
Soil Colour, 18
Soil Compaction and Building, 279
Soil Consistence, 278
Soil Development, 226
Soil Ecosystem, 137
Soil Enzyme : Dehydrogenase, 158
Soil Enzyme : Phosphatase, 158
Soil Enzyme : Urease, 158
Soil Erosion, 272
Soil Erosion and Conservation, 268
Soil Factors Impacting Tillage, 278
Soil Fertility, 13
Soil Fertility and Nutrition, 167
Soil Food Web, 93
Soil Health and Plant Health, 66
Soil Health Indicator, 60
Soil is Layered, 226
Soil Management, 13
Soil Microbes overview, 88
Soil Microbial Activity, 167
Soil Micro-organisms, 115
Soil Moisture, 135
Soil Organisms, 67
Soil Processes, 140
Soil Profiles, 18
Soil Resources, 292
Soil Sampling Plan, 174
Soil Structure, 14
Soil Temperature Regimes, 281
Soil Testing for Nutrient Content, 172
Soil Texture, 17
Soil Water and the Soil Solution, 322
Soil-Forming Factors, 44
Soil-Habitat, 56

Soils, Weathering, and Nutrients, 225
Soluble in Acids, 31
Specific Concepts, 236
Strategies for Bioremediation, 118
Strength Criteria, 221
Strip-Cropping, 300
Subsoiling, 303
Supporting Public Policy, 87
Surface Tension, 255
Synthetic Soils, 131

T

Termites, 54
Terracing of Side Slopes, 311
Test for Denitrification, 158
Test for Nitrogen Fixation, 157
Tests on Enzyme Activity in Soil, 158
Tests on Soil Microbial Processes, 155
Time Factor of Soil, 6
Timing of Phosphorus Fertilization, 202
Transformations that Reduce Plant
    Available Nitrogen, 186
Translocation of Soil, 17
Tree Clearing, 109
Types of Micro-organisms in Soil, 133
Types of Soil-Erosion, 272
Typical Values of Shear Strength, 223

U

Uniaxial Compression, 221
Uniaxial Extension, 220

Urease Inhibitors, 194
Utilisation and Recycling of Energy, 119

V

Vapour Pressure, 257
Viruses as Quasi-Organisms, 49
Viscosity, 256
Volatilisation, 188

W

Watershed Management, 290
Ways of Feeding, 89
Weathering, 227
Weeds, 71
Wet Deposition, 143
What Affects Soil Biota?, 63
What are Soil Biota, 61
What do Soil Biota do?, 62
What is the Soil?, 2
Where do Soil Biota Live?, 109
Wind Erosion, 273
Wind Strip-Cropping, 302
Wood, the Universal Raw Material, 290

Y

Yields to Expect, 282

Z

Zymogenic Soils, 130

This page left intentionally blank.